Operations Planning and Control

OPERATIONS PLANNING AND CONTROL

Joe H. Mize
Arizona State University
Tempe, Arizona

Charles R. White
Auburn University
Auburn, Alabama

George H. Brooks
Auburn University
Auburn, Alabama

PRENTICE-HALL, INC., Englewood Cliffs, New Jersey

PRENTICE-HALL INTERNATIONAL, INC., *London*
PRENTICE-HALL OF AUSTRALIA, PTY. LTD., *Sydney*
PRENTICE-HALL OF CANADA, LTD., *Toronto*
PRENTICE-HALL OF INDIA PRIVATE LTD., *New Delhi*
PRENTICE-HALL OF JAPAN, INC., *Tokyo*

Current printing (last digit)
10 9 8 7 6 5 4 3 2 1

13-637892-7

Library of Congress Catalog Card Number 78-149982
Printed in the United States of America

To
Betty, June and Hope

Preface

The objective of *Operations Planning and Control* is to present an introduction to the basic concepts of designing control systems for operating environments of many types. It is intended primarily as a textbook for introductory courses in operations planning and control (often called "production control" or "production management"). It is intended also as a reference book for practicing engineers and management scientists who are responsible for the design and management of production and inventory control systems in industry, government, the military, and service-oriented organizations. The text grew from lecture material for the course "Production Control Functions," a required course in the industrial engineering curriculum at Auburn University.

The authors have attempted to emphasize concepts rather than techniques. We say without hesitation or apology that our presentation is highly idealized. The concepts presented are those that we feel contribute most to the understanding of control system design principles, rather than those that are practical and workable with current knowledge and capabilities of system procedures and hardware. We have attempted to present a conceptual framework that will enhance the reader's ability to understand the manner in which system components interact in a complex operations planning and control system.

Although the book is quantitatively oriented, it is not mathematically sophisticated. A first course in calculus and a knowledge of elementary probability and statistics are considered the primary prerequisites. A few sections make use of elementary matrix operations, and a brief treatment of these is presented in Appendix A.

A unique feature of the text is that a new and promising method of teaching concepts of control system design is described in Appendix C. Specifically, a production system simulator is described which is capable of simulating a wide variety of hypothetical or real operating environments. The authors have used this computer-based simulator in conjuction with this text for several years. We have found it to be the most effective means of teaching this subject material.

The book is divided into three parts. Part I consists of two chapters. Chapter 1 is an introduction to the primary features of an operating environment, particularly production and inventory systems. Some of the decisions that have to be made in such a system are discussed. Chapter 2 presents several new concepts that show promise as valuable aids in designing control systems for complex operating environments. It is suggested that the first step in designing a control system is to uncover the *decision structure* of the organization: what *decisions* are required? How are they *related*?; what *information* is needed for each decision? The concept of the *Transfer Function* is borrowed from systems engineering as a general mechanism for transforming information into decisions. Individual transfer functions are combined into a total control system by conceptually "plugging them into" two data files that are common to all transfer functions; the Parameter File and the Variable Status File. Included in this system framework are the concepts of feedback and the modification of system parameters through comparison of actual results to established measures of effectiveness.

Part II consists of five chapters, one for each of the five major functions of an Operations Planning and Control System; (1) forecasting, (2) operations planning, (3) inventory planning and control, (4) operations scheduling, and (5) dispatching and progress control. All chapters in Part II have the same basic format. The major section headings are Concepts, Inputs, Transfer Functions, Feedback and Corrective Action, and Interfaces. Stated in the Concepts section are the purpose of this particular function and specific decisions that must be rendered by this part of the control system. The Inputs section identifies specific information requirements for the decisions, and their sources. The Transfer Functions section discusses and develops some of the techniques and theories that are available for executing this particular function. (We have included primarily transfer functions that are readily understood without sophisticated mathematical procedures. The instructor is encouraged to add other transfer functions at his own discretion. The modular nature of our treatment makes this especially easy and convenient.) The section on Feedback and Corrective Action discusses the concept of continuous monitoring of system performance and adjusting of control parameters. Both short-term and long-term corrective action are discussed. The Interfaces section portrays the relative position of this particular function within the total system and shows its relationships with other components, both inside and outside the control system per se. The chapters in Part II contain a generous number of flow diagrams and other graphical aids to assist the reader in understanding the complex relationships and interdependencies between system components.

Part III consists of two chapters. Chapter 8 presents a systems engineering approach to the design of a total operations planning and control system. Theoretical as well as practical considerations are treated. A case study is included to illustrate the application of the design procedure. Chapter 9 considers the future of operations planning and control systems design. An assessment is made of the applicability of formal control theory to production control.

Appendix A contains a brief presentation of elementary matrix operations that are used in parts of Chapter 4. Appendix B presents an introduction to PERT. Appendix C describes PROSIM V, a production system simulator that can be used as a teaching aid.

All exercises are grouped together following the Appendices. Exercises are included for Chapters 2 through 7. There are several sequences of exercises that carry through from one chapter to the next. This feature again emphasizes the integrated nature of operations planning and control systems.

We are indebted to our former teachers, our colleagues, and particularly to our students for the contributions they have made to the concepts presented in this book. Particular mention is due Lavon F. Jordan who contributed greatly to Chapter 9. We are also indebted to Mrs. Margaret Estes, who typed most of the manuscript. We are very grateful to our wives and families for their understanding and patience throughout the entire project.

JOE H. MIZE
CHARLES R. WHITE
GEORGE H. BROOKS

Contents

I INTRODUCTORY CONCEPTS

II OPERATIONS PLANNING AND CONTROL FUNCTIONS

Operations Planning and Control

I

Introductory Concepts

1 Production-Inventory Systems Concepts

Introduction

A large portion of modern industry consists of fabrication and assembly processes. Consider, for example, a plant that manufactures television sets, radios, and phonographs. This plant must purchase materials, and perhaps parts, convert these materials into specific components, then assemble the components into the several products offered to consumers. This greatly over-simplified process is represented in Fig. 1.1.

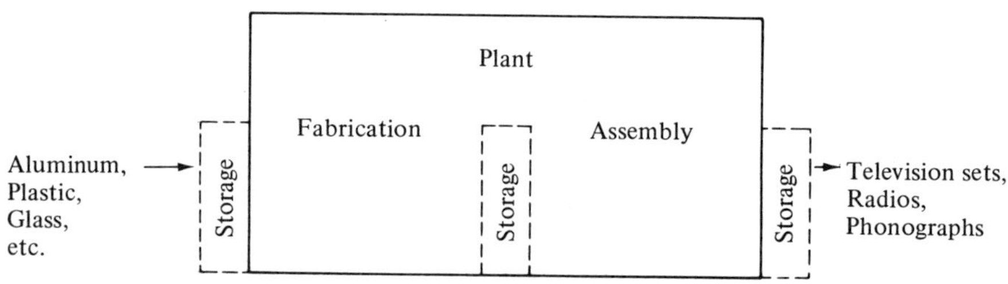

Figure 1.1 Representation of a typical Manufacturing Plant

Several observations are important at this point. First, there could be many different configurations of each of the three primary products. For example, there are many sizes and styles of television sets. Second, different products could contain several common components. It would not be surprising, for example, to find the same tube in all three products. Third, different components could contain common raw materials. For example, certain components

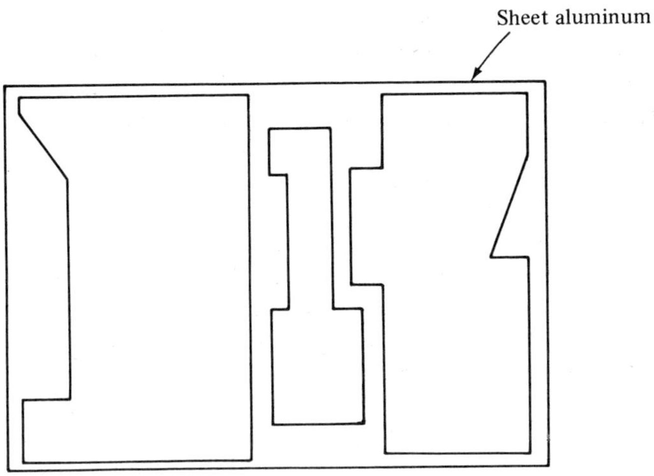

Figure. 1.2 Components for three different products stamped from same sheet of Aluminum

for each product may be stamped from the same roll of aluminum. The three components shown in Fig. 1.2 could be for different products.

Fourth, storage occurs at three stages of the manufacturing process. For the sake of convenience, we classify these three storage phases as raw material inventory, in-process inventory, and finished goods inventory. We defer to a later section the question of why inventories are necessary and here simply assert that we wish to hold inventory to a low level and yet have sufficient material and components when needed.

Our observations thus far have been of a general nature. Let us now examine in greater depth the two major divisions of the plant, fabrication and assembly. If we were to look closely at the activities in the fabrication division, we would see many machines performing many different operations on raw materials. For example, plastic molding machines convert a special powder into knobs, cases, etc. Metal forming machines convert tubing and sheet metal into brackets, frames, braces, etc. Special equipment produces electronic components, printed circuits, etc. An important feature that we would notice would be the large number of components being produced in batches. If we observed long enough, we would see certain equipment discontinue the production of one component and, perhaps after adjustments and retooling, begin producing another component. The fabricated components are then moved to intermediate storage to await assembly.

We now shift our attention to the other major division of the plant, assembly. Here components are brought together to form subassemblies and, eventually, television sets, radios, or phonographs. Work stations are arranged in a sequential fashion, at each of which the operator performs a few selected operations on each product unit moved past his station. Finished products emerge from the end of the assembly lines and are then transferred to finished goods inventory to await distribution. Again, if we observed long enough, we would see one line discontinue the production of one particular product configuration, and, after adjustment, begin assembling another product.

The scene described above is repeated thousands of times in the modern industrial world. The described activities are strikingly similar, no matter what the final product might be. Such manufacturing systems are usually difficult to control. The magnitude of the control problem can best be appreciated if we imagine ourselves suddenly thrust into the position of managing a manufacturing plant. Here are many of the decisions that would have to be made and remade through sequential time periods:

1. The number of each finished product to be made tomorrow, next week, next month. (Long range forecasting is a planning decision. See number 19, below.)
2. Which components to fabricate during this period and their batch sizes.
3. How best to use available equipment and facility capacities.
4. What inventory levels to maintain in each of the three storage categories.
5. Proper work force levels: number and types of employees, number of shifts, amount of overtime, use of subcontract support.
6. When to release materials, tools, and instructions to the shop floor.
7. How to handle rush orders and other unexpected events.
8. How to reschedule work on a dynamic basis in response to delays and other internal factors.

In addition to the above operating decisions, there are several planning decisions that must be made on an irregular basis, among which are:

9. What the product(s) will be: size, style, cost, quality, etc.
10. What materials to use in the product: use of standard components where possible, selection of least expensive material that will perform the required functions.
11. What equipment to purchase; when to replace or expand current equipment; equipment arrangement.
12. What manufacturing operations to perform on each component.
13. The sequence in which to perform the operations (technological ordering).
14. What time and cost standards are applicable.

15. Standard batch sizes for manufactured components; order quantities for raw materials and purchased parts; safety stock; reorder levels.
16. When to schedule maintenance operations.
17. When to modify or phase out a product.
18. When to introduce a new product.
19. Over-all level of operation for the planning period.

Other decisions will become apparent as we proceed further into the problems associated with manufacturing systems. This list is sufficient, however, to demonstrate the complexity of controlling such systems. The primary function of operations planning and control systems design is to provide a sound framework within which the above decisions can be satisfactorily made.

A control mechanism to aid the management in making these decisions must be developed. The basic ingredients of such a system are:

1. A current *operation plan* to specify the desired performance of the manufacturing system in terms of meaningful effectiveness criteria.
2. A *data acquisition system* to determine on a timely and accurate basis the current status of the manufacturing system.
3. An *effectiveness comparer* to compare meaningful measures of effectiveness with the effectiveness criteria.
4. A *parameter adjuster* to determine a new operation plan for the following period.

Such a control system is illustrated in Fig. 1.3. Note that the system must respond adequately to external disturbances as well as to internal events. Note also that the "operations control loop" is usually performed on a short-term periodic basis, whereas the "planning control loop" is performed on an irregular, longer-range basis.

An analogy can be drawn between such a control system and a temperature control system for a building, in which the thermostat acts as the effectiveness comparer. In both systems, we want to maintain performance within certain limits. That is, we do not want to overreact to insignificant fluctuations in system performance or to underreact to significant fluctuations.

Another useful analogy to our kind of control system is that of guiding a missile to its target. Suppose a missile is to be launched from point A to a target at point B, as shown in Fig. 1.4. A desired trajectory (shown by the solid line) is determined. As the missile progresses, it does not follow the prescribed trajectory precisely. When its path (dashed line) has deviated *significantly* from the desired path (such as at point C), a new path to the target is determined. Notice that we do not attempt to return to the originally prescribed path.

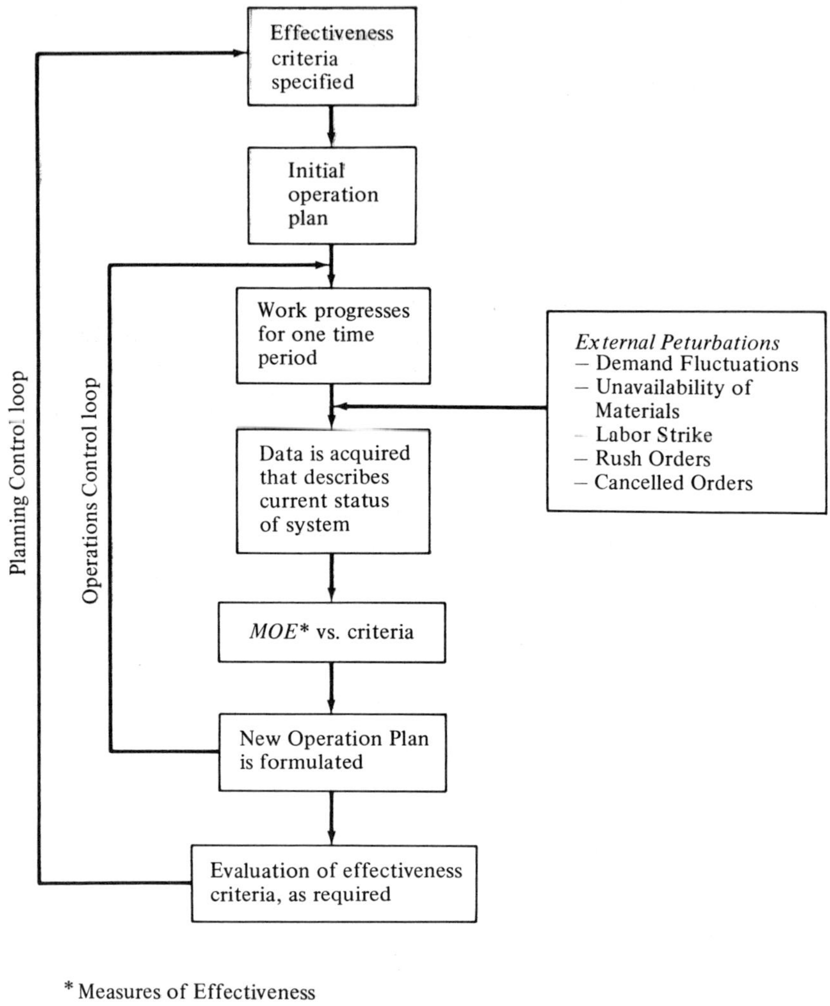

Figure 1.3 An idealized Control System for Manufacturing

Rather, we calculate the best path from the current point. As the missile proceeds on the new path, it may again deviate significantly, such as at point *D*. Again, a new path to the target from point *D* is determined. This correction process is continued until the missile "converges" on a point very near the target. Rarely will it hit the target precisely.

We must do the same sort of thing with our manufacturing system. We first determine the best operation plan, based upon expected occurrences. As production progresses, deviations from the original plan occur because of unex-

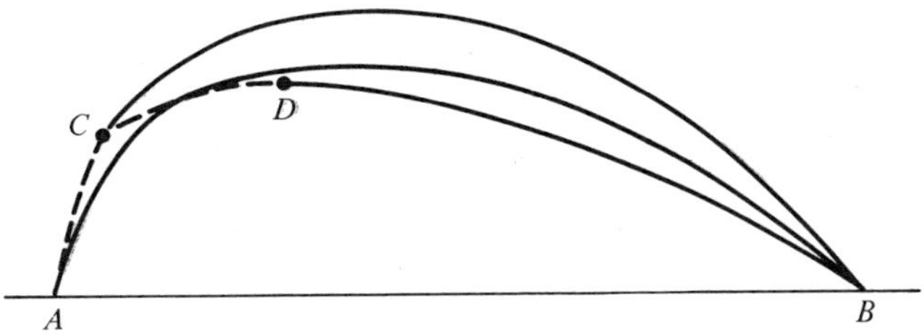

Figure 1.4 Analogy of Control System of a Guided Missile

pected occurrences. We must detect such deviations, measure them, and then formulate a revised operation plan.

Functions of Operations Planning and Control

We are interested in learning to *design* planning and control systems that will facilitate (1) the construction of the operations plan, thereby accounting for the "planning" decisions that have to be made and, (2) the evaluation and control of operations, thereby accounting for the "operating" decisions.

There are several elements of planning and control systems. We might think of these elements as subsystems, or simply as functions of operations planning and control. We briefly mention and describe these functions here and note that later an entire chapter is devoted to an in-depth analysis of each function (Chaps. 3 through 7). Each function is related to one or more of the 19 decisions discussed in the preceding section.

FORECASTING

Forecasting is an estimation of *how many* and *what* will be demanded. It is also concerned with *when* the demand will occur. This function of a control system relates to decisions 1 and 19.

OPERATIONS PLANNING

Operations planning is concerned with the allocation of available productive resources to the production requirements as determined by demand forecasting. This function relates to decisions 3, 5, and 9 through 18.

INVENTORY PLANNING AND CONTROL

Inventory planning involves the specification of inventory system procedures, such as the type of ordering system to be used, the amount to order each time, permissible safety stocks, etc. Inventory control involves the operation of the inventory system, such as the placement of production orders and purchase orders. Inventory control must be considered at three levels: raw materials, purchased parts, and finished goods. Proper consideration must also be given to product distribution. This function relates to decisions 4 and 15.

OPERATIONS SCHEDULING

Operations scheduling assigns specific work activities to specific work stations with specific start and stop times. This function relates to decisions 2 and 8.

DISPATCHING AND PROGRESS CONTROL

Dispatching is the action function of operations planning and control; it releases work to the operating divisions, provides for material and tool releases, and conveys instructions to the shop floor. This function relates to decision 6. Progress control is the feedback and corrective action function, in that it provides a means of measuring actual performance and comparing it with intended performance. Progress control determines whether there is a need for replanning. This function relates to decision 8.

INTEGRATING THE FUNCTIONS

There is, of course, considerable overlap among these five functions. The performance of certain functions depends on the previous performance of other functions, which may depend on still others. It is usually necessary, therefore, to perform certain functions essentially simultaneously.

Certain of these functions have received research attention in recent years, and analytical tools have been developed for their performance. Little progress has been made, however, in developing quantitative models for optimizing the performance of the total production system. In lieu of quantitative models, we currently must rely on logical decision rules and sensible procedures for integrating the several individual functions into a working operations planning and control system. Much of this book is devoted to the design of *integrated* planning and control systems.

Types of Production

The final configuration of any control system we design depends upon the type of production in which the particular company is engaged. Production systems are commonly divided into two broad classes: *make-to-stock* and *make-to-order*. The distinction is based upon the degree of certainty of knowledge of demand and the length of a production run. It should be noted, however, that a particular company could engage in both types of production.

An example of make-to-stock production is the manufacture of men's sport shirts. The manufacturer will produce hundreds of shirts of a certain size, style, and color on the faith that the shirts will be sold. Seasonality is, of course, a factor here. An example of make-to-order production is the manufacture of nuclear powered submarines. Regular, continuous demand cannot be assured. Furthermore, each submarine has particular requirements.

Certain characteristics distinguish the two types of production. Make-to-stock production is normally employed for standard products with a high demand volume. Other characteristics of make-to-stock production are fixed routing and non-flexible materials handling, high equipment investment, relatively low in-process and raw material inventories, and special purpose equipment. Make-to-order production has correspondingly opposite characteristics. Perhaps the most important distinction is that planning of work is relatively easy for make-to-order production, while control of work is relatively easy for make-to-stock production. Note that each characteristic of one type of production is expressed relative to the other. For example, the equipment investment may be high for both types, but it is relatively higher for the continuous type.

In general, make-to-stock production is more economical when the two necessary conditions, high volume and standardization, are met. In make-to-stock production, we attempt to forecast customer demand as accurately as possible, since we are producing in anticipation of sales rather than for a firm order. In make-to-order production, we either know in advance the precise demand, or we must be ready to accept on-the-spot orders, which are impossible to forecast and for which the customer is willing to wait a long lag time.

There are several other ways of classifying types of production. (Timms 1962, Chap. 7) for example, states two broad classifications: (1) production to order and (2) production to stock. He then specifies three subclasses within each of the two broad classifications: (1) jobbing production, (2) continuous production, and (3) intermittent production. Other writers present still other classifications. In any case, the important distinctions are how well the demand is known before the fact and how long the production runs will be.

The several functions of operations planning and control have varying degrees of pertinence to the two types of production. While it can be dangerous to generalize, and while exceptions can surely be found, the primary differences are summarized in Table 1.1. We should not overlook the importance of attempting to perform all the functions for each type of production, even though certain functions are not extremely pertinent.

Another useful classification of manufacturing industries (IBM 1960, p. 3) is:

1. *Basic Producer*—uses natural resources (ores, wood, rubber, etc.) to produce materials for other manufacturers.

2. *Converter*—changes the products of the basic producer into a variety of industrial and consumer products (metal plates, paper, plywood, etc.).

3. *Fabricator*—transforms the products of the converter into a larger variety of products (fasteners, paper goods, etc.).

Table. 1.1 Pertinence of the Functions of Operations Planning and Control to the two major types of Production

Function	Pertinence to Make-to-Stock (I)	Pertinence to Make-to-Order (II)	Discussion
Forecasting	Extreme	Little	I. Routings are fixed. Little flexibility in process. High equipment investment. Rigid operations plan. II. Usually not possible to forecast. More responsive to sudden change.
Operations Planning	Extreme	Moderate	I. With this function, much of the control is built into the production process. II. Little detailed planning is possible.
Inventory Planning and Control	Extreme	Moderate	I. A highly structured process depends upon availability of materials, components, and assemblies. II. Must maintain more kinds of materials because of uncertainty of orders. Little concern with finished goods inventory.
Operations Scheduling	Moderate	Extreme	I. Performed in operations planning function. II. Must continually fit new orders into existing workloads.
Dispatching and Progress Control	Moderate	Extreme	I. Little dispatching. Only exceptions important in progress control. II. Close control of highly variable processes is needed.

4. *Standard Assembler*—brings together the products of fabricators into finished products (washing machines, electrical appliances, etc.).
5. *Standard Assembler with Options*—assembles products according to customer specifications (automobiles, computers, commercial airplanes).
6. *Customer-Specified Assembler*—manufactures products according to the exact specifications and requirements of the customer (nuclear submarines, buildings, bridges, etc.).

The most common type of manufacturing plant is a combination of standard assembler with options, and fabricator. Most of our developments, therefore, pertain to such a company. Fabrication is regarded as similar to a "job-shop" operation, in which general purpose machines are grouped in work centers by machine type. Parts are fabricated according to their own particular routings and enter an intermediate storage stage. They are then extracted as needed by the assembly process.

Inventory Considerations

We have discussed in a general way the manner in which inventory is a necessary part of an operations planning and control system. Additional discussion is appropriate, however, because of the vital role inventory plays in production operations.

Inventories must be maintained whenever there is a lack of synchronization between the following pairs of factors:

1. The raw material supplier and our production process.
2. Consecutive stages within the production process.
3. The production process and demand for the product.

The three inventories needed are analyzed as an input-output process in Fig. 1.5. To demonstrate this concept further, suppose a certain assembly proceeds from Stage 10 to Stage 11 in a process. Suppose further that the processing rate of each stage is an independent random variable (to preclude the possibility of synchronization between two stages). If we begin each stage idle and empty, a possible result of 10 consecutive hours of operation is show in Table 1.2. We assume that the total production of Stage 10 for 1 hour is available to Stage 11 at the beginning of the following hour. Obviously, Stage 11 can process only those assemblies supplied to it by Stage 10.

Note that negative inventory indicates idle time for stage 11. This time is productive capacity lost forever. Note also that both positive and negative inventories result, even though the average hourly productions for the two stages are identical. This example shows why inventory is necessary to avoid

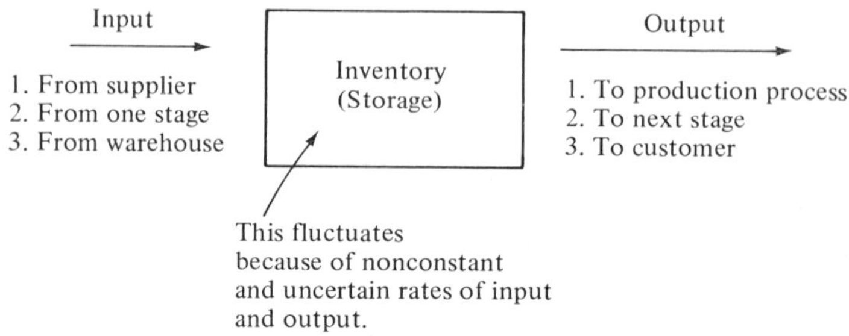

Figure 1.5 The three types of Inventory, represented as an Input/Output Process

expensive idle time. It also demonstrates the importance of the inventory function in a control system for a manufacturing process. Finally, we observe that a similar phenomenon can occur between any two consecutive components of the system.

Table 1.2 Inventory resulting from unsynchronized Production Stages

Hour (I)	Stage 10 Production $P10(I)$	Stage 11 Production $P11(I)$	Surplus (Inventory) $S(I)$
1	3	—	0
2	4	4	−1
3	2	3	1
4	4	3	0
5	3	4	0
6	4	2	1
7	2	3	2
8	3	3	1
9	2	3	1
10	4	4	−1
		2	2

In Table 1.2, the $P10(I)$s are random production rates for Stage 10. The $P11(I)$s are random production rates for Stage 11. Surplus inventory is calculated as follows:

$$S(I) = P10(I - 1) - P11(I) + \max[0, S(I - 1)]$$

Examples:

$S(1) = 0$ (initial starting conditions)
$S(2) = 3 - 4 + 0 = -1$
$S(3) = 4 - 3 + 0 = 1$
$S(4) = 2 - 3 + 1 = 0$

Even if synchronization of the above factors was possible, we still might wish to maintain inventories for purely economic reasons. For example, suppose our raw material supplier was located such that he could respond almost instanteously to an order. We could then conceivably order one unit of raw material at a time as our production process needed it. In addition to being a lot of bother, such a policy would result in a relatively large cost of placing orders, even though we would have eliminated all costs of storing the material.

Another reason to maintain inventory is that certain processes, such as the manufacture of alcoholic spirits, involve aging the product. Even though the aging process can be considered simply as another operation, the product is idle and as such possesses all the characteristics and problems of an inventory.

We maintain inventory, then, for the following reasons:

1. To compensate for lack of synchronization discussed above.
2. Economic considerations: quantity discounts, economical order quantities, production smoothing, price speculation.
3. Process aging.

We discuss specific methods of analyzing inventory problems in Chap. 5.

The Life Cycle of a Product Line

No product lasts forever. Eventually, all products must either be phased out or modified. Our operations planning and control system must be able to accommodate such changes, as well as completely new products. We must consider the effects of such changes on plant capacity, manufacturing processes, labor mix, etc. When new or modified products are introduced, the accompanying production data must be inserted into the master information system of the plant. This topic is discussed in greater depth in Chap. 3. We want to emphasize here, however, that it is as important to phase out a product in the most economical manner as it is to manufacture it economically at the peak of its life demand cycle.

Orientation of Book

In an earlier section, the book was said to focus on control systems for the type of plant that can be described as a fabricator and standard assembler with options. It should be understood, however, that the concepts we use are broadly applicable and, with little effort, can be extended to other activities. In particular, the words *operation* and *manufacturing* can have much broader interpretations than they are commonly given. Most of the concepts and methods presented in the book are certainly applicable to such military activities as engine repair and depot operations. They are also applicable to distribution systems, large-scale information processing systems (such as insurance companies), and many others. The fact remains, however, that the terminology and general orientation used is more closely associated with manufacturing operations, in the traditional meaning of the term.

This book is oriented toward the design rather than the operation of operations planning and control systems. As such, the book emphasizes design principles rather than case studies of existing control systems. As we shall quickly see in Chap. 2, the design approach taken in this book is that of first discovering the decision structure underlying a manufacturing environment. Only then do we concern ourselves with the information system and specific techniques of performing basic functions.

In addition, this book emphasizes primarily the plant. That is, the marketing, sales, and financial components of the total system are not given primary attention, although certain interfaces with these components are necessary occasionally.

Finally, the concepts are presented from the point of view of minimizing cost rather than of maximizing profit. The reason for this approach is that we can measure cost much more accurately, albeit not perfectly, than we can measure profit, especially at the subsystem level. For example, when we attempt to determine the best inventory policy, we have reasonably good measures of the cost components, but it would be very difficult to express the needed relationships in terms of profit.

We should say something more about the criterion of minimizing costs and the degree to which this result can be assured for the total operations planning and control system. It would be extremely misleading to imply that equations for all relationships can be stated, the derivatives found, set to zero, and the variables of interest solved for. This operation can be done for certain subsystems, such as optimum batch size and order quantities. For the total system, however, so many mathematically complex relationships exist that we simply have no guaranteed optimizing procedure. The best we can hope to do is to use those analytical tools available for optimizing the individual functions and then to use logical decision rules, tested by simulation studies, to obtain "good" performance from the other functions.

An Introductory Example*

To assist us in grasping the magnitude of operations planning and control systems, and later (Chap. 8) to furnish an illustrative design example, we here

*This case study draws upon the collective experience of the authors and intentionally represents a real-world sort of situation. Any resemblance, however, to a single real company is purely coincidental.

describe a quite typical, although fictional, company as a systems designer might encounter it in practice. The information presented here is typical of what might be distilled from several hours of conversation with the primary executive of such a company. As such, the information is primarily situational rather than quantitative and operational. As we will see in Chap. 8, the actual design activity requires much more detailed information about the company.

THE COMPANY

Our company, VAST, Inc.,* was founded in 1884 by an immigrant German toolmaker, Hans Schmidt. It had its origin in Toledo, Ohio, in a small plant along the Maumee River. The initial products were a line of machine tools. Because of the high technical skill of Mr. Schmidt and other family members who soon joined him, these products found ready sale to other industries then emerging in the area.

From the first, the company depended greatly upon innovation and technological advancement, and was one of the first to start a formal research program. In the early 1900's, the company started a program of diversification of product. Some of this diversification was accomplished through acquisition, but much of it was done by development of new processes and products as a result of the research program. Except for a minor set-back after World War I and during the Great Depression in the early 1930's, the company grew steadily. Throughout this time, the company pursued a vigorous research program, a policy which was instrumental in its post-Depression and post-World War II recovery and growth.

By 1965, VAST, Inc., had become one of the largest companies in the country. With sales of over a billion dollars per year, gained from a very wide variety of products, VAST was a major force in our economy. On the stock market, it was classified as a "Blue Chip," but was considered still to have growth potential. The company continued to pursue an aggressive research program, allocating approximately five percent of gross sales to this program. It was difficult for the average American to go through one day without encountering a VAST product in some form.

In 1965, the organization of the company and its philosophy of management could be described as "decentralized." For many years, management's principal criterion had been return on investment. This emphasis meant, in general, that each subordinate manager, including individual plant managers, were quite free to run their own show as long as they produced an adequate return on the investment in facilities and resources charged to them.

The company continued to maintain its headquarters, with some 10,000 employees, in Toledo. These personnel were primarily managerial, technical, and

*The current name, adopted in 1954, is an acronym for "Vereinigte Allgemeine Schmidt Technischegesellschaften," which translated means "United General Schmidt Technical Companies."

support personnel. A few of the company's 118 plants were in the Toledo area, but the balance were well spread out over the United States. In addition, VAST owned several foreign subsidiaries outright and shared ownership in several others. Sales offices, warehouses, and distribution centers could be found in almost all principal cities.

Below the usual president–board of director's level, the company was organized into 15 production departments, each with a general manager responsible to the president and to the board. Each of these production departments had full responsibility for a broad class of products, sales, accounting (with exceptions and some company constraints), research and development, and production. In addition to these 15 production departments were several service departments serving the company as a whole, such as the comptroller's department, the personnel and employee benefits department, the central service department, and the engineering department. The latter had responsibility for design and construction of new facilities, for new process development, and for furnishing in-house consulting services to the several production and service departments. *We will imagine ourselves as one of these in-house consultants, specializing in management information-decision systems design.*

THE PRODUCTION DEPARTMENT

The production department with which we will be concerned is the plastics fabrication division. This department, organized just after World War II, was a direct outcome of a new product line development arising from the basic research of another production department. The company had acquired some similar products by purchase of foreign patents in the 1930's, so that in addition to three plants that had been built since 1948, two older plants had been transferred from the control of another production department in order to place all plastics fabrication in one department.

The five plants produced four broad product lines closely related in end use and therefore in marketing but differing greatly in production methods and range of individual product diversification. The three new plants employed a newly developed and highly efficient plastic molding process, whereas the two older plants employed two different older processes generally regarded as inferior.

The plant at Akron, Ohio, produced plastic parts and assemblies by a process in which a solvent was used to dissolve the basic plastic. The process then required the evaporation of the solvent during that phase in which the piece-parts were formed. The assembly of these piece-parts was complicated by problems of solvent remaining in the parts for as long as several days. This plant was the oldest of the five and had by far the largest number of final end items.

The process employed at the Indianapolis, Indiana, plant was based upon a different process, wherein the plastic parts were formed by impact, the raw materials being in granular or dust form. The process was quite complicated, and it was felt that this type of fabrication would become obsolete within ten years. The long-range plan for this plant called for its phasing out over a five-year period and for conversion of the plant to the more modern process used in the three new plants.

Both of these older plants depended largely on companies other than VAST for their raw materials, per agreements made when the basic patents were purchased. Both, furthermore, had a strong technological tie with the production department with which they had formerly been affiliated and had also retained much of the organizational structure and operational philosophy of the older department.

The three newer plants all depended on a common basic technology, the results of basic research in the late 1930's. World War II found the company engaged in production for defense, which precluded exploitation of this new technology on a commercial basis until after the war (although pilot plant production during the war had been used for some very vital military ends.) The basic process employed in these plants made use of thermo-plastic raw materials. The preponderance of these raw materials was produced by another production department and internally transferred to the plastics fabrication division at a competitive price. A few raw materials were purchased from outside vendors.

The three plants had been built in 1948, 1959, and 1963. While all used the same fundamental process and served the same markets, there were substantial differences between the three. The 1948 plant, located in Louisville, Kentucky, in particular, had smaller equipment and more product diversification than the other two. The plant built in 1959 was located in Syracuse, New York, while the 1963 addition was a plant in Memphis, Tennessee.

Process variations between these two plants were relatively small. However, the Memphis plant had two basic product lines, the second of which was primarily for heavy industrial end uses. The Louisville plant also produced this industrial line but accounted for only about 15 percent of the volume.

All five plants had relatively common markets and shipped to a wide variety of customers. The product of the two older plants had found many end uses and was less expensive. The newer thermoplastic process had a higher cost, but the excellence of the product was such that demand was very great. The three new plants were less diversified and operated at or near capacity much of the time, while the older plants generally were quite diversified and seldom operated at over 90 capacity.

At the Toledo (corporate) level, there were two directors of production, one in charge of the two older plants and one in charge of the three newer ones. The former was designated "Director of Particulate Production," the latter, "Director of Thermo-Plastic Production." Both reported directly to the general manager of the plastics fabrication division. Each had a separate organization, including a production planning and control group.

Because of the commonality of markets and end use, the division had a single director of sales, who also reported to the general manager. His sales and marketing activities were carried on primarily through eleven district sales offices located throughout the country. Nine of these sales offices had clearly defined geographical boundaries. The sales office in Detroit, however, confined its activities to automotive sales, while the Toledo office handled only export sales. The other nine offices were located in New York City, Philadelphia, Richmond, Atlanta, Chicago, Dallas, Los Angeles, Denver, and Spokane.

Each of these sales offices, except for the Detroit office, was responsible for sales of products of all plants. However, each district sales office had three divisions, one concerned with solvent-type sales from the Akron plant, one with impact-type sales from the Indianapolis plant, and a third with thermo-plastic sales from the three newer plants. These three divisions shared physical facilities, communications, and certain other functions and were headed by a single manager. However, procedures differed in detail between them, and each group had developed certain informal channels of communication to their corresponding groups in Toledo and to their producing plant.

All the production of the solvent and impact plants, and preponderance of the production of the theromplastic plants, was shipped directly to customers by motor freight. The Los Angeles and Spokane districts did have small warehouses, where a nominal stock of the more common thermoplastic items was held because of the long shipping distances involved. The plants warehoused extensively, carrying in general about three weeks of sales' demand as an inventory level. Production was geared both for inventory and to customer order. Trade practice was such that a customer might demand immediate shipment or call for a shipment scheduled over several weeks' time, with up to about three weeks' notice before the first shipment was due. Production times for orders ranged from two to four weeks depending on the process and product involved.

The competitive situation was somewhat varied. Both of the two older product lines had several direct competitors. These competitors were old enough to have become well established, and those of them with newer plants had a competitive advantage because of lower production costs.

The thermo-plastic product lines were in a period of transition in competition. While they had always had some competition from similar products produced by other processes, they had for years enjoyed a very marked advantage based on the physical properties of their product. In 1960, in an anti-trust action, the Federal government forced VAST to license their thermo-plastic patents to "any legitimate producer." In 1962, a well-funded smaller company had started up a plant in direct competition with VAST and by 1965 had encroached heavily upon the market. While all three VAST plants were operating at or near capacity, management felt a strong need to prepare for even more vigorous competition within a very short period of time. They were particularly concerned with becoming more responsive to market trends and to the demands of individual customers.

OPERATIONS PLANNING AND CONTROL

While the details of the then existing operations planning and control system are given in Chap, 8, we summarize the organization and general function of the existing system here and give some approximate data to indicate the magnitude of the situation.

Primary responsibility for the planning of operations rested with the main office in Toledo. Each director of production had a group charged with this responsibility. In general, they obtained each month, from sales, a forecast by end item of the demand expected for each product line. Based on this forecast, their knowledge of plant capacity, projected inventories, etc., these groups sent to each plant a statement of specific production requirements for the month. The plant production control section (under the manufacturing superintendent) was then expected to procure material through plant purchasing (under the accounting superintendent) and see that production requirements were met. The Toledo office also frequently changed these requirements, according to short-term sales demands. The actual sales orders were given to the plants by the district sales offices after they obtained clearance to do so from a sales scheduling group under the director of sales in Toledo. Orders were received by mail at the plant in the order section (under the accounting superintendent). The primary short-range criterion of plant performance was daily production, in pounds and unit cost per pound shipped.

The magnitude, and to some extent the scope, of planning varied considerably between plants. The oldest plant, at Akron, produced over 4,000 different items and used 120 different raw materials. In the four main steps of their process, some 1,400 individual pieces of equipment or assembly work stations were involved. About 75 orders per day were received, which resulted in about 180 shipments each day. Plant production, in pounds, was the smallest of all five plants, however, at 1.5 million per month.

At the other extreme, the newest plant, in Memphis, was much less complex in these respects, even though it produced two basic product lines. Only 12 basic raw materials were involved in producing about 150 end items. The three steps of their process included fewer than 400 pieces of equipment and work stations. Their 50 orders per day resulted in about 70 shipments per day, but plant production approached 7.0 million pounds per month.

The other three plants ranged between these two. Each had some unique situations, which would need to be taken into account in designing an over-all system for the division.

THE SITUATION

The situation relative to operations planning and control consequently varied according to the product line and plant involved. However, each situation had developed over a period of time as a natural result of changes in markets, processes, and personnel. Management had a growing concern over what they felt to be a lack of control manifested in high and unbalanced inventories, high plant costs, high rates of machine downtime, and poor shipping performance.

In the first section of this chapter, we discussed 19 planning and operating decisions that must be made almost continuously in a manufacturing operation. It would be well for the reader to review these decisions and relate them to the situation described for VAST, Inc. Also, review the section, Functions of Operations Planning and Control, and relate the five functions to the hypothetical company described.

Imagine that we, as systems designers, have been asked to help solve the problems of lack of control, high and unbalanced inventories, high plant costs, high rates of machine downtime, and poor shipping performance. However, we must defer solving these problems and designing a new system until we have learned a great deal more about the concepts and techniques of operations planning and control systems design. We shall return to our problem in VAST, Inc., in Chap. 8, where we will see how the concepts developed in the text can be applied to a realistic design situation.

REFERENCES

IBM Reference Manual E20-8041, 1960, *General Information Manual: Management Operating System for Manufacturing Industries*. White Plains, New York.

Timms, H. L., 1962. *The Production Function in Business*. Homewood, Ill.: Richard D. Irwin, Inc.

2 Information-Decision Structure and Processes

The purpose of operations planning and control systems is to assist management in making decisions. All decisions require information. The principal purpose of information is for use in arriving at, implementing, and evaluating decisions. It is clear that information and decision are topics that should not be discussed separately; thus, the title of this chapter.

While most everyone agrees that information and decisions are very closely related, there appears to be a disagreement as to which should be considered first in designing operations planning and control systems. The authors of this text are in an apparent minority who believes that the *decision structure* of an organization should first be specified as the basis for the control system design and that the *information system* should then be designed to meet the requirements of the decision structure.

All too often, systems analysts and designers take an approach similar to the following:

1. Survey different levels of management as to their information needs or, more often, their information "wants."
2. Design a reporting system that often has the following characteristics:
    a. Produces far more information than can possibly be digested and much of which is not really pertinent to the decision to be made.
    b. Much of the information must be consolidated and processed further by the recipient.
    c. Relationships between decision points and mutual information needs are not provided for.

3. Design the system according to the *hardware* with which the designer is familiar.

It seems to us far more logical to begin the system study by uncovering the decision structure of the organization. What decisions are required? How are they related? What information is needed for each decision? Only when these questions are fully answered is it meaningful to consider specific content of management reports. That the authors are not alone in this orientation is evidenced by the very fine article by Dr. Russell Ackoff, "Management Misinformation Systems" (1967, pp. B-147-156). Others are also beginning to recognize the importance of the decision structure.

Decision Principles

The principal function of an operations planning and control system is to make decisions. These decisions in turn profoundly influence the extent to which the production system meets its objective of producing goods or services to meet customer's needs while furnishing a profit to the enterprise. It is therefore critical that the nature of decisions be explored and that the decision structure of an operations planning and control system be specified.

It is first necessary to define the term *decision* very carefully. Dictionary definitions are not very helpful for our purposes since they emphasize such terms as "verdict," "judgment," and "conclusion." We will use the following more rigorous definition of a decision:

Given that there exists a situation such that two or more distinguishable courses of action (alternatives) are possible, a decision consists of selecting one of these courses of action to the exclusion of the other or others.

With this definition, we can proceed to examine the entire decision-making process, which can be conceived of as consisting of five steps:

1. Collection of data
2. Establishment of alternatives
3. Assignment of measures of utility to each alternative in view of some criterion of success
4. Selection of an alternative (decision)
5. Implementation of the chosen alternative.

This process is entirely general and may be discerned in operating decisions, design decisions, and, for that matter, in personal decisions. To illustrate the process, consider the situation confronting a person about to leave his home for work on a day in April. He wishes to determine what, if any, outer garment he should wear.

Collection of Data. He looks out the window and finds that the sun is shining but through scattered clouds. His radio informs him that the current temperature is 43°, with a predicted high of 69°. The weather bureau cites 0.40 probability of rain. He knows that he will drive to and from work and that he has a walk of about two blocks between his parking place and his office. He does not own an umbrella.

Establishment of Alternatives. Our subject's alternatives, based on garments available, are (1) wear a topcoat (2) wear a raincoat and (3) wear no coat.

Assignment of Measures of Utility to Each Alternative in View of Some Criterion of Success. Our subject's criterion of success in this case would be personal comfort, a subjective measure. He would then determine the utility of his personal comfort for each alternative, in this example, on a somewhat intuitive basis.

Selection of an Alternative. Assuming that our work-bound man has assigned measures of utility to each alternative in such a manner that one alternative has a higher utility measure than either of the others, the decision, or selection of an alternate, is made in favor of the alternate bearing this higher utility measure. If two alternatives have the same measure of utility, and it is greater than that of the third, then some random method of selection must be employed. In this case, the flip of a fair coin would be suitable.

Implementation of the Chosen Alternative. In our example, the implementation is simple. Our man removes the chosen apparel from his closet, or if he has selected the alternative of no coat, merely goes on to work.

Viewing the decision process in this manner, we can now establish some broad principles useful in the design of the decision structure of a planning and control system. While some of these principles may seem obvious or trivial, they are often violated in the design of such systems.

Principle 1. Data is a necessary basis for decision. Without some data, it is not possible either to establish alternatives or to assign utility measures to the alternatives.

Principle 2. The data collected must be of two types, that which serves to establish the alternatives and that which may be used to establish utility measures.

Principle 3. The data collected must be directly relevant, or such that it can be made relevant by some data transformation, to the criterion of success that will be used. For example, if a cost criterion is to be used, only cost data or data that can be transformed to a cost basis is useful.

Principle 4. Assuming that alternatives have been established and measures of uitility assigned, additional data is of no use unless it affects the already assigned utilities. Note that this principle is somewhat contradictory to the idea commonly expressed by decision makers, "If I had more data, my decision would be easier." What the decision maker really needs in most cases is not more data but better methods of assigning measures of utility to the available alternatives.

Principle 5. Data collections must be completed prior to establishment of alternatives and assignment of utilities. While this principle appears to be obvious, it underlies the so frequently heard demand for timely data collection. It has therefore been the motivating principle behind the installation of many elaborate and sophisticated data collection and transmission systems.

Principle 6. The requirement for accuracy of data is a function of the techniques used to assign measures of utilities to alternatives, of the relative weight given the data in the utility measurement, and of the relative utilities assigned to the alternatives. This principle underlies sensitivity analysis of mathematical models. If a given model is relatively insensitive to a given parameter, or if the given parameter is weighted lightly, the demand for accuracy is lessened. Also, if two alternatives have (preliminary) assignments of utility of, for example, 0.95 and 0.05, the need for accuracy of data is low, and very rough estimates may suffice.

The observation of this principle in systems design frequently results in substantial savings. For example, verification in a key-punching operation may be unnecessary in many cases if it can be established that an occasional error will not affect the ultimate decisions.

Principle 7. Assuming that the five steps of the decision process are well defined and routinized for a given decision that must be made on a repetitive basis, then this entire decision process may be delegated to a lower organizational level or programmed to be performed by mechanical or electronic equipment. Note that in every such case, a design or management decision has to be made. The alternatives are to retain the decision process to the manager or to delegate the process to a subordinate or a machine. In making the decision, the system designer or manager must assign measures of utility to the alternatives. Then, the decision to delegate (or program) the decision process in question implies that the utility of the delegate or the machine making an acceptable decision is high.

This principle also underlies the so-called management by exception principle. The implication is that in every routine, delegated, or programmed decision process one of the alternatives is to refer the matter to higher (or human) authority. Then, when all the other alternatives have low utility measures, the referral alternative will have high utility and therefore be selected.

Principle 8. By elimination of low utility alternatives, most decision situations can be reduced to a selection between two alternatives or to a series of such selections. The implications of this principle of reduction to a binary process are important in simplifying and routinizing decision processes.

Principle 9. The making of a given high-level decision usually implies, or sets in motion, a large set of subordinate operations, which may in turn involve decisions but which in many cases are at best implementation steps involving no real decision as herein defined.

With these principles in mind, we can now examine the existing general decision structure which plans and controls production operations.

Generality of Management Decison Processes

While the details of each manufacturing or service enterprise vary greatly, the type of decisions involved and their bases are relatively common to all such enterprises. The reason for this commonality may be ascribed to a commonality of purpose or goal. That is, the *criteria of success by which alternatives are appraised* are relatively similar.

Profit is often viewed as the principal criterion of success. There is some validity, however, to the arguments of Drucker (1954, p. 37) and Reinfeld (1959, p. 13), which define the objective of an enterprise as "the making and holding of a customer." Profit is then viewed as a measure of the extent to which the enterprise is successful in meeting this objective.

Regardless of the specific point of view held, there is no doubt that the criteria used in every decision *must be related* to the ultimate objective of making and holding customers and to the attendant prospect of profit. Even where it seems to be the intent of a given company to forego profits in the short run, one must assume that they are doing so to maximize profit in the long run. For example, it must be assumed that the decision of railroads to discourage passenger traffic must be ultimately aimed at diverting resources to freight service, with an overall increase in profit.

It is this commonality of criteria, then, which allows us to generalize a decision structure. It is critical to note that *decisions* and *functions* are being generalized, not organizational structure or organizational activities. The view held here is that common decisions and functions exist, regardless of the title of the person or group that performs them in a given company.

A second point of commonality is that, in general, the decisions to be made involve commitment of available resources. Management has at its command resources of physical facilities, manpower, materials, and money. Money, within limits, may be converted to physical facilities, manpower, and materials. Because of this fact, it is generally believed that maximum flexibility of action can be maintained by making this money-to-other-resource conversion at the latest feasible time prior to the use of this other resource. Therefore, it is necessary to carefully control the cash flow within an enterprise.

A third common characteristic of management decisions is that they may be logically divided into three phases. The first phase determines *whether or not* some action should be taken. The second phase involves some decision relative to *quantity* or *extent*, while the third phase concerns *temporal relations*, or *timing*. For example, consider the decision to be made in the face of a forecasted decline in sales. First, should a lay off of workers take place? If so, how many should be layed off, and when should the lay off take place? For another example, the decision to buy supplies proceeds in the sequence: Do we need to buy? What quantity should be bought? When should they be bought?

Integrating several preceding notions then results in the following generalization: Management decisions are made relative to the commitment of resources and must first determine whether the commitment of resources is to be made. If resources are to be committed, further decisions must be made relative to the amount of resources to be committed and the timing. These decisions have the common criterion of profit maximization and the underlying philosophy of preserving flexibility of operation by deferring conversion of cash to more rigid resources as long as possible.

It should be noted that the first decision phase, that of whether to commit resources, may be made at a higher level and much less frequently than the other two phases. The general manager of a company may, for example, make the decision to devote a given plant to the manufacture of a specific product. By this decision, he implies that any reasonable customer demand for the product, within plant capacity, will result in the further repetitive commitment of labor, materials, and money. He usually also delegates, by implication, decisions relative to amount and timing of resource commitment to a subordinate decision level.

The foregoing is an extremely important point in the design of planning and control systems. The erroneous impression is often gained by the engineer designing the system that a given person or function makes a decision as to whether to commit recources, when in reality the decision concerns only amount and timing, or even just timing. For example, a production control dispatcher may say he makes the complete decision to release material from the stock room to production. Such is not the case; the decision to commit resources was made at the time it was decided to make the product. The decision of amount to be committed was made when the decision relative to amount of finished product to be made was reached. Only the timing decision is up to the dispatcher, and then only if the scheduler did not imply this decision when he scheduled the product into production.

The realization of this fact can lead to extensive system simplification. So-called materials authorizations, move authorizations, etc., can be seen as merely implementation devices and frequently eliminated or drastically simplified.

Throughout the design of the planning and control system, the engineer should characterize the decision phase required (commitment, quantity, or time). He should then establish whether a higher level decision, usually of the commitment phase, precludes the necessity of further decision. Then he should determine the extent to which a given decision can be routinized for action by lower organizational echelons or for programming in a device such as a computer. Further comments of this nature relative to specific decisions are made as these specific decisions are discussed throughout this text.

Concepts of Information-Decision Structure

Our primary concern in this text is to learn to design operations planning and control systems. The first step in any system design effort is to specify the objective, basic purpose, or desired performance of the system. The next step is to specify as explicitly as possible the criteria by which the performance of the system will be measured. Once these steps are accomplished, it is necessary to discover the information-decision structure by which action in the system is implemented. Basically, we must discover all those decisions that must be made and then determine what information will be necessary.

In the following section we explore certain fundamental system building blocks, called information and decision components. Then, in the next section, we employ the concept of a transfer function to convert information into decisions.

INFORMATION AND DECISION COMPONENTS

We saw in Chap. 1 that, for a production system to perform its intended purpose, many decisions are required. Decisions were classified into two groups: planning decisions and operating decisions. We were warned in the previous section of this chapter that many so-called operating decisions are not true decisions but merely administrative actions. For example, when the inventory level of a purchased item falls to its reorder point, an order is automatically placed. No true decision is involved in placing the order; the decision was made when the reorder point was established. To avoid clumsy phraseology, however, we will continue to call such administrative actions decisions, even though we should always keep in mind the proper distinction.

It would perhaps be beneficial to consider an operations control system as a directed* network of information and decision components, as illustrated in Fig. 2.1. Each decision that must be made is designated *D*, a decision component. Each information component (parameter or variable) available to the system is designated *I*. Information flow is represented by arrows. After a decision has been made, that decision is transmitted, *now as information*, either to a following decision component or to the specific operation location at which the decision will be implemented.

*For a discussion of directed and undirected networks, see Ford and Fulkerson (1962). Here we simply mean that the decisions are made in some ordered way, relative to each other or to the total system.

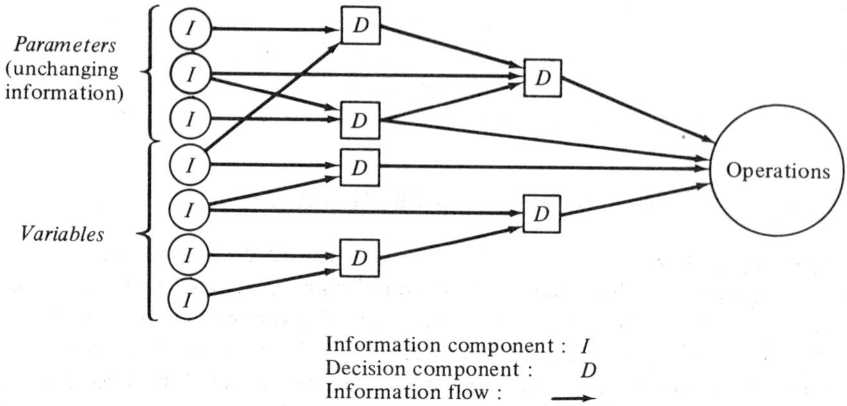

Figure 2.1 Information and Decision Components in an Operations Control System

Note that certain decisions depend upon previous decisions, plus perhaps more information components as inputs. Note also in Fig. 2.1 that an information component may be needed for more than one decision component. Also, a single decision component may require several information components. In other words, the same piece of information can sometimes be used in making different decisions, and many decisions require several different pieces of information, including perhaps the results of a previous decision.

To illustrate these concepts, let us consider the "decision" (actually an administrative action) of whether an order should be placed for a particular purchased item. This single decision requires two information components: (1) the reorder level, a parameter previously determined (see Chapter 5), and (2) the current stock on hand, a variable determined from stock records. Note that the second information component, the stock on hand, is also one of the infor-

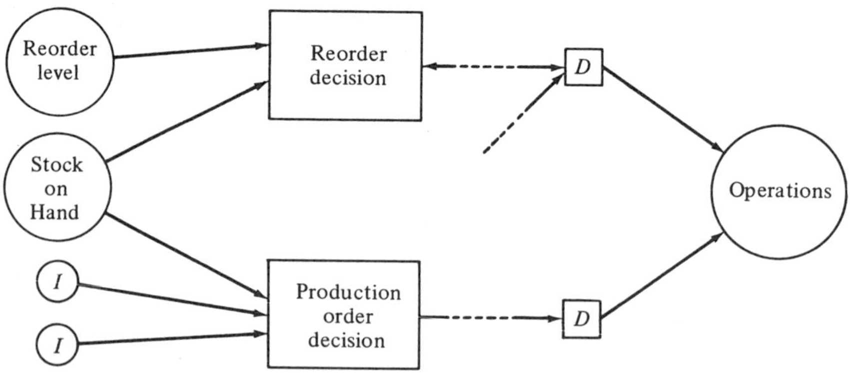

Figure 2.2 Illustration of Information and Decision Components

mation components required for making decisions regarding production orders for assemblies which use this purchased item. (Obviously, we cannot produce more assemblies than current stock on hand permits.) Perhaps these decisions are prerequisites to other decisions which must be made before reaching operations. Possible relationships are illustrated in Fig. 2.2.

From the concepts discussed above, we can state several general principles regarding the information-decision structure of a control system:

1. Decisions are always unique; there should be no opportunity for a given decision to be made in more than one organizational entity.
2. Information is not unique; it can be used in several decisions. There should, however, be no opportunity for a given information component to be originated more than once. Also, information is not perishable; when used, it is not consumed. It is still available for other uses and other decisions.
3. Information should be processed to the maximum extent possible at the same time for all information-decision subsystems.
4. At least two components of information are required for each decision. We can illustrate this perhaps surprising principle by first recognizing that any decision requires the processing of information. The simplest possible way of processing information is to scale it, i. e., multiply it by a constant. The information being processed and the scaling factor are two different information components, the former being a variable, the latter, a parameter. Thus, each decision requires at least two information components.
5. Any pair of decisions cannot require the same exact set of information components. At least one component required by one of the decisions cannot be in the set required for the other. A single set of information components (both parameters and variables) can result in only one decision.

Now that we have established the distinction between information components and decision components, it is in order to explore further their temporal relationship. Obviously information must be available at the time the decision is made. It is natural, therefore, to regard information as input to the decision process. This concept is developed in the next section.

THE TRANSFER FUNCTION CONCEPT

A classical concept in electrical engineering is that of a "transfer function." This term is used to denote the functional relationship between input and output of various electrical system components. Specifically, electrical engineers define the transfer function (Hare 1967) of a component as the ratio of its output to its input. Forrester (1961, p. 52) used the term in a broader sense to denote any mathematical relationship between the inputs and the outputs of a part (component) of a system. We use the term *transfer function* in a still broader sense to represent all decision processes, mathematical or otherwise, in a control system.

Conceptually, we may regard a decision process as consisting of three basic elements:

Input: the information available.

Output: the decision required.

Transfer Function: the process by which the input is converted to a decision.

The relationship of these elements is shown in Fig. 2.3. Although we extend this representation later, we can now make the following observations:

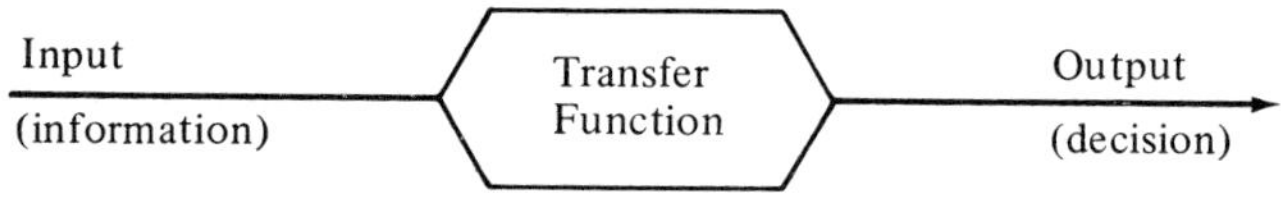

Figure 2.3 Transfer Function Concept applied to Decision Processes

1. The input may consist of new data feedback from operations, a previous decision, and parameters.
2. The output (the decision) may become input to another decision process.
3. The transfer function may be of many forms, such as:
   - a mathematical expression.
   - a linear, nonlinear, or dynamic programming model.
   - a statistical analysis procedure.
   - a tabular procedure (e. g., Gantt chart).
   - a decision rule.
   - a simulation or other computer model.
   - a heuristic procedure.
   - human judgment.
   - a combination of the above.

The transfer function concept is applicable to any decision level. We can use it to specify both planning and operating decisions. This concept agrees with our discussion of the five-step decision-making process. The transfer function assigns utility measures to the alternatives considered and then selects the alternative having largest utility measure. We also see that the three phases (commitment, quantity, and timing) of decision making are performed by the transfer function.

To give meaning to the transfer function concept, let us cast the reorder decision discussed in the previous example into this context:

Decision required	Whether to order *EOQ* units of item
Transfer Function	If ($SOH \leq RL$), order *EOQ*
Input	*SOH* and *RL*

EOQ is the economic order quantity (determined by methods discussed in Chap. 5), *SOH* is current stock on hand, and *RL* is the previously established reorder level. The transfer function in this case is in the form of a decision rule.

Later on, in Chaps. 3 through 7, we identify the individual decisions required for performing the several functions of operations planning and control. We then determine the information needed to arrive at each decision. Finally, we discuss the transfer functions available for converting the available information into the required decisions.

There is frequently more than one transfer function available for executing a particular operating decision (most of which are administrative actions rather than true decisions). In such cases, we must select the appropriate transfer function. This selection can be cast into the context of a higher level transfer function. We are literally "deciding how to decide." Such decisions are called "structural" decisions, since they determine the set of decision processes for the control system. Structural decisions are discussed further in the following section.

The transfer function representation in Fig. 2.3 is lacking in several respects. First, it does not account for repeated performance of the transfer function. Most decisions must be made repeatedly during successive periods of operation. Hopefully, the decisions become better as they are repeated. This repetitiveness leads to the second weakness of Fig. 2.3, the lack of a feedback loop. Another weakness is that information is not distinguished as to type or source. It is meaningful to distinguish two broad types of information:

1. *Parameters:* relatively constant information, such as standard processing times, overhead rates, reorder levels, mathematical constants, performance criteria, and the like.
2. *Variables:* information that is subject to frequent change. Variable information may be further classified as follows:

- *System status data:* data that describes the current status of the operations system, such as stock on hand, actual processing times, schedule delays, new orders, and other external occurrences, and the like.
- *Results of other decisions:* another transfer function may provide results required as information.

It is convenient to consider two separate data files for the two types of information. The first we call a *Variable Status File* and the other a *Parameter File.** These files must be kept up to date at all times so that the transfer function has access to the necessary information whenever it is called upon to make a decision. The files are updated through feedback data that describe the results that occurred from previous sets of decisions and through data which arises outside the system.

The extended representation of a transfer function that accommodates the above concepts is shown in Fig. 2.4. Let us mentally trace through the sequence

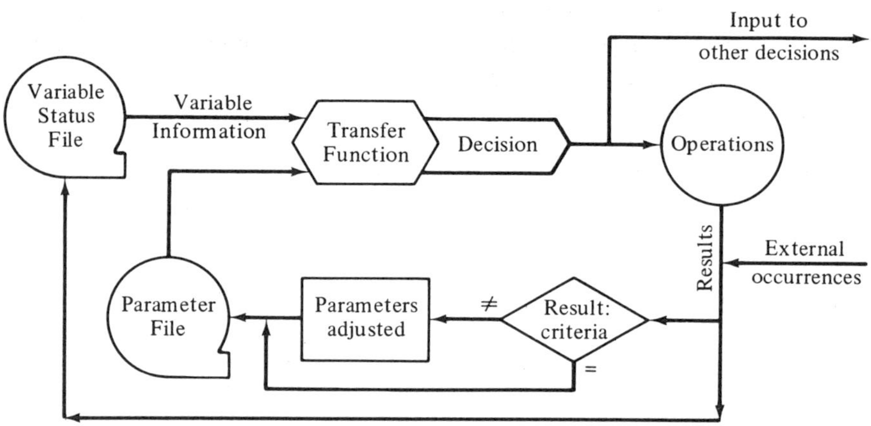

Figure 2.4 Extended Transfer Function Concept

of events that would occur upon a single execution of a transfer function. Suppose the transfer function is of the form $D_1 = a + bx + cy^2 + 1/2D_6$, where D_1 is the decision required, a, b, and c are parameters, x and y are system status variables, and D_6 is the current value of decision 6, determined in another transfer function. Now, in order to determine the new value of D_1, all the information on the right side of the equation is required. The decision is made and forwarded to operations and perhaps to another transfer function. Operations implements decision D_1, along with all other decisions that must be

*For now, we will consider a file as any device in which information can be stored in a systematic way. It may be magnetic tape, drum or disk, punched cards, or simply a manual file. This concept is explored further later in this chapter.

executed at this time. The results of the implementation, along with any effects of external occurrences, are returned as feedback. The results serve two purposes: (1) the status of all system variables is updated, and (2) the results are compared to certain performance criteria to determine whether parameter values should be adjusted. Thus, new values for x and y are entered into the variable status file and, if needed, adjusted values of a, b, and c are entered into the parameter file. These new values would then be used the next time this particular transfer function is executed. D_6 may or may not have a new value at the next execution, depending upon whether the transfer function that determines D_6 has been executed in the meantime. The execution of D_1 cannot change the value of D_6 on the same iteration* of the decision set, since a one-way precedence relationship must hold for decision dependencies. That is to say, if D_i requires D_j as input, D_j cannot require D_i as input on the same iteration. D_j may, however, use D_i as input on the following iteration.

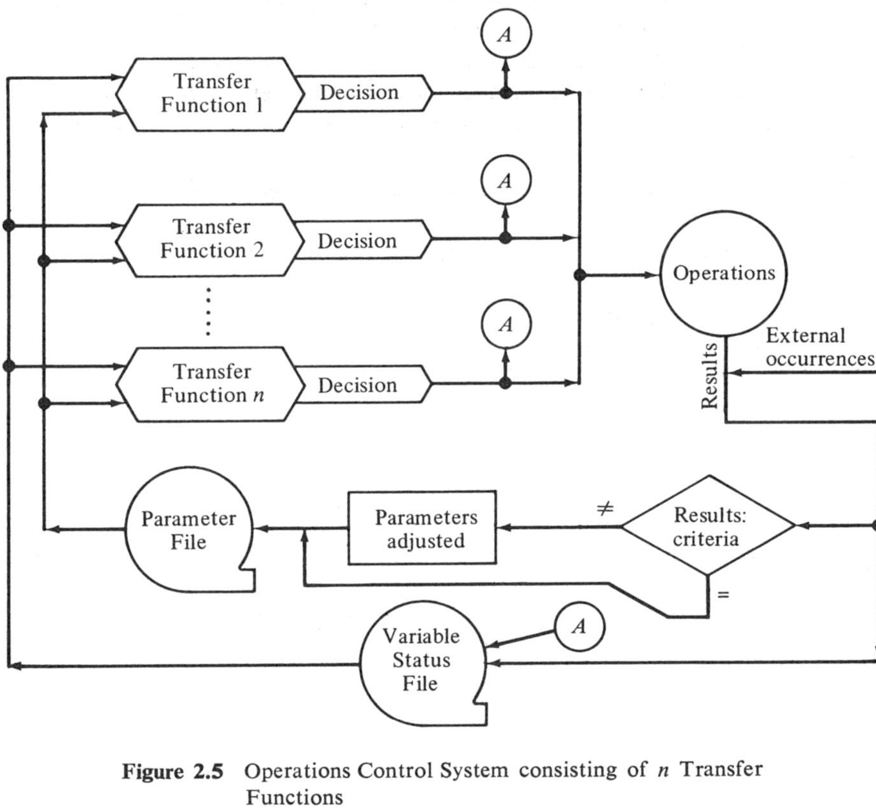

Figure 2.5 Operations Control System consisting of n Transfer Functions

*An iteration is simply one time through the total decision process. In this section, we are tactically assuming that all decisions are executed sequentially whenever any are executed. We relax this restriction in the next section.

Figure 2.4 illustrates a single transfer function. There is one such transfer function for each decision required in the operations control system. The extension of our concept to include several transfer functions is straightforward (see Fig. 2.5). Note that there is one principal feedback loop for the entire system and one each of the two types of information file.

DECISION HIERARCHY IN CONTROL SYSTEM DESIGN

In Chap. 1, we distinguished between planning and operating decisions. Then, in a previous section of this chapter, we introduced structural decisions. Planning decisions are concerned with getting ready for production; i. e., how best to meet expected demand with available or obtainable resources. Operating decisions are concerned with implementation of planning decisions. Structural decisions are concerned with the selection of transfer functions from those available. In general, then, structural decisions determine the appropriate set of transfer functions for the control system, while planning and operating decisions result from executing the transfer functions.

The transfer functions in Fig. 2.5 pertain only to planning and operating decisions. Structural decisions determine the particular set of transfer functions making up the control system and therefore are not included in the control system itself.

The major emphasis in this text is on structural decisions, since our primary concern is to design a control system rather than to operate it. We are concerned with selecting the best transfer function for each planning and operating decision that must be made. All such selections are structural decisions, and we can apply the transfer function concept to the execution of these higher-level decisions. Most transfer functions for structural decisions consist of computer simulations, in which various methods are tested, and human judgment. In fact, many operating and planning transfer functions for a particular company must be developed rather than selected.

THE INFORMATION TIMING PROBLEM

It is relatively easy to visualize the manner in which the control system of Fig. 2.5 would operate if all decisions were executed simultaneously. Reality does not afford us that luxury, however, since certain decisions must be made several times each day, while others may be made perhaps once each year. Some decisions are made on a regular time schedule, e. g., once each week, while others are made at essentially random times.

At first glance, the additional consideration of the time dimension might appear to greatly complicate the operation of the control system. Fortunately, however, all we must do is add "frequency of execution" to the specification of each decision process in the system. Thus, each decision process consists of the following specifications:

Decision required
Transfer Function
Input
Frequency of execution

The complete transfer function for the reorder decision discussed in the previous section would be:

Decision required	Whether to order EOQ units of item
Transfer Function	If ($SOH \leq RL$), order EOQ
Input	SOH and RL
Frequency of execution	Continuously

We discover in Chap. 5 that this is the transfer function of a fixed order size inventory system, one in which stock level is continously monitored and compared to a reorder level. When that level is reached, a fixed amount, EOQ, is ordered.

The primary reason we are able to include the timing problem with very little additional complexity is that we categorized information into two types (that which rarely changes and that which frequently changes) and then consolidated all of each type of information into a common information file. This process makes all information in the system immediately available to each decision process, whenever it must be executed. Dependencies between decisions are accommodated by channeling decision results directly to the variable status file. Immediate updating of all information components is provided for by a feedback loop.

Another factor contributing to the easy inclusion of the timing problem is the manner in which we relate dependent transfer functions. Certain decisions cannot be made until previous decisions have been made. By specifying that certain decisions are inputs to other decisions, we thereby establish these dependency relationships and further simplify the timing problem.

The timing problem discussed above relates to the manner in which the various transfer functions can be synchronized in view of the differing execution cycles. Another timing problem, and a very serious one, is that related to time lags in the control system. This problem is discussed in a following section.

THE CONTROL SYSTEM IN PERSPECTIVE

Much of the preceding development has necessarily been rather abstract. We have progressed through several concepts, beginning with elementary building blocks and then putting them together to form the system shown in Fig. 2.5.

It is natural to ask at this point just how these concepts are related to the fundamentals of designing operations planning and control systems. We can visualize the relationship by reviewing the set of decisions discussed on page 4 in Chap. 1, recognizing that each such decision, as well as others brought out in later chapters, requires a transfer function block in the control system shown in Fig. 2.5.

The abstract control system we have developed is truly an *integrated information-decision system.* Let us not believe, however, that such a system is easy to develop and operate in practice. There are two primary obstacles to the development and operation of such a system. First, such a system requires the decision structure of the organization to be formalized in the manner decribed in this chapter. The most difficult problem here is to specify explicitly all decision processes and the relationships between them. The second major obstacle is that the information files must be kept current. Although great strides have been made in remote data entry and in information file centralization, such systems are both expensive and time consuming in their development and maintenance. Compromises must sometimes be made. Despite the practical obstacles to the development of an ideal control system, it is important to understand the features of an ideal system in order to compromise intelligently where necessary. Significant advances are being made in overcoming both of the obstacles discussed above and perhaps the compromises will be far fewer than most managers now contemplate.

Information Availability, Handling, and Storage

The control system described in the preceding section is idealistic in that all information is assumed to be immediately available and up to date whenever it is needed for making a decision. For several fundamental reasons, this assumption rarely holds. In this section we discuss some practical considerations of the control system in regard to the availability, handling, and storage of information.

INFORMATION AVAILABILITY

Recall that decisions require information of two types, parameters and variables. All such information components are contained in the parameter file and the variable status file, as shown in Fig. 2.5. There are two general reasons why some information may not be available when needed for the execution of a transfer function. First, the information simply may not exist. Second, time delays in the control system may prevent prompt updating of the information files, particularly the variable status file.

Certainty, Risk, and Uncertainty. In decision theory terminology (Miller and Starr 1960, p. 80), decision making is said to be under certainty, risk, or uncertainty, depending upon the amount of information available concerning factors affecting the outcome of a decision. When all inputs are deterministic and completely known, the results of the transfer function can be predicted precisely. We call this *decision making under certainty.*

Inputs to many transfer functions are random variables, such as process times, lead times, demands, etc. If the probability distributions of these random variables are known, we can predict the expected value of the outcome. We call this *decision making under risk.*

Some decisions have to be made in almost total ignorance of certain of the factors affecting the outcome of the decision. For example, one of the inputs needed to decide whether to expand operations would be the national economic health for the next several years. This input is a variable, but we do not know its probability distribution. We call this *decision making under uncertainty.*

Rather significant progress has been made in dealing with decision making under risk. The entire field of statistical decision theory is directed toward this and related problems. We discover in the following chapters that many transfer functions are designed to accommodate random variables as input.

Very little progress has been made in dealing with decision making under uncertainty. A little reflection on this problem leads us to wonder how many decision situations arise in which we are totally ignorant of the inputs. We can usually relate a decision situation to some similar experience, or through logic and reasoning we can estimate reasonable boundaries (upper and lower limits) of the inputs. Decision theorists classify this kind of decision problem as *decision making under partial information.* The procedures for dealing with this decision problem are rather sophisticated and still in their early stages of development. In practice, the transfer function for this kind of decision is usually human judgment.

We do not mention again the classification of decision problems discussed above. Instead, we develop transfer functions according to the concepts presented in the previous section and employ human judgment to compensate for incomplete information.

Time Delays in the Control System. We now discuss the second general reason why certain information may not be available when needed for the execution of a transfer function. We discuss this problem in the broader context of considering all types of time delays in the control system.

We used the electrical engineering concept of a transfer function to facilitate the development of a control system consisting of decision processes and information components. We can make a further electrical engineering analogy by considering each transfer function in Fig. 2.5 as having a button (i. e., an activator) on it which, when depressed, executes the transfer function. Such a system would operate as follows: Whenever a decision needs to be made, the button is depressed immediately; the decision is made; it is implemented; results are returned; and information files are updated. All these actions occur at electronic speed.

There are some process control systems (such as those for certain chemical processes) which operate essentially like the system described above. Most production systems, however, cannot be programmed so precisely because of time lags that occur in the system. We now examine three categories of such time lags (Forrester 1961, p. 15).

1. *Time required to make a decision.* Most decisions cannot be made instantaneously by simply pushing a button for two basic reasons:
   a. The decision process (i. e., executing the transfer function) may be time consuming. For example, it takes a lot of time to manipulate a large Gantt Chart or a PERT diagram. If a transfer function involves a linear programming model, the problem has to be set up, checked, keypunched, and finally run on the computer: this process could involve several hours. Many decisions require discussion and compromise between department supervisors, and the plant manager may have to intervene.
   b. The decision maker (even an on-line digital computer) may be preoccupied with other decisions, and this decision may have to wait its turn.
2. *Time required to implement a decision.* Certain decisions require printing of instructions, updating of schedule charts, preparation of various documents, etc. These actions must then be relayed to the point of implementation, at which another delay may occur while the operations people review the decision and decide how best to implement it. All these steps require time and can cause delays.
3. *Time required to acquire feedback data.* The Variable Status File and the Parameter File are not kept current minute by minute because of these two basic reasons:
   a. Many decisions call for action that may require hours, days, or even weeks to accomplish. While the intermediate status of long lasting operations can be reported periodically, it is simply not practical to attempt to report work progress on a continuous basis.
   b. The act of recording and reporting progress is itself time consuming. Progress and accomplishment must first be measured and then recorded on the shop floor. It then must be transmitted through reporting channels to a central data collection point, where it must be recorded and processed. Only then is it ready to be inserted into the two information files.

We see, then, that there are many reasons why current information may not be immediately available or complete. These factors must be taken into account when designing a control system. More is said about this in the last section of this chapter.

Let us consider an additional practical consideration of the control system. Many on-the-spot decisions made by operators and first-line supervisors are never formally recorded. These decisions concern the minute details of implementing and executing formal decisions from the control system. The decision process is usually not very complex, and the transfer function (performed mentally) often involves nothing more than looking up certain facts in handbooks or operation manuals and performing a few relatively simple calculations.

Our previously discussed concepts apply even to the lowest level of decision making. Both information files are still needed for input to the decision process. Handbooks, engineering drawings, etc., constitute the Parameter File. The variable status file may exist exclusively in the memories of workers and perhaps in handwritten records. The results are compared to mental criteria, and mental parameters are adjusted as experience is accumulated. The Variable Status File is updated by word of mouth, notes on route sheets, quality charts, etc.

INFORMATION HANDLING

The electrical system analogy of our control system suggests that information flows at electronic speed between the components of the system. Some advanced control systems do transmit much information in this manner. Most companies still convey information in the form of written documents, such as route sheets, schedules, drawings, operation manuals, purchase orders, quality reports, production reports, etc. Often multiple copies of written documents are required, and a large variety of duplicating equipment is available commercially for this purpose.

Information is conveyed physically in a variety of ways: dispatchers, plant mail service, telephone, face to face, teletype, air tubes, postal service, etc. Most of these methods are time consuming and contribute to the time lag problem discussed in the previous section.

Modern developments in data processing permit us to approach the information handling problem in a new way. It is now possible to install an elaborate remote access computer system with input/output terminals distributed wherever we wish. Decisions (i. e., the results of transfer functions) can be relayed to operating foremen and personnel via typewriter-like terminals, television-like screens (cathode ray tubes), a high speed printer, or any combination of these devices. Process control information can be transmitted directly to automatically controlled machines. Production information (i. e., the results of implementing the decisions) can be relayed back over the same channels to the central data collector. Here, the results are compared to established performance criteria, and the data files are updated.

We see, then, that we now can come nearer to the electrical system analogy than was possible a short time ago. Further advances in this area can be expected. A system such as we have described is discussed further in Chap. 7.

INFORMATION STORAGE

Closely related to the information handling problem is the problem of information storage. Surely, information is in inactive storage a far greater portion of the time than it is being moved or otherwise processed. Much information is still stored in written records such as engineering drawings, bills of material, operations manuals, handbooks, catalogs, professional literature, vendor literature, cte. A surprising amount of information is stored in the memories of operators and supervisors. Literally millions of business forms have been designed to record and store information.

A great advance in information storage was made when punched cards were introduced. Some very good information systems were designed around the punched card and associated unit record equipment. Microfilm was added to the punched card for the purpose of filing and quickly retrieving engineering drawings, parts lists, fingerprints, etc. More recent developments include magnetic tapes, drums, and disk files. Drum and disk files are particularly important since they permit the concept of random access to be introduced. All information can be reached in just a fraction of a second in a random access file. Because this is an important concept to understand, we discuss the random access concept in terms of a magnetic disk file, shown in Fig 2.6.

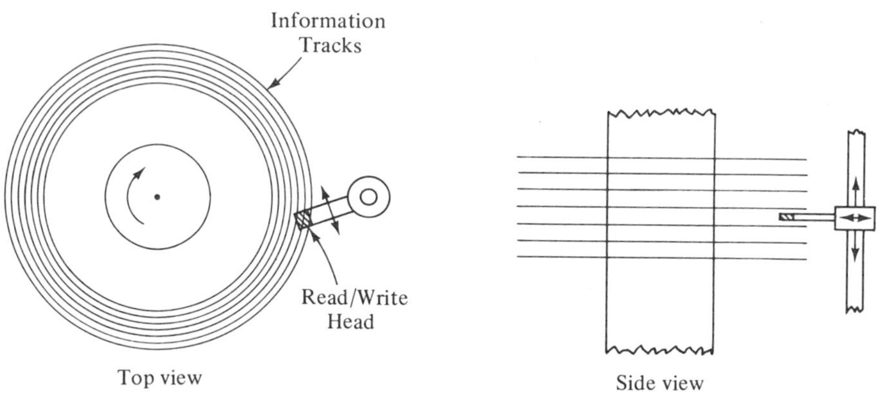

Figure 2.6 Random Access Disk File

The file consists of several disks similar in appearance to oversized phonograph records. The disks are coated on both sides with ferrous oxide, which can be magnetized by an electrical impulse from the read/write head. The read/write head can be moved in and out to read from or write on either side of the disk. Since the disks are rotating (at very fast speeds), the entire surface of each disk is always within a single revolution of the read/write head. Information is arranged on each side of the disk in circular tracks (usually, there are 100 tracks on each side). Position on the track is determined by the angle relative to the center of the rotating shaft. If a particular piece of information is wanted, it is necessary to know only the disk number upon which it is written, the track number, and the angular position. These instructions are conveyed to the appropriate read/write head, which then locates the information on the disk in a manner very similar to that of a juke box selecting a record after the customer punches a button.

To really appreciate the importance of the random access concept, consider the problem of finding a particular piece of information on a magnetic tape. If the information is located near the end of the tape, almost the entire length of the tape has to pass the tape read/write head before reaching the desired location. There is simply no other way to get to the desired information. The far greater speed of random access files is obvious.

On the Design of Information Systems

All the factors discussed in the preceding section make the designing of a good information system a very important and difficult undertaking. An information system should be designed for the entire management system, not just for the operations planning and control function. Thus an information system should include the functions of accounting, marketing, personnel, engineering, quality control, maintenance, safety, public relations, etc.

We will not pursue the topic of designing total information systems, since such a broad topic cannot be done justice as a secondary part of a text. Several texts are available that take a general pragmatic approach to designing information systems. Few texts emphasize the decision structure as the foundation upon which a sound information system may be built. The decision structure concept is just as applicable to the other management functions as it is to the production function.

By emphasizing the decision structure of the control system (especially establishing relationships between decision processes), we are in a better position to know precisely when each information component is needed. This input is invaluable to the design of the accompanying information system. A formal decision structure also gives us better insight into the practical problems discussed in the previous section: information availability, handling, and storage. By knowing the formal structure, we can accommodate better these imperfections and reduce their impact. Otherwise, we would be admitting that we are at the mercy of the imperfections.

At the beginning of this chapter, we stated that information and decision should not be treated separately. How then can we develop the concepts of an operations planning and control system without developing at the same time the accompanying information system? We overcome this difficulty by developing information needs and dependencies as they relate to the transfer functions in the operations planning and control system. We develop this system in a very general way, so that it can be plugged into any well-designed total management information system that has been designed according to the total decision structure concept.

We make extensive use of the random access concept of information storage and the remote computing concept of information availability. We do not, however, discuss the details of these systems, such as coding, addressing, etc. In other words, our developments are completely general and unrelated to specific programming methods or data processing equipment.

The design of information systems remains largely an individual undertaking for each company. There are certain principles and guidelines available (e. g., McCarthy, McCarthy and Humes 1966; Hare 1967) to assist in such an undertaking. There are also numerous consulting firms to provide technical assistance in system development and numerous professional and trade journals which contain articles concerned with management information systems.

REFERENCES

Ackoff, R. L. 1967. Management Misinformation Systems, *Management Science* 14: B-147-156.

Drucker, P. F. 1954. *The Practice of Management*. New York: Harper and Brothers Publishers.

Ford, L. R., Jr.; and Fulkerson, D. R. 1962. *Flows in Networks*. Princeton, N. J.: Princeton University Press.

Forrester, J. W. 1961. *Industrial Dynamics*. Cambridge, Mass.: The M. I. T. Press.

Hare, Van Court, Jr. 1967. *Systems Analysis: A Diagnostic Approach*. New York: Harcourt, Brace, and World, Inc.

McCarthy, E. J.; McCarthy, J. A.; and Humes, D. 1966. *Integrated Data Processing Systems*. New York: John Wiley and Sons, Inc.

Miller, D. W.; and Starr, M. K. 1960. *Executive Decisions and Operations Research*. Englewood Cliffs, N. J.: Prentice-Hall, Inc.

Reinfeld, N. V. 1959. *Production Control*. Englewood Cliffs, N. J.: Prentice-Hall, Inc.

II Operations Planning and Control Functions

Introduction

Authorities in the field of operations planning and control disagree significantly as to the specific functions that constitute an operations (or production) planning and control system. Some argue, for example, that the system should include the purchasing function, since raw materials must be readily available for plant operations. A similar argument is often made for such functions as product design, sales and distribution, and quality control, etc. On the other hand, other authorities argue that the function of forecasting should not be included, since it merely provides an input to operations.

The disagreement arises from the fact that it is very difficult to treat all functions of the firm in a single framework. These functions have been divided among two academic disciplines in our universities: (1) schools of business or industrial administration and (2) schools of industrial engineering. Traditionally, the former have dealt primarily with marketing, sales, finance, etc., while the latter have concentrated primarily on production problems directly related to plant operations. In such a situation, considerable overlap is bound to occur; one result is lack of a clear delineation of propriety regarding the functions.

We certainly make no attempt here to resolve this conflict, nor do we specify that this treatment is directed to one or the other of the disciplines. Let us simply make it clear that this text is concerned primarily with those functions directly related to the operations of the firm. The authors hope that the text will provide a new and useful approach to the design of operations planning and control systems, an approach equally appealing to scholars and practitioners in both disciplines.

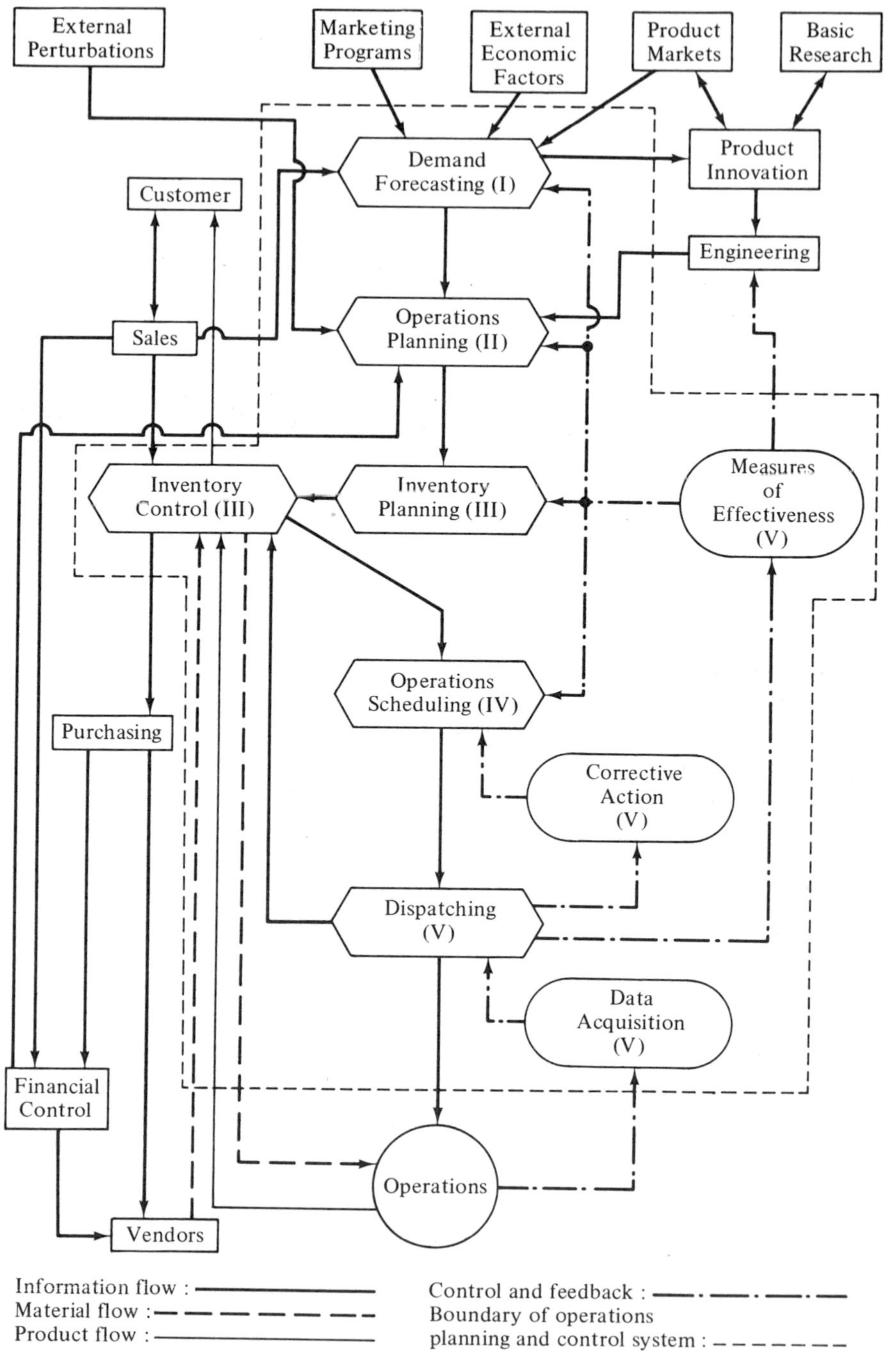

Figure II. 1 Master Flow Diagram: Operations Planning and Control System

Both disciplines are becoming increasingly aware of the importance of the many interfaces among the major functions of the firm. Our approach specifies those functions that we consider to belong to an operations planning and control system and discusses important interfaces at appropriate points in the development.

The set of operations planning and control functions we have chosen to include are: Demand Forecasting (I), Operations Planning (II), Inventory Planning and Control (III), Operations Scheduling (IV), and Dispatching and progress control (V). These functions, explained briefly in Chapter 1, are shown in Fig. II. 1, enclosed within the dashed-line boundary of the operations planning and control system. Progress control includes the three blocks entitled Data Acquisition, Corrective Action, and Measures of Effectiveness. Important interfaces with the control system are shown outside the dashed-line boundary.

Discussion of Master Flow Diagram

Figure II. 1 is an overview of an integrated operations planning and control system. The details of performance of the planning and control functions are presented in subsequent chapters. An intricate flow diagram, such as Fig. II. 1, is conceptually difficult to understand in its entirety. We could claim, in fact, that the primary objective of the book is to explain fully this flow diagram. The authors strongly suggest, therefore, that Fig. II.1 be referred to very frequently as the reader progresses through the book. Many of the component blocks shown in this master flow diagram are later expanded into detailed flow diagrams.

DEMAND FORECASTING (I)

Inputs. The most tangible input to the demand forecasting function comes from Product Markets in the form of demand histories of existing products and, especially, trade information. Demand trends can be determined quantitatively both for one's own company and for the entire industry. Special trade information can either modify or explain the demand trends. Another input comes from Marketing Programs. An attempt must be made to anticipate variations in the demand pattern that result from advertising and promotion schemes. A third important input is External Economic Factors, which represent the general economic health of the nation (and possibly of other nations). The state of the national economy at any point in time depends upon the collective effects of such factors as the political situation, the military situation, the extent of government involvement in large projects (space, welfare, road building, other construction), labor conditions, stock market trends, and literally millions of other factors. It is very difficult for a company to evaluate quantitatively the total effect of the national economy upon its own operations. Finally, actual demand comes from Sales and is compared to forecasted demand.

Outputs. The primary output of the demand forecasting function is a statement of expected demand quantities for the several products over some planning period. This output is conveyed to the Operations Planning function. Another important but often overlooked output is the recognition of declining demand for a product. Any indication of declining demand is conveyed to Product Innovation for possible modification of the product. When it becomes evident that a product line will soon cease to be sufficiently profitable, a decision must be made by higher management to phase it out. This decision then has implications in the Operations Planning function in terms of changes in capacity.

OPERATIONS PLANNING (II)

Inputs. The primary input to the operations planning function is from Demand Forecasting. The demand forecast provides the basis for determining the activity level at which the company should plan to operate over the planning period. Another large and important set of input data comes from Engineering and concerns new products, modifications to existing products, or modifications to the production process. This data includes manufacturing sequences, bills of material, standard operation times, set-up times, standard process manuals, and the like. A third input, from Financial Control, concerns monetary constraints and budget limitations. A fourth input, called External Perturbations, includes emergency orders, cancelled orders, labor strikes, unavailability of resources, and other problems from the world at large.

Outputs. There are two broad categories of outputs from the operations planning function. One deals with long-range planning of plant expansion and new facility acquisition. In this category, consideration is given to the total life cycle of a product line and the impact on production and resource capacities of proposed new or modified products. The second category, concerned with relatively short-term (usually 1 year or less) planning, consists primarily of the allocation of available resources to production requirements. Machine groupings, facility layout, and assembly line balancing are performed. Make or buy decisions are made, as are decisions regarding work force size, skill mix, manpower leveling, number of shifts, approximate time phasing of production runs, and the like. This information is conveyed to Inventory Planning and Control and Operations Scheduling.

INVENTORY PLANNING AND CONTROL (III)

The inventory planning and control function is shown in Fig. II.1 as two blocks: Inventory Planning and Inventory Control. Inventory Planning determines the material requirements (components, parts, raw materials, assemblies, supplies, etc.) necessary to satisfy the operations plan. Inventory Control determines the proper inventory levels, reorder points, safety stocks*, and the like.

*Safety stocks are needed as protection against unusually large demand during a short period of time.

Input to Inventory Planning. The primary input to inventory planning is the approved operations plan from Operations Planning. Specifically, the operations plan must provide a time-phased outline of the plant's activities in terms of production for the market.

Outputs of Inventory Planning. The operations plan is extended into the net, time-phased requirements for components, parts, raw materials, assemblies, and supplies. These requirements are forwarded to Inventory Control.

Inputs to Inventory Control. Primary are the time-phased requirements from the Inventory Planning function. Other important information inputs are sales orders from Sales and material requisitions from Dispatching. In addition to information input and output, we are concerned also with the receipt and issue of materials and products. Material is received from the Vendor and product is received from Operations.

Outputs of Inventory Control. Order quantities, reorder points, safety stock of raw materials and purchased parts, manufacturing batch sizes, and safety stock of assemblies, fabricated parts, and finished product are determined. When stocks on hand of raw materials and purchased parts reach their reorder level, a purchase order is sent to Purchasing. When stocks on hand of assemblies, fabricated parts, and finished products reach their reorder level, a production order is sent to Operations Scheduling. Material and product outputs include finished product sent to Customer and materials sent to Operations. Note that subassemblies and fabricated components go from Operations to Material Control as products; however, when they are brought back into the production process, they return to Operations as materials.

OPERATIONS SCHEDULING (IV)

Inputs. Primary input to the scheduling function are requests for fabrication and assembly from Inventory Control. Another extremely important input comes from Corrective Action in the form of updated priorities and schedule adjustments of production orders currently in process.

Outputs. Detailed operation sequences for individual work activities, as well as start and stop times for all operations, are released to Dispatching. New work is assigned to operating facilities, with proper consideration given to work already in process and to priority assignments. Schedule conflicts on production facilities are resolved. This function is at the heart of the entire planning and control system, since it is here that compromises must be made between economic batch sizes, due dates, resource constraints, manpower leveling, and facility utilization. Here also corrections are made for nonstandard process behavior.

DISPATCHING AND PROGRESS CONTROL (V)

Since this function is shown in Fig. II.1 as four blocks, we discuss each block in turn.

Dispatching. This function is responsible for initializing production: i.e., for releasing work orders to Operations at the appropriate time. Material is requisitioned from Materials Control, and arrangements are made for changing the production process over to the new production item.

Data Acquisition. As work progresses, data is acquired from different processing points on the shop floor. Specific data required are the progress of work in production, the status of work awaiting processing, the status of production facilities, and the availability of required workers. Material usage is relayed back to Inventory Control. It should be noted that much of the data is also useful for accounting functions, quality control, personnel, etc.

Corrective Action. Short-term corrective action is performed here. Specific decisions are made regarding operating problems. These decisions include expediting critical jobs, determining new priorities, balancing work loads between work centers, handling personnel problems, problems with product quality, and equipment breakdowns, etc. These decisions are conveyed to Operations Scheduling and are included in the next execution of the scheduling transfer function.

Measures of Effectiveness. Long-term corrective action is performed here. Control parameters are maintained to which measures of plant performance are related. Actual performance is compared with planned performance for such measures of effectiveness as production output, standard and premium labor cost, investment in inventory, hiring and layoff cost, amount of scrap, facility utilization, schedule slippage, etc. Particular attention is given to the identification of cause-effect relationships so that we can anticipate problems and try to prevent them. Information from this function constitutes the principal form of long-range system feedback and is used for recalculation of pertinent governing parameters.

INTERFACES

The purposes of several of the interfaces shown in Fig. II.1 were discussed in the preceding sections; others are self-explanatory. Explained below are certain important interfaces whose functions are important to the operations planning and control system but whose purposes may not have been clear from the flow diagram.

Operations. Production orders reach the shop floor through the Dispatching function. Materials and components are issued by Inventory Control. Production is accomplished, products are returned to Inventory Control, and progress information is fed back through the Data Acquisition function.

Basic Research. Basic research originates ideas for new products essentially independently of market considerations. An example would be the development of nylon by the du Pont research group. This development opened the way for hundreds of commercial uses. However, many research ideas prove fruitless.

Product Innovation. Ideas for new products and modifications of existing products are consolidated by this function, whether the ideas originate in Product Markets, Basic Research, an independent inventor, or a member of the product engineering group.

Engineering. Specific configurations of all models in the product line are finalized here, often in consultation with Product Innovation. The manufacturing engineering functions, including the determination of manufacturing operations, raw material and purchased parts requirements, standard processing times, set-up times, etc., are also performed here. Standard operating procedures for performing the operations are prepared. The complete file for each new or modified product is conveyed to Operations Planning. Important feedback, including cost information from the production process, quality control information, labor usage, material usage, scrap rates, etc., comes from Measures of Effectiveness.

Purchasing. Purchase orders are received from Inventory Control. A vendor is selected, and the order is placed. Material costs are conveyed to Financial Control.

Financial Control. This function determines and reports the effects of the company's activities upon its financial position. Standard costs are developed from production records, material costs, line labor, and overhead rates. Labor charges are accumulated against work orders, accounts, and departments. Reports are produced which cover variances between actual and standard costs of material, labor, and overhead. Budget constraints are transmitted to Operations Planning. Vendors are paid for raw materials. Production costs are received from Operations, and sales dollars are received from Sales.

Data Requirements for Transfer Functions

The discussion in the preceding section emphasizes the need for large quantities of data in a system such as that pictured in Fig. II. 1. Data must be gathered at its origin, entered into the system, transmitted through the system, processed, stored, and disseminated. Its ultimate use will be in one or more transfer functions that yield decisions. This concept was discussed in Chap. 2 (see Fig. 2.5).

We recall also from Chap. 2 that there are two broad types of data files in our system: a Parameter File and a Variable Status File. It is convenient to use an elementary coding scheme to identify the two files for each function of operations planning and control. The following codes are used throughout this text:

Parameter File

Function		*Code*
I.	Forecasting	P. I
II.	Operations Planning and Control	P. II
III.	Inventory Planning and Control	P. III
IV.	Operations Scheduling	P. IV
V.	Dispatching and Progress Control	P. V

Variable Status File

Function		*Code*
I.	Forecasting	VS. I
II.	Operations Planning and Control	VS. II
III.	Inventory Planning and Control	VS. III
IV.	Operations Scheduling	VS. IV
V.	Dispatching and Progress Control	VS. V

Conceptually, imagine that we have two random access disk files, each with five disks, as shown in Fig. II. 2. These two files are part of the total control system, as shown in Fig. 2.5. We use the term *disk file* in a very loose sense; we include in it all available information, some of which would be in written form. For example, engineering drawings and process manuals are included in our conceptual disk file. By using this simplification, we can appreciate the concept of all information centralized in one set of files and readily available for the execution of the several transfer functions.

Let us consider again the master flow diagram in Fig. II. 1. Our job is to learn to design the best set of transfer functions for a particular operating environment that will execute the five operations planning and control functions.

Numerous transfer functions have been developed for each of the operations planning and control functions. For example, some of the transfer functions available for forecasting are subjective opinion, regression analysis, weighted average, etc.

A large percentage of the technical articles published in such journals as *Management Science*, *Operations Research*, and *Journal of Industrial Engineering* are presentations of new and improved transfer functions. The person who designs an operations planning and control system thus has a large and growing inventory of transfer functions from which to choose. His job is to understand these transfer functions, select those that seem most appropriate for his system, make adjustments where required, and most importantly, tie them all together into an integrated control system.

Each of the five chapters in Part II deals with one of the operations planning and control functions. In each chapter, we discuss the basic nature of the particular function, the decisions required, the information needed for making those decisions, and some of the transfer functions available for converting the

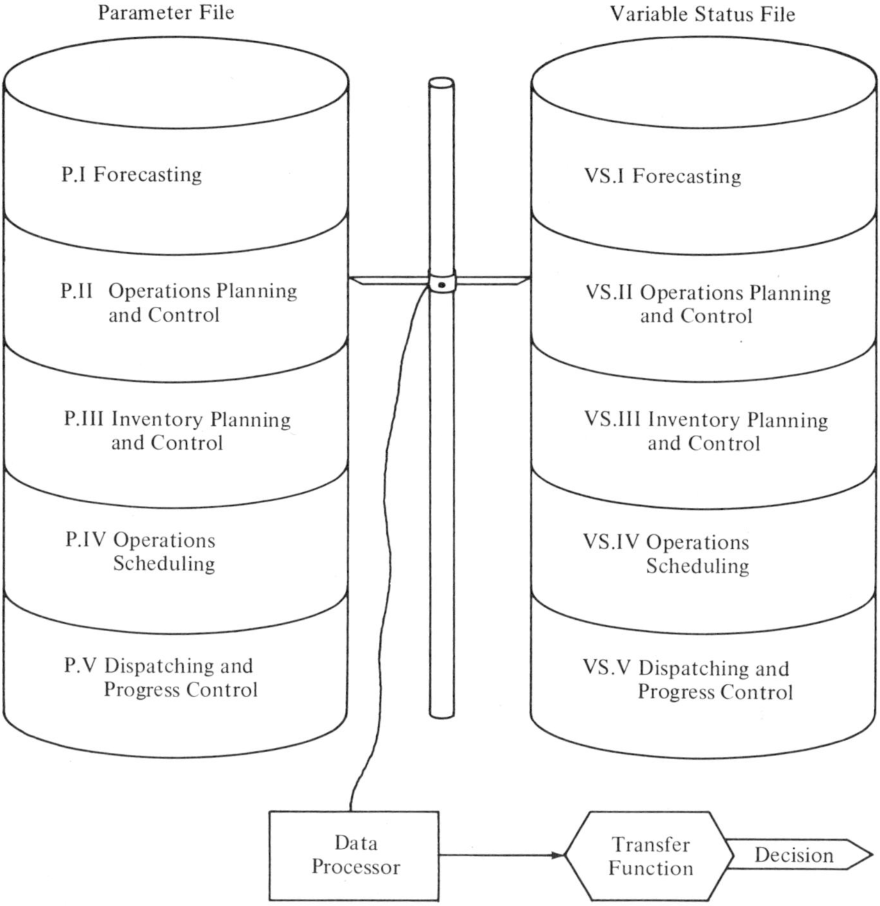

Figure II. 2 Data Files for Operations Planning and Control System

information to the required decisions. Finally, we discuss the interrelationships among functions by showing dependencies, using the codes presented previously.

Once again the authors encourage the reader to refer frequently to Fig. II.1 to maintain perspective and over-all orientation.

3 Demand Forecasting

Concepts

A demand forecast is an estimate of future customer demands for products or services. It provides the basis for establishing the over-all activity level of operations. Forecasting provides guidelines on *what* our product should be, *how many* are likely to be demanded by our customers, and *when* these demands are likely to occur.

PURPOSE OF FORECASTING

Demand forecasting provides the major input to the other functions in the operations planning and control system. The other functions convert this forecast into material requirements, parts lists, manpower requirements, schedules, and other decisions.

All planning must begin with an estimate, or forecast, of the amount of business our company can expect during the planning period. The means by which the estimate is arrived at may be completely subjective or unscientific, but the fact remains that all other planning of the company's activities depends upon an estimate of business volume.

Forecasting is not required directly in producing or selling the product. The requirement for a forecast should be explicit before the forecast is developed. The accuracy requirements of a forecast should be determined comparing the relative cost of an error in the forecast to the cost of providing a forecast of a given precision. This is simply to say that highly sophisticated and costly forecasting procedures should not be employed unless the increased accuracy of the resulting forecasts save the company (through fewer stockouts and lower inventory cost) at least as much as the additional cost of the sophisticated forecasting procedure.

It is important to understand that a forecast is not a sales goal. In establishing sales goals for our products, we can and should be highly optimistic. For example, a company may set as a goal a 20 percent increase in sales. Our forecast, however, should attempt to project in hard, cold facts what we think demand will actually be. We would certainly not expand our production facilities 20 percent on the basis of a sales goal. On the other hand, the forecast should not be constrained to manufacturing capability, since this would give no information relative to expansion needs. Rather, the basic forecast should estimate true sales demand as closely as possibly. Other factors will be incorporated in later refinements.

TYPES OF FORECASTS

Different levels of management planning require different types of estimates about future company activities. The board of directors would be interested in an estimate of total dollar volume over several years. The production manager, on the other hand, would be more interested in an estimate of the number of units of each product to be sold over some short time span.

Four broad categories of forecasts can be identified in most organizations:

Market Forecast. This forecast, made to guide research and development efforts and to plan long-range expansion of facilities, covers from one to twenty years. Since it sets the course the company will follow, this forecast is extremely important and should be prepared with great care.

Financial Forecast. From this forecast, which estimates future profits, the cash flow and capital requirements are determined. The budget may be estimated for 1 month to 2 years; however, for most companies the projection is for 1 year.

Sales Forecast. This forecast, made for short-term sales, is used to plan sales campaigns and other market strategies. It may be for periods of 1 month to a year, but a quarterly estimate is generally the most useful.

Production Forecast. This forecast estimates the demand in units of each product we sell. An estimate is made for each time period (usually a week or a month) in the planning period (usually one quarter or one year). The individual time period estimates are cumulated over the planning period for a total estimate. From the total estimate, we determine a long-range operations plan, which includes such decisions as the number of shifts of operation, over-all size of work force, additional equipment requirements, and amount of subcontract support. The estimates for the individual time periods are used to generate specific production orders and material requirements, which, in turn, form the basis for detailed schedules, men and machine assignments, and other short-term decisions.

In this chapter only production forecasts are treated. The other three types of forecasts are obviously not performed within the boundaries of the operations planning and control system shown in Fig. II.1.

FORECASTING DECISIONS

We can now explicitly state the decisions the forecasting function is expected to render.

Demand Trend. The forecasting system must project a demand trend for each product that shows how much we can expect to sell during each time period. This trend is based upon some sort of an analysis of the demand history. Any indication of declining demand is conveyed to product innovation for possible modification of the product. All products eventually lose their customer appeal and are phased out. While the decision to phase out a product is made by higher management, it usually enters the control system through the forecasting function.

Other Decisions. Several other decisions may enter the control system through the forecasting function. These decisions interact with the derived demand trend and thereby influence the final forecast.

For example, the subjective opinions of salesmen are sometimes an important input that should be considered. Salesmen are usually quite close to their customers and can react to intangible factors. Another input that could affect the forecast is the expected impact of promotional programs. A sales campaign often results in a surge in demand, followed by a sharp decline, and finally a return to normal.

External factors that can influence the forecast are competition (such as a new product or a price reduction of a competing product), political factors, and the general economic health of the industry.

Combined Forecast. The forecasting function must combine all the "other factors" mentioned above with the demand trend to provide an estimate of the number of units of each of our products that will be sold during each time period in our planning period. The cumulative demands are used to develop our operations plan, and the periodic demands are used for detailed production scheduling. These decisions are input to Operations Planning.

We should remember that a forecast is only an estimate and that, therefore, forecasted demand and actual demand for a time period will rarely agree. It is often necessary to revise our forecasts in consideration of recent demand values. Most forecasting systems are constructed such that a new production forecast is generated for each operating period (usually each week or each month). This process is illustrated in Fig. 3.1, in which the planning period is one year* and the operating period is one week.

*We will use a planning period of one year in this text for illustrative purposes, although the planning period for many firms is considerably shorter.

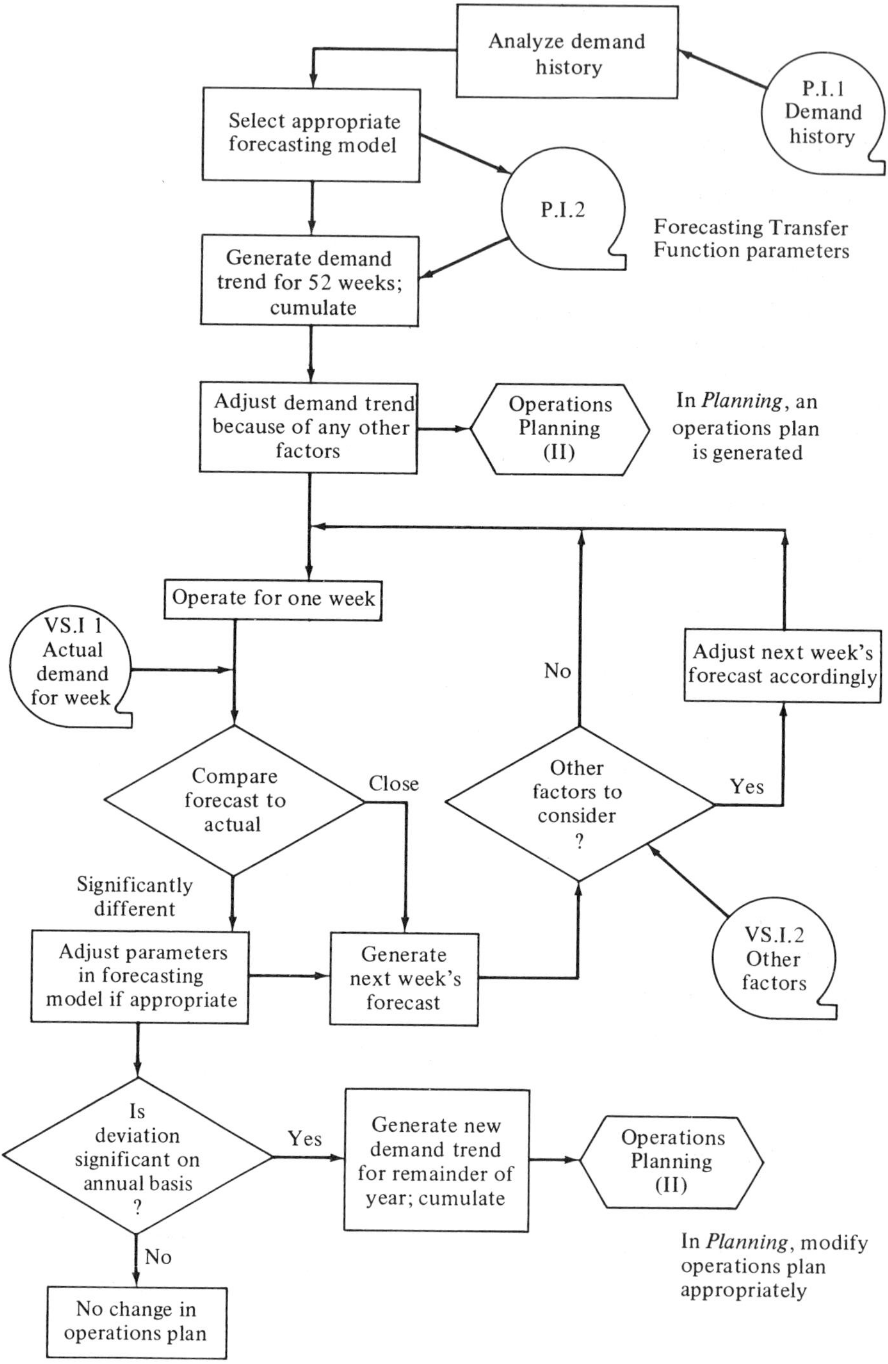

Figure 3.1 Flow Diagram of General Forecasting Procedure

Note in Fig. 3.1 the block labeled "Select Appropriate Forecasting Model." Later in this chapter we discuss several of the forecasting models that are available. The models we discuss consider only the demand trend. The effects of other factors (such as sales campaigns, etc.) must be considered as part of the forecasting control system, as shown in Fig. 3.1. The models themselves do not inherently accommodate the other factors.

Inputs

We recall from our development of the transfer function concept in Chap. 2 that decisions are arrived at by processing information. The decisions that must be rendered by the forecasting function require both types of information, parameters and variables. In this section we examine the specific information requirements for forecasting and where this information comes from. Only that information required for production forecasts is discussed here.

PARAMETER FILE FOR FORECASTING (P. I.)

There are two primary types of parameter inputs to forecasting: (1) the past demand history of each product and (2) parameters (numerical constants, control limits, equations, etc.) used directly in the transfer function. We designate the demand history as P. I. 1 and the transfer function parameters as P. I. 2.*

Demand History (P. I. 1). The most basic input to forecasting is the past history of each product. If possible, these histories should be actual demand records rather than sales records. The records should reflect units of product rather than dollar volume.

Table 3.1 contains twelve months of demand records for a product sold by three distributors. We analyze this data in a later section. The purpose here is to demonstrate what a typical set of demand data might look like. For most products, of course, more than one year of demand history is available.

Demand records may be summarized in several different ways, depending upon the size of the company, the type of distribution system used, the nature of the total information system of the company, and the requirements of the particular transfer function being used for forecasting. One way to summarize demand records is by individual customer or type of customer. If we have distributors in different geographical locations, demand records of each product should be kept for each distributor. By summarizing demand records in ways such as these, our forecasting control system can react quickly to subjective inputs. For example, suppose that we summarize our demand records by type of customer and that we learn the federal government will accelerate its road building program in two months. We can estimate the resulting effect of this action on our production system by adjusting one part of our forecast upward.

*These codes are further breakdowns of the information files conceptualized in Fig. II. 2.

Table 3.1 Demand Records for a Product

Month	Distributor *A*	Distributor *B*	Distributor *C*	Total Demand
January	500	430	110	1,040
February	510	380	160	1,050
March	480	420	150	1,050
April	600	370	100	1,070
May	600	410	130	1,140
June	660	380	70	1,110
July	590	440	120	1,150
August	700	380	80	1,160
September	680	420	80	1,180
October	740	370	120	1,230
November	790	410	40	1,240
December	760	390	90	1,240

We can also classify our demand records into regular orders and emergency orders, if emergency orders play a significant part in our system. We can generate a probability distribution of time between emergency orders and another probability distribution of size of emergency orders. These distributions provide a basis for anticipating emergency orders over our planning period and for including this much extra activity in our over-all operations plan.

Forecasting Transfer Function Parameters (*P. I. 2*). This category of information input consists of the parameters used directly in the decision-making process. The parameters are determined from an in-depth analysis of demand histories. Suppose our analysis of past demand indicates that we can predict next month's demand by averaging the last two months' actual demand and adding a constant, say 16 units. If, however, our last forecast was more than 10 units higher than the actual demand for last month, then next month's forecast will be the average of the last two months' actual demand plus a smaller constant, e. g., 8 units.

Let us enumerate the parameters in this elementary example. The first parameter is the number 2, the value we divide into the sum of the last two months' actual demand. Two other parameters, 16 and 8, are the values we add to our two months average, depending on whether our last forecast was more than 10 units higher than the actual demand for last month. Still another parameter is 10.

The particular parameters needed for the forecasting function depend entirely on the specific transfer function selected. We develop a few of the possible transfer functions in a later section of this chapter.

VARIABLE STATUS FILE FOR FORECASTING (VS. I)

Variable information is that whose value is likely to change very frequently. For the forecasting function, it is convenient to separate variable input into two categories: actual demand (VS. I. 1) and other factors (VS. I. 2).

Actual Demand (*VS. I. 1*) This input is the actual customer demand for one time period (week, month, etc.). We hope that our forecasted demand (determined by the transfer function) will be very close to actual demand. If the two values are relatively close, our forecasting procedure is correct, and we may make next period's forecast using the same procedure. If the two values are sufficiently different, then we must adjust the transfer function parameters (P. I. 2) in our forecasting procedure before making next period's forecast. This procedure is shown in Fig. 3.1. Emergency orders as well as regular orders are included in this category.

Other Factors (*VS. I. 2*). This category of input consists of those other decisions, including subjective opinions of salesmen, sales campaigns, new competition, price changes, economic fluctuations in the economy, and political considerations, that may affect the volume of actual demand. Again, the manner in which these inputs influence the forecast is shown in Fig. 3.1.

UPDATING THE FILES

All values in the two files must be current whenever the forecasting transfer function is executed. The values in the Parameter File do not change very often. The Demand History (P. I. 1) receives an additional value after each time period, but the parameters describing it do not change unless we suspect that the underlying cause system has changed. This concept is explored in greater depth in the next section. Similarly, the transfer function parameters (P. I. 2) remain unchanged unless it is determined that the present values of the parameters are not predicting demand closely enough.

The values in the Variable Status File are subject to frequent changes. Actual demand (VS. I. 1) usually changes each time period. This value comes from Sales. It is important that this value be entered into the system as promptly as possible. Recent advances in communications make it possible to collect demand data from geographically dispersed sales regions on a daily (or even shorter) basis, if such a fast response is needed and can be justified economically. In such sophisticated systems, sales records are generated automatically as a byproduct of sales transactions. Actual demand is usually entered into the system on a regular time basis. Other factors (VS. I. 2) are entered into the system as they occur. How fast they should be entered depends upon the urgency of the situation.

Remember that we have been considering our files to be magnetic disk files because this facilitates our conceptualization of the information-decision processes within our control system. In reality, a forecasting file may consist only of sales slips, or it may be on punched cards or some other storage medium. The kind of file actually used depends upon the information processing system the company uses.

Transfer Functions

We have come to what might be considered the most important section of this chapter: the development of the specific decision processes, which we call transfer functions, that actually generate forecasts. The transfer functions presented in this section estimate the demand during a particular time period for a particular product, given only the Demand History (P. I. 1) of that product.

All of the methods presented in this chapter fall under the heading "time series analysis," a topic about which many books have been written (See Brown 1963 and Wiener 1950). Only those methods most easily and economically used are presented. The type of forecasting presented in this section is called technological forecasting, since no attempt is made to identify underlying causes. Forecasting based upon known cause and effect relationships is fundamental forecasting.

DISCUSSION OF CAUSE SYSTEMS

When we base our forecasts entirely upon past demand records, we are assuming that the underlying cause system will continue to operate unchanged. A *cause system* is the entire group of factors, or variables, that collectively generates the demand pattern of a product. Consider the demand values that were given in Table 3.1. It is helpful to plot these values on a graph in order to analyze demand trend behavior .

Figure 3.2, which follows, shows the demand histories of three distributors as well as the history of the total demand for the product. Each point on the total demand curve is simply the sum of the demands of the three distributors for that month.

Notice the demand pattern of Distributor *B*. This is basically a level pattern, neither increasing nor decreasing. This is not to say that each month's demand will be identical; indeed, it is obvious that such is not the case. We are simply saying that the trend of demand is level for Distributor *B*. We know that the cause system will generate demands which in some months are higher than in other months. As long as the cause system remains unchanged, the trend will continue as shown.

Distributor *C* shows a downward trend, while Distributor *A* exhibits a rapidly increasing trend. Each distributor has its own cause system. It is entirely possible that some factors included in the cause system of one distributor are also included in the cause system of another distributor. There are literally thousands of factors that interact simultaneously to cause customers to behave as they do at a particular time. The cumulative result of these interactions are the demand values that occur.

Notice that the cause system for total demand is simply the union of the cause systems for the three distributors. There is an additive relationship between the one cause system for total demand and the three cause systems for the distributors.

Consider again the demand pattern for Distributor *B*. If we were to predict a demand value for the next month (January of the next year), our best estimate

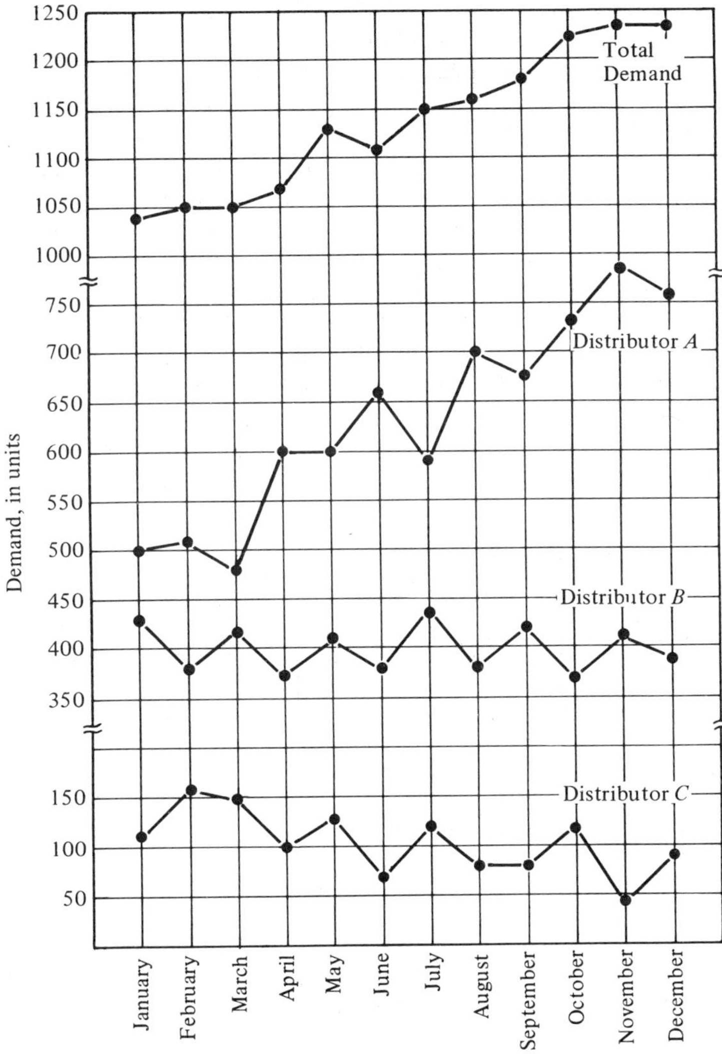

Figure 3.2 Graphical Representation of Data in Table 3.1

would simply be the average demand, or 400 units. How good is this prediction? Judging from the last 12 months' demands, demand for any month will not be below 380 nor above 440 units. In statistical terminology, demand for Distributor B is a random variable. There are easy formulas for computing the mean and standard deviations of this random variable. Furthermore, we can plot the demand values in a histogram to determine what type of probability distribution best describes this random variable. It is not uncommon for the probability distribution to be very nearly normal.*

If we now combine the two concepts just developed, we see that a cause system generates a demand value for a time period. The demand value is a random variable whose probability distribution can be specified approximately. These concepts form the basis of some of the transfer functions developed later.

A disturbing problem in forecasting is that cause systems change quite frequently. Consequently, one of the most important tasks of a forecasting system is to detect when a significant change occurs in the cause system. As an illustration of a changing cause system, consider the demand data of a product listed in Table 3.2 and graphed in Fig. 3.3. In Fig. 3.3, we see that the cause system changed twice during the 36-month interval. A possible explanation might be: during the first eight months, the product was being introduced, and retailers were rapidly building up their inventories. The growth rate during the second eight months reflects an expanding export market for the item. The demand trend appears to have stabilized during the final 20 months.

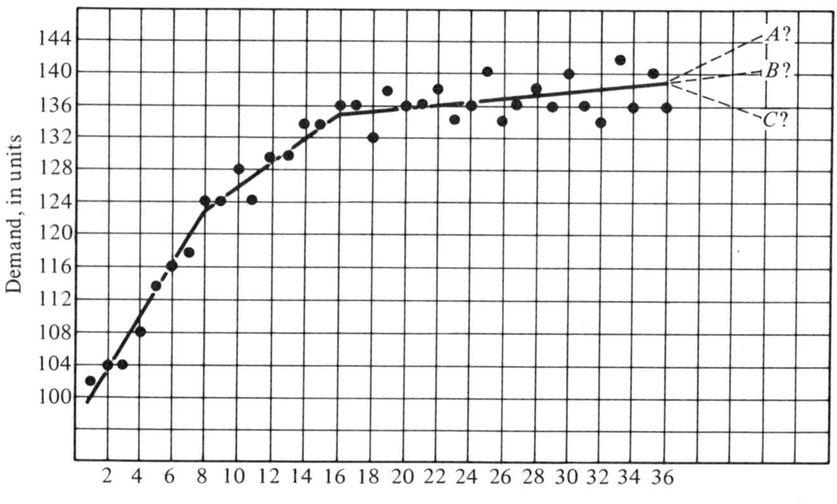

Figure 3.3 Graphical Representation of Data in Table 3.2

The problem that a forecaster must concern himself with is *when* the cause system will change. Beginning in month 37, will the trend follow lines *A*, *B*, or *C* on the graph in Fig. 3.3? We cannot outguess the future. While it may be possible to identify a cause system for past time periods, it is impossible to know exactly the kind of demand trend a cause system will generate in the future. It

*Normal is represented by the bell-shaped probability distribution curve.

Table 3.2 36 Months of Demand Data for a Product

Month	Demand	Month	Demand	Month	Demand
1	102	13	130	25	140
2	104	14	134	26	134
3	104	15	134	27	136
4	108	16	136	28	138
5	114	17	136	29	136
6	116	18	132	30	140
7	118	19	138	31	136
8	124	20	136	32	134
9	124	21	136	33	142
10	128	22	138	34	136
11	124	23	134	35	140
12	130	24	136	36	136

is important, therefore, that our forecasting system be responsive to significant changes in the cause system but not overreact to random fluctuations. We cannot develop an ideal forecasting system that will behave perfectly all the time; we can, however, develop meaningful control measures to keep us in the right ball park. We discuss one such method in one of the transfer functions developed below. All forecasting methods are intended to predict accurately future demands for our product. Many elaborate forecasting methods have been developed and implemented by various companies. Undoubtedly, some companies spend more money maintaining sophisticated forecasting procedures than is saved by the difference in accuracy between the sophisticated procedure and a less sophisticated procedure. Presented in the following sections are a few of the less sophisticated forecasting transfer functions that have been found useful for production forecasting.

LAST PERIOD DEMAND

Probably the simplest forecasting transfer function is one in which next period's forecast is made equal to the previous period's actual demand. This method requires no calculations and only one piece of data for the forecast. If we plotted the forecasted values on the same chart with the actual demand values, our forecast pattern would be identical to the demand pattern and lag behind it by exactly one time period. This plotting is done in Fig. 3.4 for Distributor *A* in Table 3.1, whose demand pattern was originally plotted in Figure 3.2.

The forecast error is the difference between actual demand and forecasted demand for a time period. Column 4 of Table 3.3 shows this difference for each of the eleven months for which we have both values.

The accuracy of a forecasting procedure is determined by how well it performs over time. Notice in Table 3.3 that for some months the forecast is too high, while in other months it is too low. We thus encounter a cancelling effect of the forecast error if the algebraic deviations are summed. We see in column 5 that the sum of algebraic deviations is 260 units. We can obtain a better measure of total forecast error simply by keeping a cumulative sum of the absolute deviations of the forecast from actual demand. This value, shown in

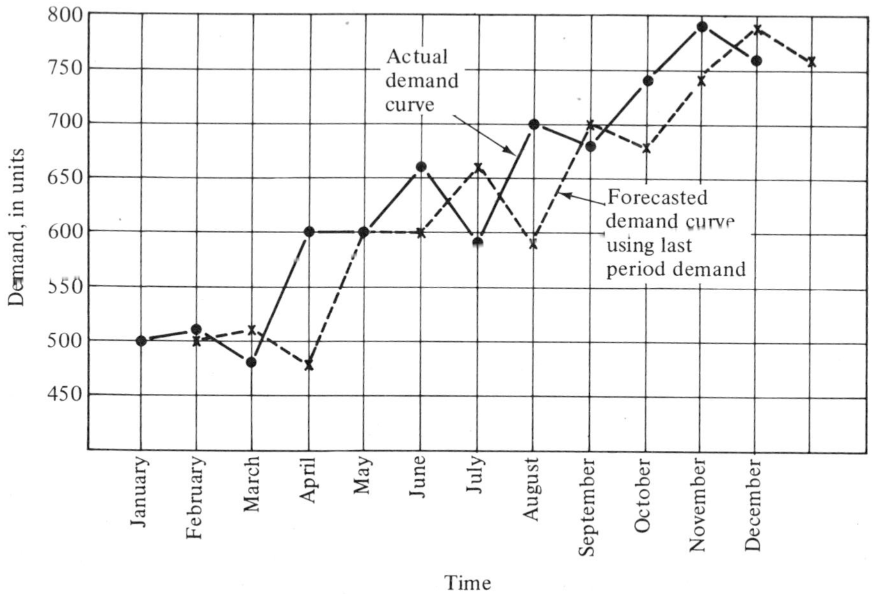

Figure 3.4 Actual and Forecasted Demand Using Last Period Demand Method

Table 3.3 Forecast Errors Using Last Period Demand as Transfer Function

1	2	3	4	5	6
Month	Actual Demand*	Forecast	Algebraic Deviations	Sum of Algebraic Deviations	Sum of Absolute Deviations
February	510	500	10	10	10
March	480	510	−30	−20	40
April	600	480	120	100	160
May	600	600	0	100	160
June	660	600	60	160	220
July	590	660	−70	90	290
August	700	590	110	200	400
September	680	700	−20	180	420
October	740	680	60	240	480
November	790	740	50	290	530
December	760	790	−30	260	560

*Demand Values Are for Distributor *A* in Table 3.1.

column 6, is 560 units for our example. Therefore, if we recognize that it is possible to obtain zero as the sum of algebraic deviations, even when the individual deviations are large, it is easy to see that the sum of absolute deviations is a better measure of the performance of the forecasting function than is the sum of algebraic deviations.

MOVING AVERAGE

Another commonly used forecasting technique is the moving average method. This method, a logical extension of the one discussed in the preceding section, generates next period's forecast by averaging the actual demand values for the last n time periods. Mathematically, we have

$$\hat{x}_t = \frac{\sum_{i=1}^{n} x_{t-i}}{n} \tag{3.1}$$

where $\hat{x}_t$ = forecasted demand for period t,

x_{t-i} = actual demand for ith period preceding period t, and

n = number of time periods to include in the moving average.

A forecast generated by this method is generally too small if the demand pattern has an upward trend and too large if a downward trend exists. If n is too small, reaction to actual demand values may be exaggerated. If n is too large, reaction is too damped.

The choice of a value for n is arbitrary and should be determined by experimentation (perhaps in a computer simulation model).

A variation of the above procedure is the weighted moving average, in which more weight is given to more recent data. The forecasting transfer function is

$$\hat{x}_t = \frac{\sum_{i=1}^{n} c_i x_{t-i}}{n} \tag{3.2}$$

where c_i = weight given to ith actual value, and

$$\sum_{i=1}^{n} c_i = n.$$

For certain demand patterns, this method partially overcomes some of the weaknesses of the standard moving average method.

The choice of values for n and the weighting coefficients are arbitrary and can be determined by experimenting with several combinations.

Let us now cast the forecasting transfer function given in Eq. 3.2 into the context of the concepts shown in Fig. 2.4. Parameters c_i and n (P. I. 2) are maintained in the forecasting Parameter File. The x_{t-i} $(i = 1, \ldots, n)$ values are actual demand values (VS. I. 1) for the previous n periods and are maintained in the forecasting Variable Status File. Equation 3.2 is the transfer function itself. The value $\hat{x}_t$ is the decision (forecasted demand for period t) that results from the processing of information by the transfer function. This process is illustrated in Fig. 3.5 for the transfer function shown in Eq. 3.2.

Note in Fig. 3.5 that a sales campaign is indicated as an external occurrence. This external factor will surely cause a disturbance in the stability of our system. The sales campaign is entered into the Variable Status File as VS. I. 2, according to our information coding scheme developed in an earlier section. Note also that sales in Fig. 3.5 is equivalent to operations in Fig. 2.4. For all other functions, operations is the prime recipient of decisions.

It would be highly instructive at this point to compare closely Figs. 2.4, 3.1, and 3.5. In particular, the reader is encouraged to compare the bottom portion of Fig. 3.1 with Fig. 3.5. The direct correspondence between the two figures should be recognized.

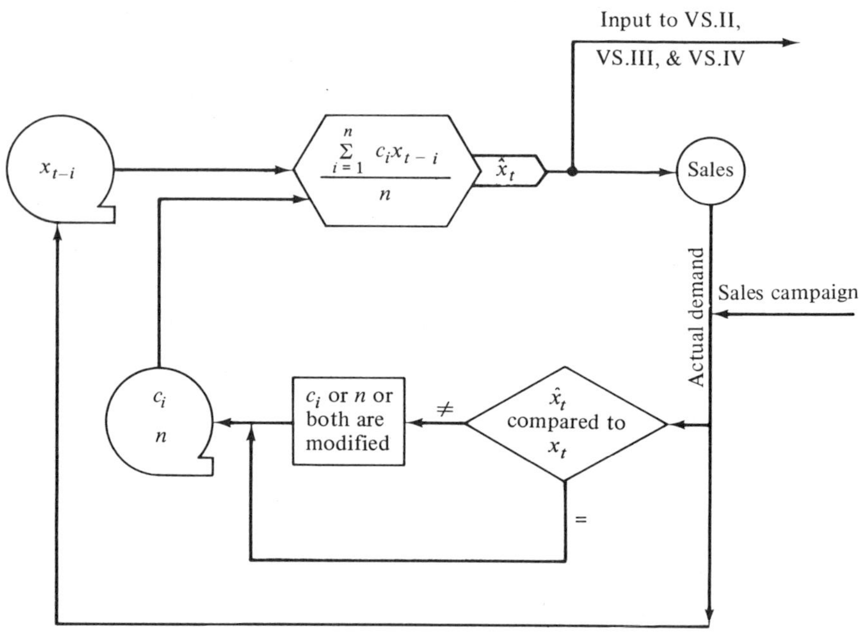

= Denotes approximately equal
≠ Denotes significantly different

Figure 3.5 The Transfer Function Concept for Weighted Moving Average Method of Forecasting

EXPONENTIALLY WEIGHTED MOVING AVERAGE

The exponentially weighted moving average (EWMA)* is a forecasting method which overcomes many of the disadvantages of the ordinary moving average and weighted moving average methods. Like the weighted moving average method, the EWMA method gives more weight to newer data; however, the EWMA method is much more efficient and is more readily adapted to computer application.

EWMA assigns weights to the demand values of previous periods in inverse proportion to their age. It does this in a very ingenious manner in which only three pieces of data are required to generate next period's forecast: (1) last period's forecast, (2) last period's actual demand, and (3) a smoothing constant, which determines the relative amount of weight given to recent demand values.

The rationale behind EWMA is that we would like our forecasting method to track the demand pattern through its general trends (i. e., its "ups and downs") without overreacting to purely random fluctuations. Suppose, for example, that for a particular product we forecasted a demand of 100 units. When actual demand became known, it was only 95 units. We must now forecast demand for the following month. Our question now becomes: How much of the difference between 100 units and 95 units can be attributed to an actual shift in the demand pattern and how much can be attributed to purely random causes? If our next forecast is for 100 units, we would be assuming that all of the difference should be attributed to chance and, therefore, that the demand pattern has not shifted at all. If our next forecast is for 95 units, we would be assuming that the entire difference should be attributed to a shift in the demand pattern and none to chance causes. (The latter assumption is the basis for the last period demand forecasting method.)

EWMA attributes part of the difference between actual demand and forecasted demand to a trend shift and the remainder to chance causes. In our present example, suppose we decide that 20 percent of the difference should be attributed to a trend shift and the remaining 80 percent to chance causes. Our new forecast, then, would be determined by reducing the previous forecast by 20 percent of the observed difference of 5 units. Since our previous forecast was 100 units and 20 percent of 5 units is 1 unit, the new forecast would be 99 units. After the actual demand for the new period became known, we would determine the forecast for the following period by again computing 20 percent of actual demand minus forecasted demand and adding that quantity (which may be positive or negative) to the previous forecast, or 99. The 20 percent value is called the smoothing constant and is denoted α.

We can generalize this procedure as follows:

*Also called "exponential smoothing" in many texts and articles on forecasting.

$$\underset{\text{forecast}}{\text{next}} = \underset{\text{forecast}}{\text{previous}} + \underset{\text{constant}}{\text{smoothing}} \times \underset{\text{error}}{\text{forecast}}$$

$$\hat{x}_t = \hat{x}_{t-1} + \alpha(x_{t-1} - \hat{x}_{t-1}) \qquad (0 \leq \alpha \leq 1) \tag{3.3}$$

This expression can be rearranged algebraically to obtain an equivalent expression more convenient for computation.

$$\hat{x}_t = \alpha x_{t-1} + (1 - \alpha)\hat{x}_{t-1} \tag{3.4}$$

In order to use Eq. 3.4 as a transfer function, two further tasks must be accomplished. We must first determine the particular value of α, the smoothing constant, that we should use. We must then devise a way of getting started, since our forecasting equation involves a previous forecast.

Determining the Smoothing Constant. In developing the rationale for EWMA, we discussed the manner in which the forecast error (difference between actual demand and forecasted demand) might be divided between random causes and an actual shift in the demand pattern. The smoothing constant α was defined to be the proportion of total forecast error that could be attributed to a shift. Our problem, then, is to determine this proportion over many time periods.

There are numerical procedures for determining an optimal value of α for a particular demand series. The reader is advised to see Brown (1963) and Winters (1960) for some of these procedures.. Another way to determine α is simply to experiment (usually by computer simulation) with different values of α until a "good" value is found. Each value can be used with the actual demand series. An evaluation similar to that shown in Table 3.3 would be made for each α tried. The value having the lowest sum of absolute deviations (column 6 in Table 3.3) would then be used in the forecasting transfer functions. In practice, values of α between 0.05 and 0.30 are usually used.

Getting Started. For the first forecast, we need a forecasted value $\hat{x}_{t-1}$ for the previous time period. We obviously do not have such a value. If we reflect on the basic nature of the EWMA transfer function, however, we realize that no matter what initial $\hat{x}_{t-1}$ value we use, the method will eventually track its way to very near the true demand pattern. The fact remains that the better our initial estimate of $\hat{x}_{t-1}$, the quicker the method will correct itself and become an accurate forecasting procedure.

One fairly good method of estimating the initial $\hat{x}_{t-1}$ is simply to use x_{t-1}, last period's actual demand. This results in the first forecast $\hat{x}_t$ being equal to the last period's actual demand. The method is quite satisfactory if we have reason to believe that last period's actual demand is typical in this region of the demand pattern.

Another possible method is to determine the average demand for a certain number of immediately preceding time periods—say the past five periods—and then use this average as $\hat{x}_{t-1}$ for the first forecast.

Example Using EWMA. To demonstrate the mechanics of EWMA, we here apply it to the demand pattern of Distributor B in Fig. 3.2. We use two different smoothing constants and also obtain a forecast using the last period demand method. We then compare these three results by determining the sum of the absolute deviations, as in Table 3.3.

For the EWMA method, we use an initial $\hat{x}_{t-1}$ value of 430, which is also x_{t-1} since our first forecast is for February. We use smoothing constants of 0.1 and 0.2.

Equation 3.4 is used to generate the EWMA forecasts shown in columns 3 and 6 of Table 3.4. Example calculations for March and April using $\alpha = 0.1$ in Eq. 3.4 are as follows:

$$\begin{aligned} \hat{x}_3 &= (0.1)(x_2) + (0.9)(\hat{x}_2) \\ &= (0.1)(380) + (0.9)(430) = 425 \\ \hat{x}_4 &= (0.1)(x_3) + (0.9)(\hat{x}_3) \\ &= (0.1)(420) + (0.9)(425) = 424 \end{aligned}$$

The forecast obtained with the last period demand method is shown in column 9.

Columns 5, 8, and 11 in Table 3.4 provide a means of comparing the three forecasting transfer functions. Both EWMA transfer functions performed considerably better than the last period demand method. At first glance it would appear that $\alpha = 0.2$ performed better than $\alpha = 0.1$ This is true, however, only because we chose $\hat{x}_{t-1} = 430$. We can see from Fig. 3.2 that the true average is 400 units. The larger smoothing constant reacts faster to a change in demand level and thus converged close to the true average much faster than the smaller smoothing constant. This "closing" can be shown graphically by plotting the forecasted values in columns 3 and 6 of Table 3.4 along with the actual demand in column 1. The reader is urged to do this.

Suppose that earlier demand records had led us to choose $\hat{x}_{t-1} = 400$, instead of 430. How would our two smoothing constants perform with this choice? The results are shown in Table 3.5. We see that in this case $\alpha = 0.1$ is the better smoothing constant.

Also included in Table 3.5 are the results of using a constant forecast value of 400 units. This is equivalent to using $\alpha = 0.0$ in Eq. 3.4. This forecast was better than either EWMA forecast. This fact shouldn't surprise us if we remember the rationale behind the smoothing constant. We said earlier that α represents that portion of the forecast error that should be attributed to a shift in the demand pattern. It is obvious from Fig. 3.2 that the demand pattern for

Table 3.4 Comparison of EWMA to Last Period Demand, Using Initial $\hat{x}_{t-1}$ Value of 430

1	2	3	4	5	6	7	8	9	10	11
		$\alpha = 0.1$			$\alpha = 0.2$			Last Period Demand		
Month	Actual Demand	Forecast	Absolute Deviation	Sum of Absolute Deviation	Forecast	Absolute Deviation	Sum of Absolute Deviation	Forecast	Absolute Deviation	Sum of Absolute Deviation
January	430									
February	380	430	50	50	430	50	50	430	50	50
March	420	425	5	55	420	0	50	380	40	90
April	370	424	54	109	420	50	100	420	50	140
May	410	419	9	118	410	0	100	370	30	170
June	380	418	38	156	410	30	130	410	30	200
July	440	414	26	182	404	36	166	380	60	260
August	380	416	36	218	411	31	197	440	60	320
September	420	410	10	228	405	15	212	380	40	360
October	370	411	41	269	408	38	250	420	50	410
November	410	406	4	273	400	10	260	370	40	450
December	390	406	16	289	402	12	272	410	20	470

Table 3.5 Comparison of EWMA to Constant Forecast, Using Initial $\hat{x}^{t}_{-1}$ Value of 400

1	2	3	4	5	6	7	8	9	10	11
		$\alpha = 0.1$			$\alpha = 0.2$			Constant Forecast		
Month	Actual Demand	Forecast	Absolute Deviation	Sum of Absolute Deviations	Forecast	Absolute Deviation	Sum of Absolute Deviations	Forecast	Absolute Deviation	Sum of Absolute Deviations
January	430									
February	380	403	23	23	406	26	26	400	20	20
March	420	401	19	42	401	19	45	400	20	40
April	370	403	33	75	405	35	80	400	30	70
May	410	400	10	85	398	12	92	400	10	80
June	380	401	21	106	400	20	112	400	20	100
July	440	339	41	147	396	44	156	400	40	140
August	380	403	23	170	405	25	181	400	20	160
September	420	401	19	189	400	20	201	400	20	180
October	370	403	33	222	404	34	235	400	30	210
November	410	400	10	232	397	13	248	400	10	220
December	390	401	11	243	400	10	258	400	10	230

Distributor B has a constant average, with only random fluctuations around the average. As long as this demand pattern continues, a constant forecast of 400 units should be used.

Limitations of EWMA. The preceding example demonstrated that EWMA should not be used when the demand pattern contains only random fluctuations around a constant average. It is equally easy to demonstrate that EWMA should not be used with a demand pattern that contains an upward or downward trend. If we look closely at Fig. 3.2, we see that Eq. 3.4 would generate consistently low forecasts for Distributor A and for the Total Demand pattern and that it would generate consistently high forecasts for Distributor C. Finally, Eq. 3.4 is not appropriate for demand trends that contain seasonal patterns. While we have not yet encountered a cyclical demand pattern, it is easy to see that forecasts generated by Eq. 3.4 would be too low while demand is increasing and too high while demand is decreasing. We discuss the manner in which trends and cycles should be handled in the two following sections.

You may wonder just how EWMA can be used. The elementary form of EWMA, given by Eq. 3.4, can be used alone as a forecasting transfer function only for demand patterns whose average values shift moderately at unpredictable (i. e., erratic) time intervals. Since relatively few demand patterns are of this nature, Eq. 3.4 has rather limited application as a complete forecasting transfer function. In the majority of cases, as we see in a later section, it is used in combination with other transfer functions.

REGRESSION ANALYSIS

When a demand pattern is consistently increasing or decreasing, we can develop our forecasting transfer function using regression analysis, the details of which may be found in most standard statistics texts (Bowker and Lieberman 1959, Miller and Freund 1965, and Wine 1964). Consider, for example, the demand trend of Distributor A in Fig. 3.2. The demand for Distributor A is steadily increasing at what appears to be a constant rate.

The rationale for using regression analysis for forecasting is that the total set of factors that generated the cause system in the past will continue to operate in the future. Thus, forecasting becomes a matter of determining the general trend line and extrapolating this line into the future. Figure 3.6 shows the original twelve demand values for Distributor A in Fig. 3.2, a straight line that best describes this set of values, and a dashed line extrapolated as the forecasted trend line. While we realize that our extrapolated line will not agree precisely with future demand values, it is expected that deviations of actual demand from the extrapolated line will be the result of random fluctuations from the line. Remember, it is the *trend* that we hope to predict accurately, not each individual demand value.

It is possible to use regression analysis to fit a line to a nonlinear trend. The

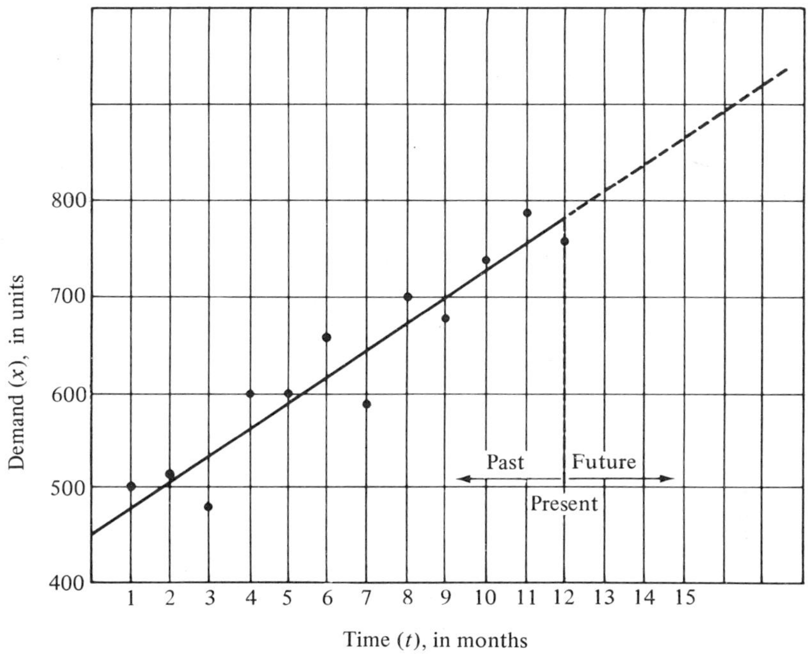

Figure 3.6 Illustration of Regression Analysis in an Upward Trend

details on how to do it are contained in standard statistics texts. We confine our discussion here to linear demand trends.

The linear line in Fig. 3.6 is determined by the method of least squares. This method fits a line to a set of observations such that the sum of the squared vertical distances of the observations from the line is minimized.

The basic equation for a straight line that expresses demand (x) as a function of time (t) is:

$$\hat{x}_t = a + bt \tag{3.5}$$

where a is the intersection of the line with the vertical axis when $t = 0$ and b is the slope of the line. Our problem is to determine the values of a and b such that the least squares criterion is satisfied. The following equations for a and b are derived in most statistics texts:

$$b = \frac{n \sum_{i=1}^{n} t_i x_i - \sum_{i=1}^{n} t_i \sum_{i=1}^{n} x_i}{n \sum_{i=1}^{n} t_i^2 - \left(\sum_{i=1}^{n} t_i \right)^2} \tag{3.6}$$

$$a = \bar{x} - b\bar{t} \tag{3.7}$$

where n is the number of periods of demand data that we include in the calculations.

A basic assumption in regression analysis is that for a given point in time, demand is a normally distributed random variable whose mean is the x coordinate on the regression line at that point in time. Thus, if we know the standard deviation of the distribution, we can make statements about the reliability of our forecasts. An estimate of the standard deviation is provided by the following formula:

$$s_{x|t} = \sqrt{\frac{\sum_{i=1}^{n} (x_i - \bar{x})^2 - \dfrac{\left[\sum_{i=1}^{n} (t_i - \bar{t})(x_i - \bar{x})\right]^2}{\sum_{i=1}^{n} (t_i - \bar{t})^2}}{n - 2}} \tag{3.8}$$

Equations 3.6, 3.7 and 3.8 are more easily used if the data are organized into a table for the calculations. Table 3.6 contains the needed calculations for the data points shown in Fig. 3.6. The column totals from Table 3.6 give the following values for a, b and $s_{x|t}$:

$$b = \frac{(12)(5338) - (78)(760)}{7800 - 6084} = 2.78$$

$$a = 63.33 - (2.78)(6.5) = 45.26$$

$$s_{x|t} = \sqrt{\frac{1198.8 - \dfrac{(391)^2}{143}}{10}} = 3.6$$

We see from the second column of Table 3.6 that the demand values were divided by 10 to simplify the calculations. Coding such as this is perfectly appropriate, but we must remember to uncode the results:

$$b = 27.8$$

$$a = 452.6$$

$$s_{x|t} = 36$$

We use $a = 452.6$ and $b = 27.8$ as parameters in our forecasting transfer function. Substituting these values into Eq. 3.5, we obtain

$$\hat{x}_t = 452.6 + 27.8t$$

Table 3.6 Table for Determining Regression Line Parameters and Standard Deviation about the Line

Month t	Coded Demand x	tx	t^2	$(t_i - \bar{t})$	$(x_i - \bar{x})$	$(t_i - \bar{t})(x_i - \bar{x})$	$(x_i - \bar{x})^2$	$(t_i - \bar{t})^2$
1	50	50	1	−5.5	−13.3	73.1	176.9	30.25
2	51	102	4	−4.5	−12.3	55.3	151.3	20.25
3	48	144	9	−3.5	−15.3	53.5	234	12.25
4	60	240	16	−2.5	− 3.3	8.2	10.9	6.25
5	60	300	25	−1.5	− 3.3	4.9	10.9	2.25
6	66	396	36	− .5	2.7	− 1.4	7.3	.25
7	59	413	49	.5	− 4.3	− 2.1	18.5	.25
8	70	560	64	1.5	6.7	10.0	44.9	2.25
9	68	612	81	2.5	4.7	11.7	22	6.25
10	74	740	100	3.5	10.7	37.4	114.4	12.25
11	79	869	121	4.5	15.7	70.6	246.4	20.25
12	76	912	144	5.5	12.7	69.8	161.3	30.25
78	760	5338	650			391	1198.8	143.0

$\bar{t} = 78/12 = 6.5$
$\bar{x} = 760/12 = 63.33$

as our forecasting transfer function, which says that demand is increasing by an average of 27.8 units per month. To estimate the future demand trend, we extend this line into the future.

We said earlier that demand at any point in time is a normally distributed random variable. The particular demand value that actually occurs at a point in time is, therefore, only one of many such values that could have occurred. The possible demand values at a particular time value constitute a normal distribution, with the mean equal to the regression line value at that point and a standard deviation whose value is determined by Eq. 3.8. It is assumed that the standard deviation is constant for the entire regression line.

Consider, for example, month number 8 in Fig. 3.6. Actual demand was 700 units. Our least squares transfer function gives us the regression line value for $t = 8$:

$$\hat{x}_8 = 452.6 + 27.8(8) = 675$$

According to our normal curve assumptions, $x_8 = 700$ is a value from a normal distribution having a mean of 675 and a standard deviation of 36. Such a distribution is shown in Fig. 3.7.

Shown on the abscissa in Fig. 3.7 are values 1, 2, and 3 standard deviations above and below the mean. It is shown in statistics texts that approximately 68 percent of the values of a normal distribution fall within the range $\mu \pm 1\sigma$;

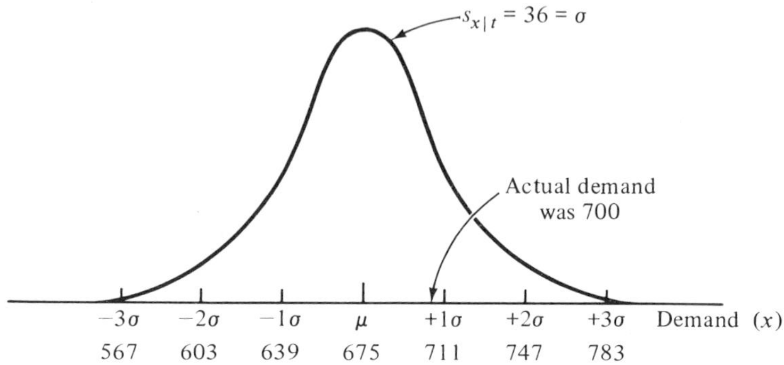

Figure 3.7 Distribution of Demand for Time Period 8 in Figure 3.6

95 percent of the values fall within $\mu \pm 2\sigma$; and 99.7 percent of the values fall within $\mu \pm 3\sigma$. Thus, we are not at all surprised that actual demand for period 8 was 700 units, since this figure is within one standard deviation of the average value. This concept is particularly useful when applied to future time periods, since it provides a way of estimating the reliability of our forecasts. We must be very careful, however, in how we use this approach.

We could construct a normal distribution curve for the other 11 time periods shown in Fig. 3.6. Only the mean would change (since we assume that the standard deviation remains constant). In each case, the actual demand value would be one of the many values that could have occurred.

Let us now discuss how we can use our least squares transfer function to estimate future demand. Assuming that the cause system that generated past demand values (from which we obtained the parameters in our transfer function) remains unchanged, we simply substitute future values of t into our least squares transfer function to obtain estimates of average demand during future time periods. The values of t must be consistent with those from which the regression line was derived. For example, the estimate of demand for the first month of next year, month 13, would be determined by substituting $t = 13$ into our least squares transfer function:

$$\hat{x}_{13} = 452.6 + 27.8(13) = 814$$

This does not mean that we expect demand during time period 13 to be exactly 814 units; rather, we expect demand for $t = 13$ to be some value from a normal distribution whose mean is 814 and whose standard deviation is 36.

If we determined demand values for the first six months of the forecast period, we would obtain the dashed line in Fig. 3.8, which is simply an extrapolation of the solid regression line. Also shown on Fig. 3.8 are upper and lower control limits.* These limits were constucted at approximately 2 stand-

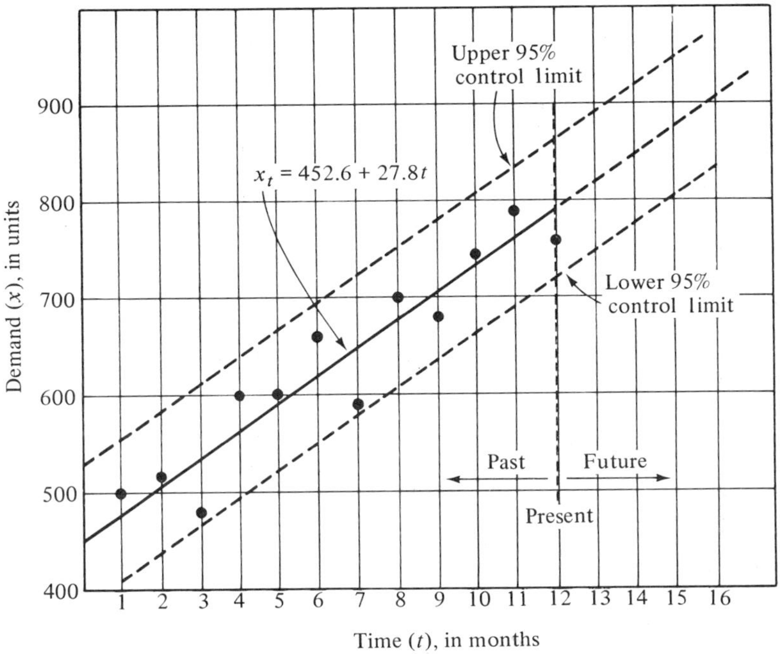

Figure 3.8 Projection of Regression Line with 95% Control Limits

ard deviations above and below the trend line. As discussed earlier, approximately 95% of the demand values should fall between the two limits.

To extrapolate the control limits into the forecast region is useful for two reasons. First, when the actual demand values occur, they can be compared to the control limits to determine whether the demand is a value that we would reasonably expect. If a demand value falls outside the control limits, we would have reason to wonder if the cause system has changed.

Second, the control limits can provide guidelines for production planning. Assuming that the cause system does not change, we can be 95% certain that demand for month 13 will not be less than 742 units nor more than 886 units. This information provides a framework within which we can plan the use of our productive resources.

We now determine how well our derived least squares forecasting function would have performed if it had been used during the twelve-month period shown in Fig. 3.8. The performance is indicated in Table 3.7. We see that the sum of absolute deviations is 333, which is considerably better than the 560

*These control limits are not the same as confidence limits on regression lines. Our control limits simply provide general guidelines as to the amount of variation we can reasonably expect in our cause system.

Table 3.7 Forecast Errors Using Regression Analysis

Month	Actual Demand*	Forecast	Absolute Deviation	Sum of Absolute Deviations
1	500	480	20	20
2	510	508	2	22
3	480	534	54	76
4	600	564	36	112
5	600	592	8	120
6	660	619	41	161
7	590	647	57	218
8	700	675	25	243
9	680	703	23	266
10	740	731	9	275
11	790	758	32	307
12	760	786	26	333

*Demand Values Are for Distributor *A* in Table 3.1

value we obtained using last period demand as a transfer function for the same data (see Table 3.3).

Limitations of Regression Analysis. There are some serious limitations of regression analysis as a forecasting method. First, we must assume that the cause system will continue without change into the future. As we saw in Fig. 3.3, however, many cause systems are subject to frequent change. It is true that our control limits will indicate when a change occurs in the cause system. Once we detect a change, however, we do not have enough data points to calculate our new slope.

A second limitation relates to the basic assumptions underlying all regression analyses. A constant standard deviation is assumed. Intuition tells us that this assumption is not usually true, but for relatively short forecasting intervals this assumption is usually not particularly harmful. Another assumption is that the deviations of demand from the regression line are normally distributed. This belief usually leads us to the further assumption that demand itself is normally distributed, as depicted in Fig. 3.7. It is also assumed in regression analysis that no extrapolation will be made outside the range of observed values. However, since the only purpose of forecasting is to project the demand pattern into the future, it is absolutely necessary that we violate this last assumption. To justify this projection, we further assume that the cause system will remain unchanged in the future.

A third limitation of regression analysis as a forecasting method is that it requires the storage and processing of large amounts of data. This problem can become significant when thousands of items are marketed.

Despite the limitations just discussed, however, regression analysis remains a valuable tool to the forecaster when used in full recognition of its restrictions.

In many situations, the most practical way to project the trend is "by eye" on a piece of graph paper. We can usually come surprisingly close to the line that would be obtained using Eq. 3.6 and 3.7.

Table 3.8 Demand Data with Seasonal Cycle

Month	Demand	Month	Demand	Month	Demand
1	102	13	104	25	103
2	115	14	124	26	112
3	128	15	124	27	126
4	121	16	132	28	112
5	118	17	125	29	126
6	110	18	103	30	112
7	82	19	92	31	94
8	68	20	74	32	81
9	75	21	68	33	69
10	75	22	77	34	66
11	80	23	82	35	82
12	97	24	104	36	93

ADJUSTING FOR CYCLES

Many demand patterns exhibit some sort of cyclical behavior, such as a seasonal pattern. Consider, for example, the historical demand data listed in Table 3.8 and shown graphically in Fig. 3.9. This demand pattern shows a strong seasonal cycle, with highs occurring in the third or fourth month each year and lows occurring in the eighth or ninth month each year. If we assume that this pattern will continue, we can project the future demand curve shown by the dashed line extension in Fig. 3.9.

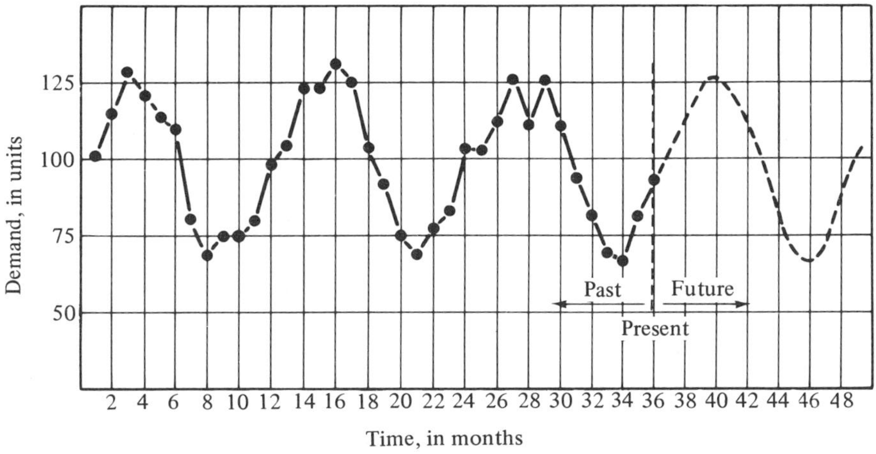

Figure 3.9. Demand Pattern with Seasonal Cycle, from Table 3.8

It is possible to develop a transfer function for this situation by fitting an equation to the points. Procedures for developing this equation, however, are beyond the scope of this text. Such sophistication is rarely needed or justified for demand forecasting purposes.

Although several procedures have been developed for forecasting demand when the underlying cause system contains a seasonal component, we discuss only two of these methods. The first method (see Biegel 1963) simply fits one straight line segment to that portion of the demand curve having an upward trend and another straight line segment to that portion having a downward trend. Two regression equations can be determined formally if this much accuracy is required. The lines shown in Fig. 3.10 were fitted "by eye." We compare the accuracy of this method to the following method after we develop the underlying logic of the latter.

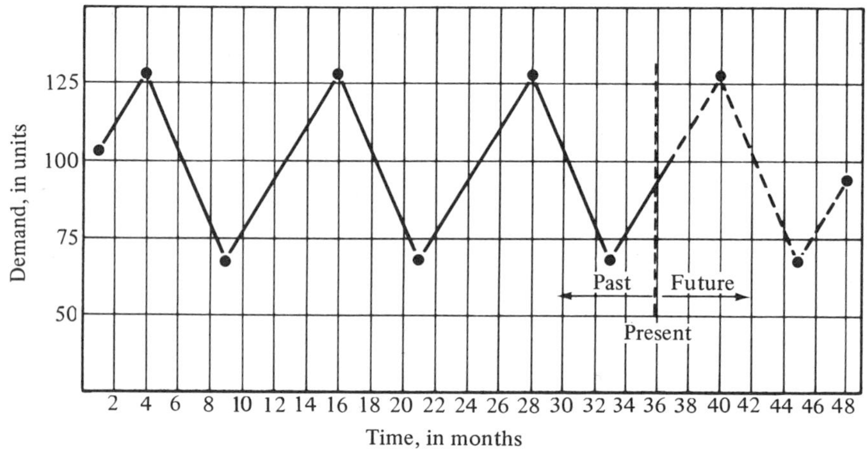

Figure 3.10 Use of Straight Line Segments in Seasonal Demand Patterns

Note that it is possible to construct control limits on the graph, as was done in Fig. 3.8. An estimate of $s_{x|t}$ would be obtained from those points on a downward trend and another from those points on an upward trend. These two values would then be averaged for our estimate of $s_{x|t}$ to be used in calculating control limits.

The second method of seasonal forecasting we examine makes use of monthly indexes to account for the effect of the cycle.* The rationale for this method is as follows: the demand pattern has some over-all average value, and if demand did not vary from season to season, each demand value would simply

*Our discussion here assumes a forecasting period of one month. However, other time periods, such as a week or a quarter, could just as well be used.

be a random deviation from the over-all average. We can remove the effect of the seasonal component in our demand pattern by determining a factor (which we call an index) for each month that, when divided into the seasonal value for that month, will result in a value very near the over-all mean. We determine such an index for each month. Then, when we forecast the demand for a particular month, we multiply the over-all mean by the index for that month.

Table 3.9 demonstrates the calculations for the demand pattern listed in Table 3.8 for the index method. The procedure for performing the calculations is:

Step 1. For each month, sum columns 1, 2, and 3, and enter sum in column 4.

Step 2. Determine monthly average by dividing column 4 by 3 for each month. Enter monthly average in column 5.

Step 3. Find over-all average by summing column 4 and dividing by 36.

$$\bar{x} = 3556/36 = 98.8$$

Step 4. Calculate monthly indexes, I_i, by dividing column 5 for each month by the over all average. Enter result in column 6.

Step 5. Determine adjusted monthly demand (the cycle has been removed) for year 3 by dividing monthly demand for year 3 in column 3 by the index in column 6. Enter results in column 7.

$$x'_{i3} = x_{i3}/I_i$$

Step 5 in the procedure and column 7 in Table 3.9 show the third year's demand data with the effects of the seasonal cycle removed. The values in column 7 are assumed to be random deviations from the over-all mean. These values can be used to estimate the standard deviation of the demand values. The standard deviation, in turn, can be used to establish control limits.

The standard deviation is:

$$s = \sqrt{\frac{\sum_{i=1}^{n} (x'_{i3})^2 - \frac{\left(\sum_{i=1}^{n} x'_{i3}\right)^2}{n}}{n-1}} \tag{3.9}$$

For the values in column 7, $s = 5.55$. We can establish 95% control limits by constructing one dashed line $[(2)(5.55)] = 11.1$ units above the projected demand trend in Fig. 3.9 and another line 11.1 units below the projected trend. Note that we could obtain a better estimate of the standard deviation by computing adjusted demand values for the first two years also and then using all 36 values in computing s.

Table 3.9 Calculations for Monthly Indexes for Seasonal Forecasting

	1	2	3	4	5	6	7
Month	Demand Data			Monthly Sum	Average $\sum_j x_{ij}/3$	Index I_i	Adjusted Demand for Year 3 x'_{i3}
(i)	Year 1	Year 2	Year 3				
1	102	104	103	309	103.00	1.04	98.9
2	115	124	112	351	117.00	1.19	94.0
3	128	124	126	378	126.00	1.27	99.1
4	121	132	112	365	121.67	1.23	91.2
5	118	125	126	369	123.00	1.25	101.0
6	110	103	112	325	108.33	1.10	102.0
7	82	92	94	268	89.33	.90	104.5
8	68	74	81	223	74.33	.75	108.0
9	75	68	69	212	70.67	.72	96.0
10	75	77	66	218	72.67	.74	89.0
11	80	82	82	244	81.33	.82	100.0
12	97	104	93	294	98.00	.99	94.0

$\sum x = 3556$

It should be clear that this procedure can be used with more than three years' data. As a matter of fact, in most cases at least five years demand data is needed to obtain good results. Most real world demand patterns are not as "pretty" as that shown in Fig. 3.9.

We use the results of the above procedure to specify parameters in our forecasting transfer function. Recalling the rationale underlying the index method, we determine the forecasted demand for a particular month by multiplying that month's index by the over-all average. Symbolically, we have

$$\hat{x}_t = (I_i)(\bar{x}) \tag{3.10}$$

where t is the ith month in a calendar year. For example, $\hat{x}_{17}$ is the forecast for the 5th month in a calendar year; therefore, I_5 would be used in Eq. 3.10.

This transfer function is applied to the third year and the results are shown in Table 3.10. Also shown in Table 3.10 are the results of the previous forecasting method discussed, that using straight line segments. We see that, although the index method is very slightly better, both methods perform extremely well for this case. The reason, of course, is that the demand pattern is exceptionally regular.

It is important to recognize that the index method as described in this section works only for cyclical demand patterns that do not contain an upward or downward trend component.

Table 3.10 Comparison of Index Method and Line Segment Method for Data in Fig. 3.10

		Index Method			Line Segment Method		
Month	Year 3 Actual	Forecast	Absolute Deviation	Sum of Absolute Deviation	Forecast	Absolute Deviation	Sum of Absolute Deviation
1	103	103	0	0	102	1	1
2	112	117	5	5	110	2	3
3	126	115	11	16	119	7	10
4	112	121	9	25	128	16	26
5	126	123	3	28	115	11	37
6	112	109	3	31	103	9	46
7	94	89	5	36	92	2	48
8	81	74	7	43	79	2	50
9	69	71	2	45	67	2	52
10	66	73	7	52	76	10	62
11	82	81	1	53	85	3	65
12	93	98	5	58	93	0	65

COMBINATION METHODS

Often a demand pattern exhibits characteristics which suggest the use of two or more of the forecasting transfer functions we have discussed previously. For example, a demand pattern can contain a trend component (upward or downward) as well as a seasonal component. The transfer function needed for this situation would be a combination of regression analysis and the index method.

Consider the historical demand data listed in Table 3.11 and shown graphically in Fig. 3.11. If we assume that this pattern will continue, we can project our future demand curve in the direction shown by the dashed line in Fig. 3.11.

Table 3.11 Demand Data with Seasonal Cycle and Upward Trend

Month	Demand	Month	Demand	Month	Demand
1	80	13	112	25	133
2	83	14	108	26	120
3	79	15	93	27	115
4	70	16	93	28	120
5	75	17	95	29	104
6	65	18	80	30	103
7	65	19	85	31	103
8	57	20	74	32	100
9	80	21	82	33	104
10	92	22	117	34	119
11	93	23	123	35	143
12	110	24	132	36	148

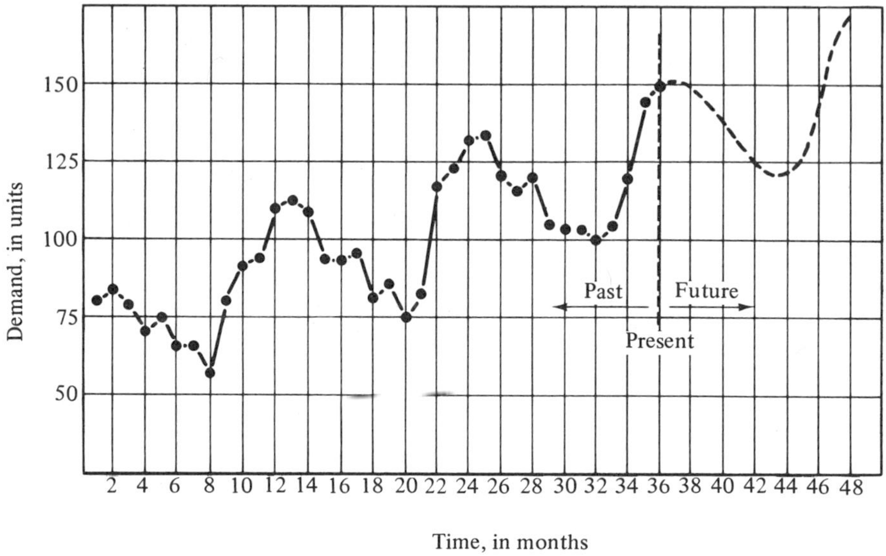

Figure 3.11 Demand Curve for Data of Table 3.11

It is possible to fit straight line segments to the data points in order to extend the demand pattern into the future. However, this project is left as an exercise for the reader.

We now develop a forecasting transfer function that combines regression analysis with the index method. The rationale for this method is a combination of those for the two individual methods. We first remove the effects of the trend component and then those of the cyclical component. These steps give us the parameters used in our forecasting transfer function to account for the combined effects of cycle and trend.

We determine the over-all trend using regression analysis. Equations 3.6 and 3.7 are used to estimate b and a, respectively, in the least squares regression equation, $x_t = a + bt$. We use the latter equation to adjust all demand values so that the trend is removed. This adjustment leaves us with a demand pattern similar to the one in Fig. 3.9. It still contains a seasonal component and random fluctuations, but the long term trend has a zero slope. Once we have removed the trend, we then use the index method to remove the cycle. The two steps result in the parameter needed for our forecasting transfer function:

$$\hat{x}_t = bt + I_i \bar{x} \tag{3.11}$$

The procedure for determining the parameters for Eq. 3.11 is:

Step 1. Determine b and a using Eq. 3.6 and 3.7, respectively. This can be done best by constructing a table such as the first four columns of Table 3.6. The results are:

$$b = 1.59 \quad ; \quad a = 69$$

Table 3.12 Calculations for Monthly Indexes after Trend Has Been Removed

	1	2	3	4	5	6	7
	Demand Data						
Month (i)	Year 1	Year 2	Year 3	Monthly Sum	Average $\sum_j x'_{ij}/3$	Index I_i	Adjusted Demand for Year 3 x''_{i3}
1	78	91	93	262	87.3	1.26	69.3
2	80	86	79	245	81.7	1.17	70.0
3	74	69	72	215	71.7	1.03	69.5
4	64	68	76	208	69.9	1.01	69.3
5	67	68	58	193	64.3	0.92	70.0
6	55	51	55	161	53.7	0.77	69.7
7	54	55	54	163	54.3	0.78	69.5
8	44	42	49	135	45.0	0.65	69.2
9	66	49	52	167	55.7	0.80	69.7
10	76	82	65	223	74.4	1.07	69.5
11	76	86	87	249	83.0	1.19	69.7
12	91	94	91	276	92.0	1.32	69.7
				$\sum x = 2497$			

Table 3.13 Comparison of Regression-Index Method and Regression-Only Method for Data in Fig. 3.11

		Regression-Index			Regression-Only		
Month	Year 3 Actual	Forecast	Absolute Deviation	Sum of Absolute Deviations	Forecast	Absolute Deviation	Sum of Absolute Deviations
1	133	128	5	5	109	24	24
2	120	122	2	7	110	10	34
3	115	115	0	7	112	3	37
4	120	114	6	13	113	7	44
5	104	110	6	19	115	11	55
6	103	102	1	20	117	14	69
7	103	103	0	20	118	15	84
8	100	96	4	24	120	20	104
9	104	108	4	28	121	17	121
10	119	129	10	38	123	4	125
11	143	139	4	42	125	18	143
12	148	149	1	43	126	22	165

Step 2. Determine adjusted demand values for years one, two, and three.

$$x'_{ij} = x_{ij} - bt$$

Enter results in columns 1, 2, and 3 of Table 3.12. The trend effect has been removed.

Step 3. Determine monthly indexes exactly as was done in Table 3.9. Also calculate double adjusted demand values for year three to estimate the standard deviation.

$$x''_{i3} = x'_{i3}/I_i$$

Determine the standard deviation using x''_{i3} rather than x_{i3} in Eq. 3.9.

We now use the obtained values of b, I_i, and $\bar{x}$ as parameters in our forecasting transfer function:

$$\hat{x}_t = 1.59t + (I_i)(69.5)$$

We can construct control limits on our demand curve as we have done in previous examples.

The above transfer function is used to obtain forecasted demand values for year three. The results are shown in the third column of Table 3.13. Also shown in Table 3.13 are the forecasts for year 3 obtained via a single regression line.

$$\hat{x}_t = 69 + 1.59t$$

By comparing the sum of absolute deviations values of the two methods, we can see that the accuracy of the forecast is improved significantly by adjusting for the cyclical component.

Table 3.14 Calculation for Forecasts Shown in Table 3.13

	1	2	3	4	5	6
i	t	$1.59t$	$1.59t + 69$	I_i	$(I_i)(69.5)$	$\hat{x}_t = (2)+(5)$
1	25	40	109	1.26	88	128
2	26	41	110	1.17	81	122
3	27	43	112	1.03	72	115
4	28	44	113	1.01	70	114
5	29	46	115	0.92	64	110
6	30	48	117	0.77	54	102
7	31	49	118	0.78	54	103
8	32	51	120	0.65	45	96
9	33	52	121	0.80	56	108
10	34	54	123	1.07	75	129
11	35	56	125	1.19	83	139
12	36	57	126	1.32	92	149

Table 3.14 is included to demonstrate a convenient computing method for obtaining forecasts for the two methods.

Feedback and Corrective Action

We discussed in the previous section several forecasting transfer functions available for use in an operations planning and control system. Many other forecasting methods, particularly variations or combinations of those presented previously, can be devised. The choice of a transfer functon depends on the demand pattern and the forecasting accuracy required.

Any forecasting method can be described in a manner similar to that in Fig. 3.5. Regardless of the particular transfer function selected, no method can predict demand precisely or prevent the cause system from changing. It is imperative, therefore, that we continuously evaluate the performance of the transfer function. In this section we discuss the feedback and corrective action loop shown in Fig. 3.5.

SOURCES OF FEEDBACK

Our main source of feedback for forecasting is Sales (see Fig. II. 1). Actual demand values for each of our products are conveyed to forecasting after the completion of each time period (day, week, month, etc.). We can imagine these data entering our two files shown in Fig. II. 2. These data become part of our Demand History and thus are entered as P. I. 1, according to our elementary coding system. They also are entered as VS. I. 1, since it may be used in a transfer function, such as the moving average method, in which last week's demand is a variable.

Another important source of feedback comes from External Perturbations, such as rush orders, cancelled orders, national or regional economic factors, and competition. These occurrences, together with information concerning marketing programs, enter our files as VS. I. 2.

SHORT-TERM CORRECTIVE ACTION

Recall that a forecaset is an estimate of the expected demand trend, not an individual demand value. Accordingly, we do not expect actual demand data to agree precisely with our forecast. There is natural variation in a demand pattern, even when the forecast is good. After each time period we must determine whether the forecast error (i. e., actual demand minus forecasted demand) can be attributed to natural variation or to some assignable cause outside the basic cause system.

In several of our previous examples, control limits were constructed on our forecast to monitor the performance of our forecasting system. As long as demand falls within the control limits, without several consecutive values appearing on one side of the trend line, our transfer function is performing satisfactorily.

Whenever a value falls outside the control limits, we attempt to identify the cause of the extreme variation. Once we have identified the cause, we ask whether it is to be considered a permanent component of our cause system. If so, we then must adjust our forecast, and all subsequent planning based upon the forecast, to accommodate the new situation. The actual operations plan may not be affected.

We normally do not perform an extensive amount of short-term corrective action in forecasting because several time periods are required to accurately evaluate a dynamic demand pattern. We have provided for expected fluctuations in demand with safety stocks of both raw materials and products. We must, of course, react promptly to serious short-term occurrences, such as the cancellation of a large order or the outbreak of an armed conflict that affects our operations.

LONG-TERM CORRECTIVE ACTION

Any time we suspect a change in our basic cause system, we must consider the possibility of adjusting the parameters in the forecasting transfer function. Whenever we observe a "run" (several consecutive demand values) above or below the projected demand curve, we might suspect that the average of the cause system has shifted, even if no values have yet fallen outside the control limits. Depending upon the extent of the shift, we may or may not have to perform our operations planning function again. This possibility will be discussed in the next chapter.

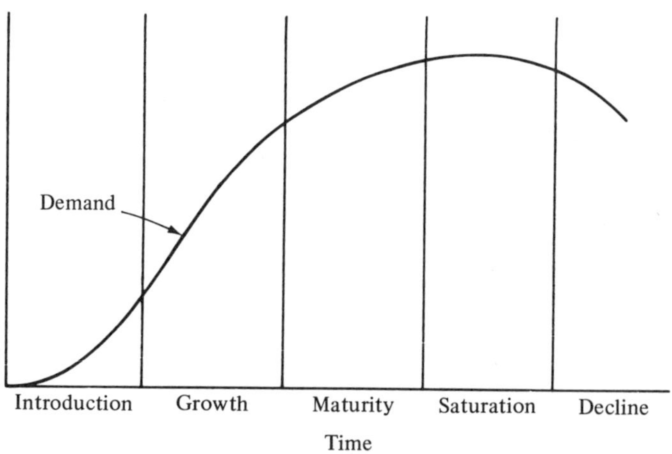

Figure 3.12 Life Cycle of a Product

Another factor that might cause us to adjust the parameters in the forecasting transfer function is the influence of the life cycle of the product. Essentially every product has a beginning and an end. The phases of a product's life distinguish themselves in a pattern common to most products. This pattern, shown in Fig. 3.12, has five phases. The actual duration of each phase varies considerably among different products.

For forecasting purposes, we must know which phase of the life cycle we are now in. For example, if we are at the end of the Growth phase, a straight line extrapolation of previous demand values would indicate a higher demand than will actually occur.

Note that on a short-term basis (e. g., week to week) we do not let ourselves be led too far astray from the basic life cycle. We see our demand pattern changing and adjust the parameters in our transfer function accordingly. However, at the beginning of each planning period (usually each year), we must estimate demand for the entire planning period so that we can determine the number of shifts to operate, whether more men and machines will be needed, etc. We discuss the details of these decisions in the next chapter. Our planning forecast, then, might be significantly high or low when projected a complete planning period into the future. It is important, therefore, that we continuously attempt to evaluate our status on the life cycle so that gross forecasting errors may be avoided.

The exponentially weighted moving average (EWMA) method of forecasting has some interesting characteristics with regard to corrective action for forecast control. The smoothing constant, α, automatically provides short-term corrective action by adjusting the forecast up or down, depending on recent demand values. If an analysis showed that the value of α should be adjusted, this change would be long-term corrective action.

In general, long-term corrective action involves the adjustment of parameters in the transfer function. This action can be prompted by a change in the cause system or by the appearance of poor results with a particular set of parameter values.

Interfaces

Forecasting interacts with other functions of the enterprise. Many of these interactions have been discussed previously. The purpose of this section is to consolidate our earlier discussions and to show where forecasting fits into the total control system. Figure 3.13 is included to facilitate our discussion of these interfaces with forecasting. Again, the dashed-line boundary separates elements of the operations planning and control system from those outside the system.

Product Markets provide the demand history for existing products. If our analysis indicates a declining demand, Product Innovation is notified for possible

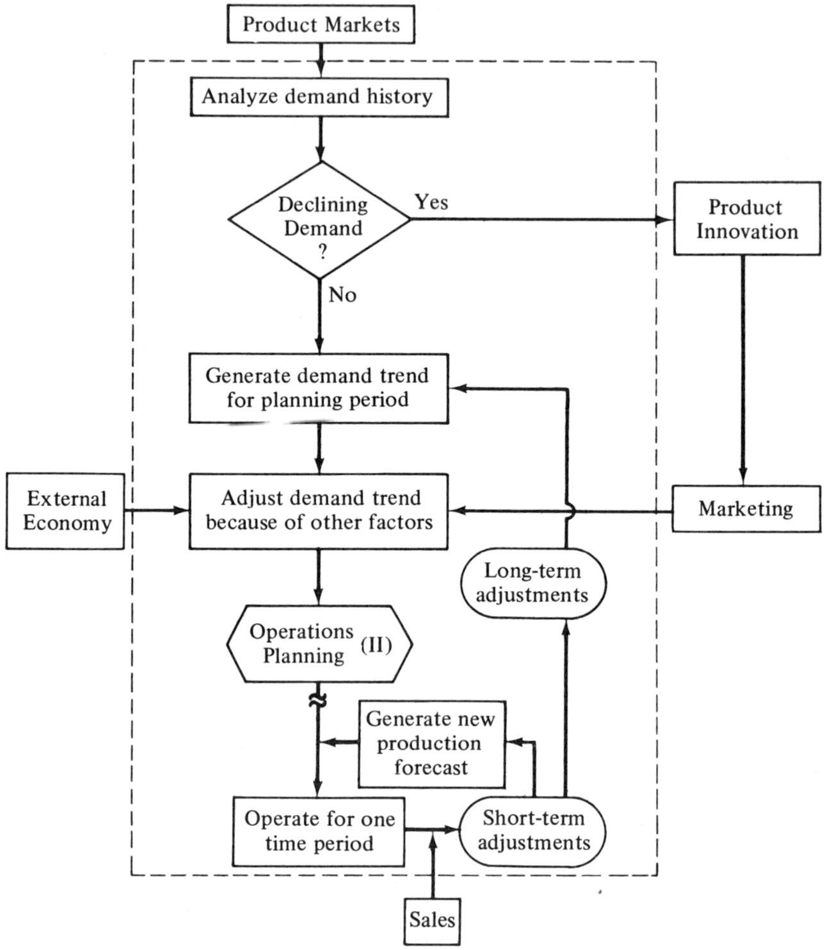

Figure 3.13 Interfaces for Forecasting Function

modification of the product. Ideas for modifications as well as for new products are forwarded to Marketing, where marketing strategies are devised.

The demand trend is generated for the planning period. Adjustments are made to the trend because of marketing programs and consideration of factors in the External Economy. The adjusted demand trend is sent to Operations Planning, where the operations plan (usually for one year) is devised.

We operate for one time period (such as a week or a month), after which actual demand is obtained from Sales. Any necessary short-term adjustments are then made and a new production forecast is generated. Actual demand is also evaluated over several time periods to determine whether changes in the forecasting transfer function are required. If long-term adjustments are sufficiently large, a new operations plan for the remainder of the planning period may be required.

REFERENCES

Biegel, J. E. 1963. *Production Control: A Quantitative Approach.* Englewood Cliffs, N. J.: Prentice-Hall, Inc.

Bowker, A. H.; and Lieberman, G. J. 1959. *Engineering Statistics.* Englewood Cliffs, N. J.: Prentice-Hall, Inc.

Brown, R. E. 1963. *Smoothing, Forecasting, and Prediction of Discrete Time Series.* Englewood Cliffs, N. J.: Prentice-Hall, Inc.

Miller, I.; and Freund, J. E. 1965. *Probability and Statistics for Engineers.* Englewood Cliffs, N. J.: Prentice-Hall, Inc.

Wiener, Norbert. 1950. *Extrapolation, Interpolation and Smoothing of Time Series.* New York: John Wiley and Sons.

Wine, R. L. 1964. *Statistics for Scientists and Engineers.* Englewood Cliffs, N. J.: Prentice-Hall, Inc.

Winters, R. P. 1960. Forecasting Sales by the Exponentially Weighted Moving Averages, *Management Science* 6: 324.

SOURCES OF ADDITIONAL TRANSFER FUNCTIONS

Listed here are several additional sources of forecasting transfer functions. The list is not intended as an exhaustive bibliography on the forecasting function, but as a representative sample of the forecasting techniques available.

Boyraktar, B. 1968. Adaptive Forecasting with General Exponential Smoothing and a Related Experimental Design for Sensitivity Analysis. Paper presented at the Joint National ORSA-TIMS meeting (San Francisco).

Brenner, J. L.; D'Esopo, D. A.; and Fowler, A. G. 1968. Difference Equations in Forecasting Formulas, *Management Sciences* 15: 141.

Collier, J. D.; and McCarthy, T. J. 1968. Short Range Forecasting for Inventory Management with Probability Tree Analysis and Monte Carlo Simulation. *Production and Inventory Management* 9: 4.

Eisenhut, P. S. 1969. Effect of Forecast Errors on an Inventory Model. *Production and Inventory Management* 10: 63.

Harrison, P. J.; and Davies, O. L. 1964. The Use of Cumulative Sum (CUSUM) Techniques for the Control of Routine Forecasts of Product Demand. *Operations Research* 12: 325.

Reed, R. R., Jr.; and Roberts, S. D. 1969. The Development of a Self Adaptive Forecasting Technique, *Proceedings of the* 1969 *AIIE National Conference* (Houston), 371.

Torfin, G. P.; and Hoffmann, T. R. 1968. Simulation Tests of Some Forecasting Techniques. *Production and Inventory Management* 9: 71.

IBM Reference Manual E20-0032-0. *Management Operating System: Forecasting, Exponential Smoothing, Detail.* White Plains, N. Y.

4 Operations Planning

Concepts

The demand forecasting function provides us with a refined estimate of how many units of each product are likely to be demanded by our customers and when during the planning period the demands are expected to occur. Operations planning must now convert the demand forecasts into a complete production program.

The demand forecast for each item is in terms of units of product. We receive a forecasted value for each time period and also a cumulative forecast for the entire planning period. The demand values for units of product must be translated into statements of resource requirements. Consider, for example, a forecasted annual demand of 42,316 type *A* radios. Each such radio requires a certain number of transistors, plastic knobs, etc., and a certain amount of sheet metal, plastic molding material, etc. Each such component requires a certain amount of labor and machine time. Many different products often require common components. Operations planning must establish a production program in which all resources (men, machines, and materials) are coordinated for the maximum benefit to the company.

PURPOSE OF OPERATIONS PLANNING

The purpose of operations planning is to assure that all resources needed to produce the required items (determined by the forecast) are at the right place at the right time and in the needed quantities and, furthermore, that waste of resources (idle time and overly large inventories of materials and product, etc.) is minimized. All this must be accomplished within the over-all constraints (such as budgetary limitations) and policies (such as steady employment practices) imposed by higher management. These tasks must also be accomplished with existing or obtainable resources, with consideration given such factors as previous commitment of resources, seasonality of demand, scrap and quality factors, lead times of purchased items, in-transit times (such as shipping to warehouse), quantities required in-process, and production lead times.

The operations plan provides only the general framework within which specific activities are to be performed. The operations plan does not specify that a particular activity must be performed on a particular facility at a particular time. Details such as these are determined in Operations Scheduling. The operations plan may, however, specify that during certain time periods a particular facility must be operated an extra shift in order to handle all the activities assigned to it during that period.

We see then that the operations plan allocates available resources to the several production requirements, but in a general way. The plan contains reasonable flexibility, so that detailed task assignments can be made within the general framework.

TYPES OF PLANNING

Just as different types of forecasts (Chap. 3) are required for different time spans, different categories of planning are also identified for different time spans.

Long-Range Planning. Long-range planning is concerned with maintaining the proper product line through research and development and with providing adequate facilities for the company's activities. The long-range plan, which includes planning for plant construction and expansion, facility modernization programs, and location of new facilities, usually covers from 5 to 20 years.

Intermediate Planning. Intermediate planning, concerned with allocating financial resources to the needs of the company, such as the acquisition of capital equipment, the building of pilot plants for new products, minor construction programs, and equipment programs, usually covers from 1 to 5 years.

Short-Range Planning. Short-range planning establishes production programs that allocate existing or due-in resources to current production requirements. This type of plan, which usually covers from 3 months to 2 years, establishes the over-all activity level for the company by making such decisions as the number of shifts of operation, work force size, additional equipment requirements, material requirements, and amount of subcontract support. It may also indicate the need for improved facility arrangements and for a new balance on the assembly line.

We concentrate here on short-range planning, since intermediate and long-range planning are not performed within the boundaries of the operations planning and control system shown in Fig. II.1.

BASIC PROBLEM IN OPERATIONS PLANNING

Imagine a situation in which demand is constant for each time period in the year. Imagine also that we have a constant work force, that our equipment is capable of producing at a constant rate, and that all our raw materials can be obtained immediately in whatever quantities we want. In such an idealized situation, no planning is required. We simply establish our production rate at exactly the demand rate, push the start button, and let the company make money (at a known rate, incidentally).

Modern industrial operations are far more complex, however. All the factors considered constant in the idealized example are, in fact, variables whose values fluctuate in an unpredictable manner. For example, we examined in Chap. 3 how demand may fluctuate. In addition to random fluctuations, demand can be generally increasing or decreasing and can contain cyclical components.

The assumption of an ideal work force is certainly invalid. In today's highly mobile and competitive labor market, continuous effort is required to maintain an adequate number of competent workers. With an increasing demand pattern, we must be prepared to expand our facilities and work force as needed.

Production rates of equipment also are rarely constant. Breakdowns and preventative maintenance interrupt production. Product quality usually varies, as does operator performance. Idle time often occurs as a result of a bottleneck at an earlier machine.

Finally, we consider the variability introduced by the acquisition of raw materials. Lead times vary. Quality of incoming product varies. A supplier may occasionally be out of the raw material we want or unable to ship a full order at one time.

The random factors just discussed are related in rather complex ways. The value of one factor at a particular time affects the values of other factors. Furthermore, the way in which we attempt to deal with one factor has a direct or indirect effect on the way we deal with certain other factors.

The presence of the random factors contributes to the complexity of the operations planning function. The greater the variation among the values of the factors, the greater the complexity of the operations planning function. Our basic problem is to devise a least-cost production program in the face of these many uncertainties. Some extreme policies (Buffa 1968) we could adopt are:

1. Use inventory to accommodate the uncertain and unsynchronized factors of production and demand. Maintain a constant work force and production rate.
2. Vary the work force in direct response to demand fluctuations. Maintain very little inventory.
3. Maintain a constant work force and little inventory; work overtime to increase the production rate.

4. Maintain a constant work force and little inventory; subcontract work during peak demand periods.
5. Maintain a constant work force and little inventory; backorder or simply forego some of the demand during peak periods.

There are hidden implications in many of these extreme policies. Policy number two, for example, assumes that we have sufficient equipment for the largest number of operators during the peak demand periods. During other periods this equipment would be idle. Policy number one requires a great amount of storage space, which has a poor over-all utilization factor.

We see, then, that rarely would one of the extreme policies be a satisfactory solution to the problem of operations planning. Ideally, we want the best overall solution, the one that results in least total cost to the company.

OPERATIONS PLANNING DECISIONS

We now can state explicitly the decisions that the operations planning function is expected to render. (Frequent reference to Fig. 4.1 will facilitate the reader's understanding of this section.)

Product Requirements. For each product we market, we must determine the number of units that must be manufactured for (but not necessarily during) each time period (1 month) in our planning period (1 year). The number of units of a product that must be manufactured for a time period may be different from the forecasted demand for that time period because of (1) existing stock on hand, (2) outstanding production orders for the product scheduled for completion during future time periods, and (3) desired ending inventory for that time period. A time profile of product requirements is obtained by determining cumulative period requirements.

Component Requirements. We must now determine the number of units of each component (raw materials, purchased parts, fabricated parts, and subassemblies) needed for the manufacturing quantities determined in the previous section for each product. Since several products may contain common components, the quantities for each component are summarized. Again, the number of units of a component that must be fabricated or purchased for a time period may be different from the number actually needed for the manufacturing quantities because of (1) existing uncommitted stock on hand, (2) outstanding production and purchase orders, and (3) desired ending inventory for that time period. A time profile of purchased and fabricated components is obtained by determining cumulative period requirements.

Equipment Requirements. The product and fabricated component time profiles provide the basis for calculating equipment requirements. Units are converted to processing times and summarized by work center and machine type. A time profile of equipment requirements is determined.

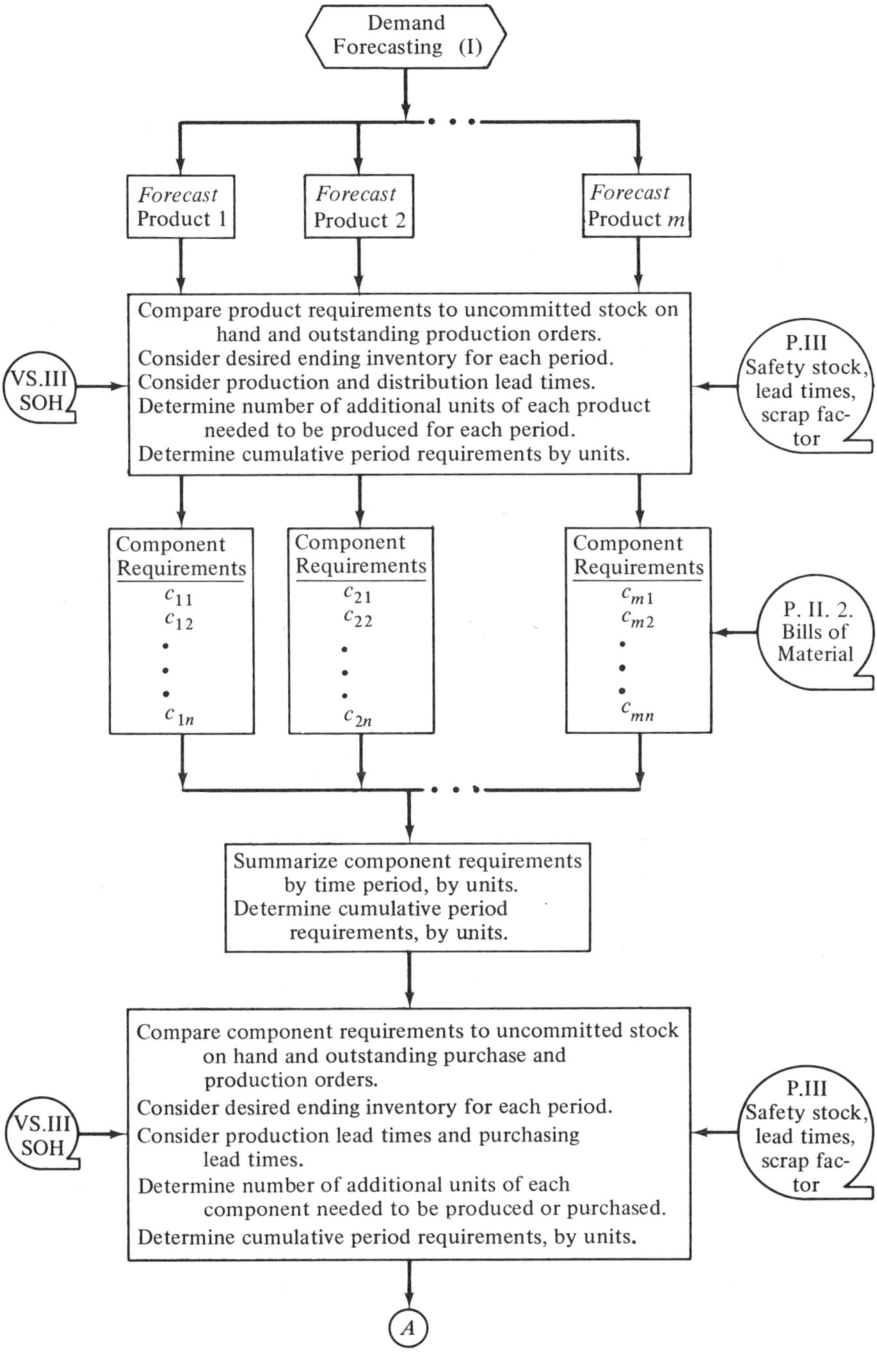

Figure 4.1 Overview of the Operations Planning Function with pertinent Information Inputs

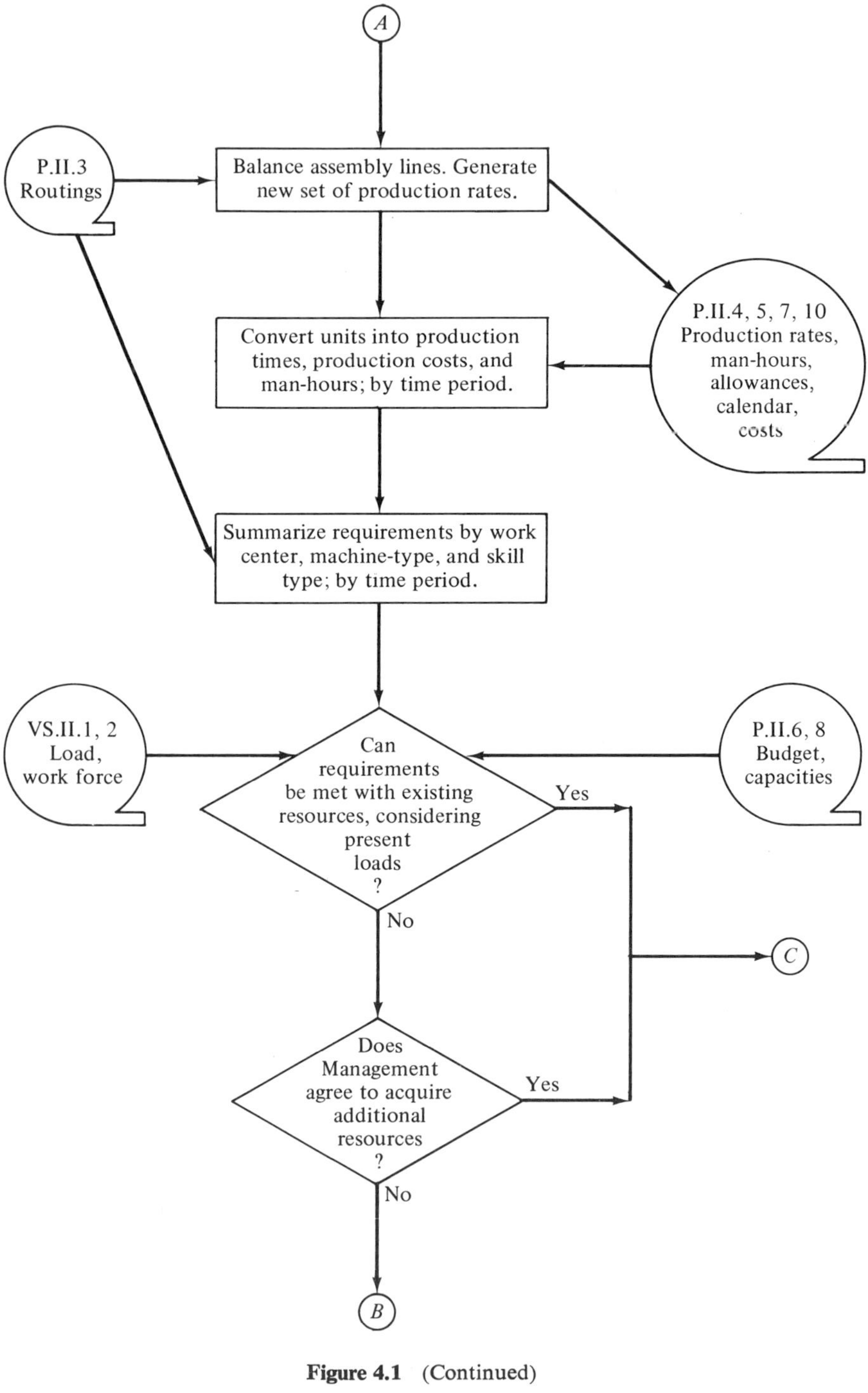

Figure 4.1 (Continued)

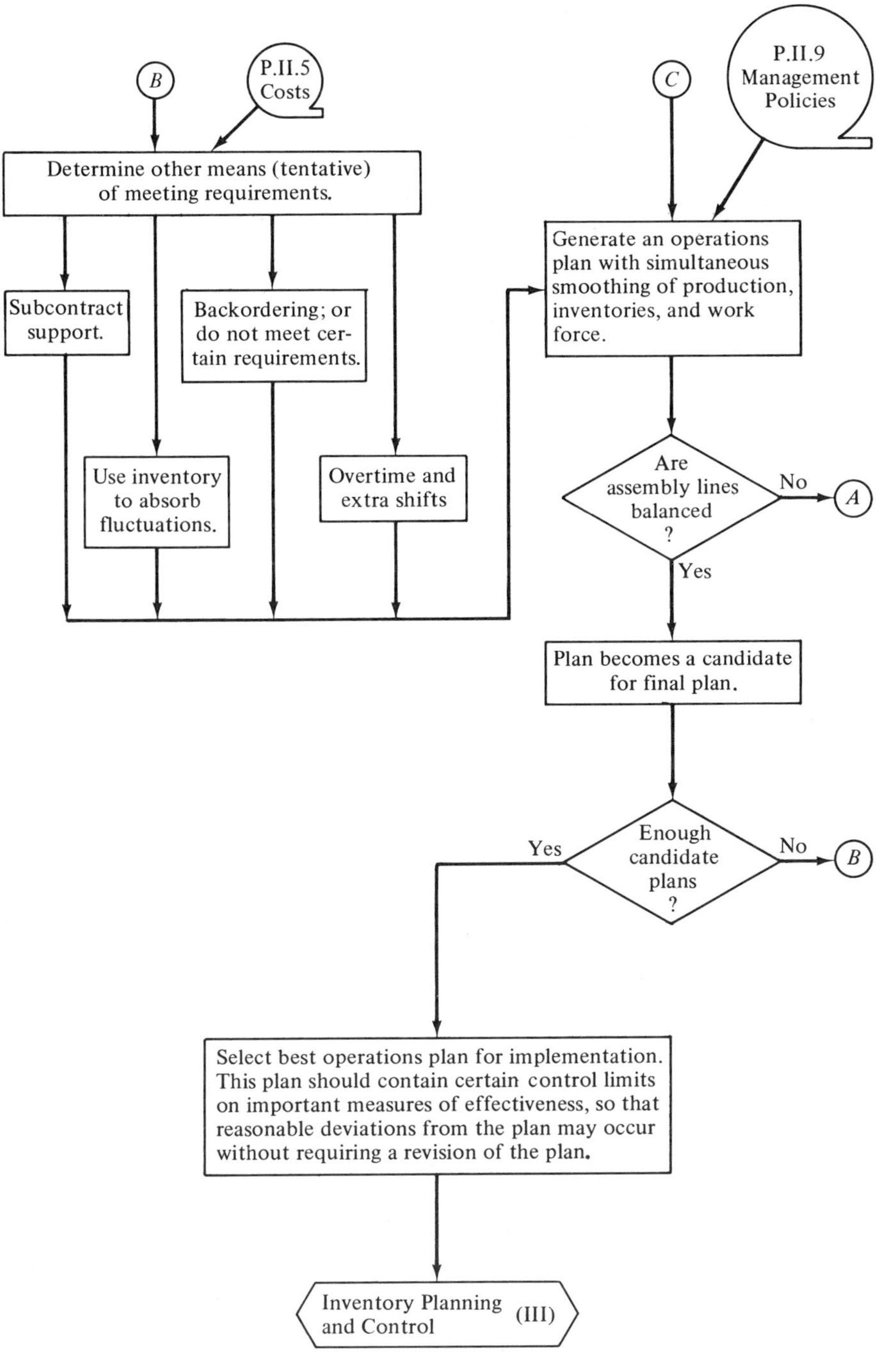

Figure 4.1 (Concluded)

Labor Requirements. A time profile of labor requirements is determined from the product and fabricated component time profiles. Units are converted to man-hours and are summarized by work center and skill type.

Consolidated Operations Plan. The previous four sections have been concerned with requirements. We must now consider how these requirements are to be satisfied. We must allocate our available resources of labor, materials, and equipment to the several production requirements, and we must do so over the entire planning period. When available resources are insufficient for the requirements, we must acquire additional resources (usually by hiring more men or purchasing more equipment), use subcontract support, or not satisfy all the requirements. Resources and requirements are sometime out of phase during the planning period, which results in our inability to satisfy requirements during the peak demand periods, even though our total resources are sufficient for total requirements over the entire planning period. A common solution to this problem is to produce at a rather constant rate during the planning period, letting inventory absorb demand fluctuations. Another solution might be to work extra shifts or overtime during the high demand periods.

A *consolidated operations plan*, then, is a time-phased plan specifying for each time period number of shifts, work force size (and skill mix), production rates, use of inventory for smoothing, use of subcontract support, and even use of backordering, where this is feasible.

Line Balancing and Facilities Arrangement. The new operations plan may very well have upset the balance of our assembly lines. Facilities in the fabrication areas may also need to be rearranged. Any additional equipment would certainly affect these two factors.

The procedure for arriving at the decisions discussed above is circular in nature. For example, the consolidated operations plan is selected from several alternative plans, each determined by changing certain planning parameters and by considering different combinations of planning strategies.

Figure 4.1 represents the total process of arriving at the planning decisions discussed in this section. The steps in Fig. 4.1 are executed at the beginning of the planning period in order to establish a consolidated operations plan for the entire period. Several things may necessitate revising the plan during the planning period. Demand patterns may change significantly (either upward or downward), thus causing errors in our forecasted demand values. Planned production quantities may not be realized because of equipment breakdowns, poor quality records or poor operator performance. Programmed additions to the labor force and equipment capacity may not be realized because of unavailable workers and equipment. Raw material supplies may fall short of expectation. Finally, financially attractive subcontract support may become available during the period.

Whenever any combination of the above factors causes a significant deviation from the consolidated operations plan, it is necessary to revise the plan. The steps in Fig. 4.1 would be executed again using current information from the Parameter File and the Variable Status File.

Inputs

The operations planning function requires information from several sources. Many of the general information requirements are shown in Fig. 4.1. In this section we examine the specific information requirements for operations planning and the sources of this information.*

Several transfer functions must be linked together to arrive at all decisions required of the operations planning function. These transfer functions require information specifically identified to the operations planning function (i.e., P.II and VS.II) and certain information components that are specifically identified to other functions (such as VS.I, VS.III, and P.III).

PARAMETER FILE FOR PLANNING (P.II)

P.II.1 Forecasted demand values, by product and by time period. Source: Demand Forecasting.

P.II.2 Bill of materials for each product and each subassembly. Source: Engineering.

P.II.3 Standard and alternative routings of fabricated parts. Source: Engineering.

P.II.4 Standard processing, set up, and maintenance times. Source: Engineering.

P.II.5 Standard manufacturing costs, overhead rates and shift change cost. Source: Engineering and accounting.

P.II.6 Equipment capacities. Source: Engineering.

P.II.7 Operations calendar for the planning period. Source: management's policy on vacations, holidays, etc.

P.II.8 Operations budget. Source: Financial Control.

P.II.9 Management policies. Source: higher management.

P.II.10 Union regulations, including worker allowances. Source: Personnel.

In addition to the above parameters specifically identified to operations planning, our transfer functions require the following information components from the Materials Planning and Control Parameter File (P.III): safety stocks for

*Frequent reference to Fig. 4.1 will indicate how the information is used, while Fig. II.1 will provide the source of the information.

finished goods and components; manufacturing and purchasing lead times; the scrap material factor; set-up, carrying, out of stock, and material costs.

VARIABLE STATUS FILE FOR PLANNING (VS.II)

VS.II.1 Current load on work centers, by machine type. Source: Progress Control.

VS.II.2 Current work force inventory, by skill type. Source: Personnel department.

VS.II.3 Actual production rates and quantities. Source: Progress Control.

VS.II.4 External perturbations, such as cancelled orders, rush orders, external economic factors.

In addition to the above variable information, our transfer functions require information components from two other sources. From the Demand Forecasting Variable Status File (VS.I) we need actual demand as it becomes known and information concerning newly planned sales campaigns. From the Inventory Planning and Control Variable Status File (VS.III) we need data on current stock on hand, availablility of raw materials, and cost of materials.

UPDATING THE FILES

All the values of the parameters and variables must be kept current so that, whenever an operations plan is made or revised, the transfer functions have the correct values available. Perhaps the most difficult values to keep current are those provided by Engineering. Products, manufacturing processes, bills of materials, manufacturing procedures, etc., are continuously changing. These changes, in turn, affect processing times, costs, and production rates.

The values in the Variable Status File are kept current through a good data acquisition system and a good progress control system, both of which are discussed in detail in Chap. 7.

Transfer Functions

If we examine Fig. 4.1 closely, we see that many decisions must be made in the planning function and that many of these decisions are sequential. Some of the transfer functions presented in this section make more than one of the planning decisions, but none make all the decisions. Our task here, therefore, is not to select one transfer function from those presented, but to develop a series of transfer functions that will make all the required decisions and that are mutually compatible.

Each transfer function developed in this chapter is but one method of executing the particular set of decisions. Other transfer functions might be equally good. The only requirement in selecting a set of transfer functions is that they are mutually compatible in the execution of the sequential decision processes.

LINE BALANCING

Although line balancing is not the first planning decision in Fig. 4.1, it is essentially a self-contained decision process that is rather easy to discuss at any point in the over-all development.

Basic Concepts. Most products are assembled from several components, some of which are assembled from still other components. An electronic tube for a radio or television set is assembled from several small components. It was discovered during the early years of industrialization that a product could be assembled much faster and much cheaper by dividing the total job into individual tasks and then assigning these tasks to different operators. The product is moved past each stationary operator, who performs the set of tasks (or elements) on each unit of product as it moves past his station. It is desired to assign equal amounts of work to each operator, so that little or no idle time exists in the assembly line. The assignment of work elements to work stations is known as assembly line balancing, or simply, line balancing.

We can now define certain terms needed in our development (Kilbridge and Wester 1961). A *work element* is the largest unit of work which cannot be split between two or more operators without creating unnecessary interference between the operators. An *operation* is a set of work elements assigned to one work station. A *work station* is an area adjacent to the assembly line where a given amount of work (an operation) is performed. We usually think of a work station as being manned by one operator, but this is not necessarily the case. The *cycle time* is the time the product is available at each work station. *Balance delay* is the total amount of idle time on the line resulting from an unequal division of work stations.

We consider here only the case in which a conveyer moves past fixed work stations at a constant speed. Other types of assembly lines also exist, but the balancing of such lines presents special problems. Figure 4.2 represents the type of assembly line we are concerned with. It is possible, of course, to arrange work stations on both sides of the line, but the concepts can be developed more easily if we consider consecutive work stations on only one side of the line. Conceptually, we can imagine the components required in the assembly to flow to the line opposite the work stations at which the components enter the assembly process.

It should be recognized that in a large plant many assembly lines are in operation simultaneously. There may be exactly one assembly line for each product. It is also possible for there to be several identical assembly lines producing the same product. It is not uncommon for one basic assembly line, with certain modifications, to be able to assemble several different but similar

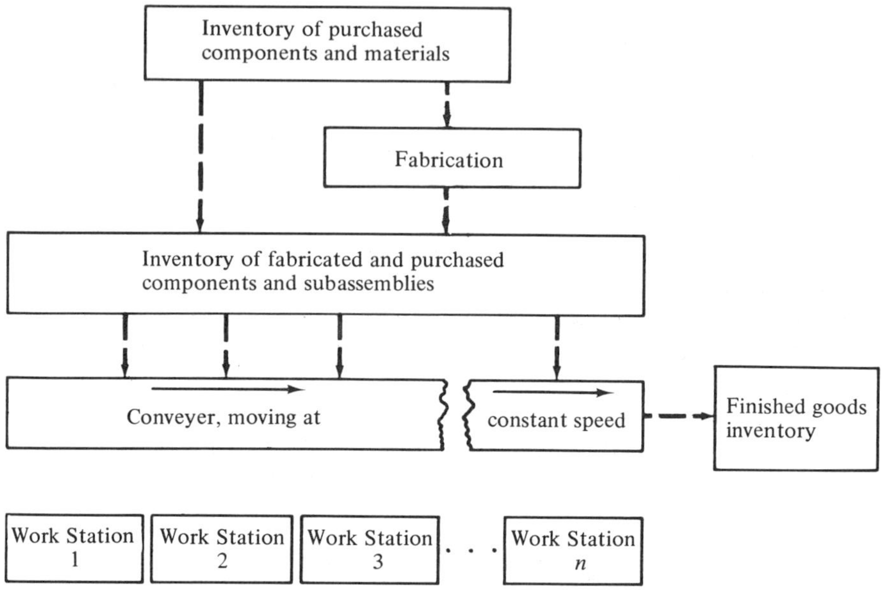

Figure 4.2 Representation of a Constant Speed Assembly Line

products. In such a case, each modification would require its own line balance.

Let us now consider an example of a product to be assembled from several components. Suppose that the industrial engineering section has broken down the assembly tasks into the 10 basic work elements and duration times shown in Table 4.1.

Table 4.1 Example of Work Elements and Duration Times

Work Element	1	2	3	4	5	6	7	8	9	10
Duration Time	5	10	5	2	7	5	10	2	5	7

Several of these work elements can be grouped together to form an operation. One work station is then established for each operation. The work stations are arranged sequentially such that a conveyer moves past at a constant speed, as in Fig. 4.2. *Our objective is to group the work elements such that we get close to an equal amount of work done at each work station.*

Consider the question of cycle time; i.e., how long will the object be available to each work station for performing the elements making up the operation at each work station? The minimum cycle time must be 10 minutes, since work elements 2 and 7 each require 10 minutes to perform. Remember that a work element cannot be divided into smaller work components and therefore must be performed completely at one work station.

The maximum cycle time would be the sum of the duration times, 58 minutes in this case. This would be the cycle time if all elements were performed as a single operation at only one work station. There would be no "line" in such an arrangement and therefore no problem of line balancing. Including all elements in a single operation is usually not feasible, however, and the maximum cycle time is rarely used in line balancing.*

There are often practical considerations that narrow the range of possible cycle times we can select. For example, there is a limit on how many different work elements an operator can perform well. This limit, in turn, limits the number of elements we can assign to one work station. There is also the practical problem of feeding components to the line. It would be virtually impossible, for example, to feed all the components of an automobile to one work station.

The principal consideration in selecting a cycle time is the production rate required. This factor is illustrated in the following example.

Suppose we group the elements shown in Table 4.1 into operations using a cycle time of 10 minutes. For now, we assume that the 10 elements can be grouped in any way we wish. Note that, using a cycle time of 10 minutes, we must have at least six work stations, since the sum of the element times is 58 minutes. In general, the minimum number of work stations possible is the total element duration time divided by the cycle time. We must always round up when fractions occur. Therefore, we must group the 10 work elements into at least six operations such that no operation exceeds the cycle time of 10 minutes. A possible solution is shown in Table 4.2. There are many other combinations equally good, but none can be better than the balance shown in Table 4.2. There is simply no other way to arrange the work elements, using a cycle time of 10 minutes, to obtain a smaller balance delay value than the 2 minutes that resulted in this solution.

Table 4.2 Ten Elements Grouped into Six Operations; Cycle Time Equals 10 Minutes

Operation	Work Elements	Σ Duration Times
1	2	10
2	7	10
3	1, 3	10
4	6, 9	10
5	4, 5	9
6	8, 10	9
		58

Let us now try a cycle time of 15 minutes, rather than 10. In this case the minimum number of work stations is $58/15 = 3.87 = 4$. We must group the

*Imagine an automobile being put together at one work station.

10 elements into four operations. We want each operation time to be as close to, without exceeding, 15 minutes as possible. An optimum solution is shown in Table 4.3. Again, there are many other combinations equally good, but none better. Again, the balance delay (unavoidable idle time) is 2 minutes.

Table 4.3 Ten Elements Grouped into Four Operations; Cycle Time Equals 15 Minutes

Operation	Work Elements	Σ Duration Times
1	1, 2	15
2	3, 7	15
3	5, 10	14
4	4, 6, 8, 9	14
		58

We could obtain different solutions using different cycle times. The reader can verify that a cycle time of 12 minutes would result in five work stations, again with 2 minutes balance delay. There are two solutions in which no idle time appears: (1) combine all 10 elements into one operation or (2) combine 5 elements into each of two operations, with a cycle time of 29 minutes. Normally, however, practical considerations would eliminate these two solutions.

Considering only the two solutions given in Tables 4.2 and 4.3, let us now attempt to determine which is best. Since they are equally good in terms of balance delay, we must make this determination based upon the relative production rates of the two solutions and the demand for the product.

With a cycle time of 10 minutes, the line will assemble 6 units per hour, or 48 units per 8-hour shift per day. Assuming 250 working days per year, the line will produce (250)(48) = 12,000 units per year on a one-shift basis. By similar reasoning, the solution with a 15-minute cycle time will produce 8,000 units per year on a one-shift basis. The selection of a cycle time clearly depends upon the demand for the product. Table 4.4 shows some possible demand values and the corresponding selection of cycle time, number of shifts, and number of lines. In some cases alternate solutions are given for the same demand value. Note in the last case that two lines are suggested, one with a cycle time of 10 minutes and one with a cycle time of 15 minutes.

In the preceding discussion we assumed that the elements could be grouped in any combination desired so long as the sum of the element duration times did not exceed the cycle time. In most actual assembly operations, however, certain work elements must be performed before others. This requirement puts additional restrictions on how we can combine the elements into operations. Since certain elements must precede certain others, the resulting relationship is called a *precedence relationship*. These relationships can be shown in a precedence diagram, the topic of the next section.

Table 4.4 Possible Solutions for Varying Demand Values

	Possible Solutions		
Annual Demand	Cycle Time	Number of Shifts	Number of Lines
16,000	15	2	1
	15	1	2
24,000	10	2	1
	10	1	2
	15	3	1
	15	1	3
360,000	10	3	10
20,000	10, 15	1	1, 1

Precedence Diagram. For illustrative purposes, let us assume that the following precedence restrictions must be observed for the work elements in Table 4-1:

Element	Precedes	Element
1	>>	2
1	>>	3
2	>>	4
3	>>	4
4	>>	5
4	>>	6
5	>>	8
6	>>	7
7	>>	9
8	>>	9
9	>>	10
10	>>	0

These relationships can be expressed as a network, or precedence diagram, as shown in Fig. 4.3. As we assign work elements to operations, we must now consider the permissible orderings as well as the cycle time. For example, element 2 cannot be assigned to an operation until element 1 has been assigned.

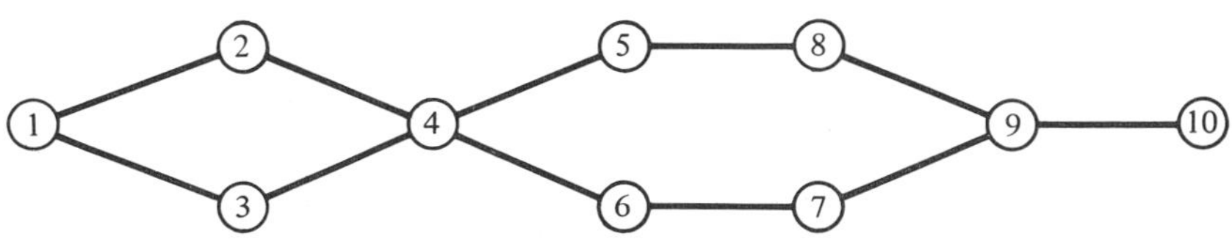

Figure 4.3 Precedence Diagram for ten Work Elements

Element 4 cannot be assigned until elements 2 and 3 have both been assigned. It *is* permissible to assign elements 2 and 4, 3 and 4, and even 2, 3, and 4 to the same operation.

The possible permutations (orderings) of the 10 elements can be shown in a tree diagram (Fig. 4.4). Note that the orderings below element 4 on the left major branch are identical to those below element 4 on the right major branch. Careful study of the tree diagram also reveals other interesting relationships. There are 12 possible orderings of the 10 elements. We can combine any number of adjacent elements within one of the orderings to make up an operation. We must not, of course, exceed the cycle time.

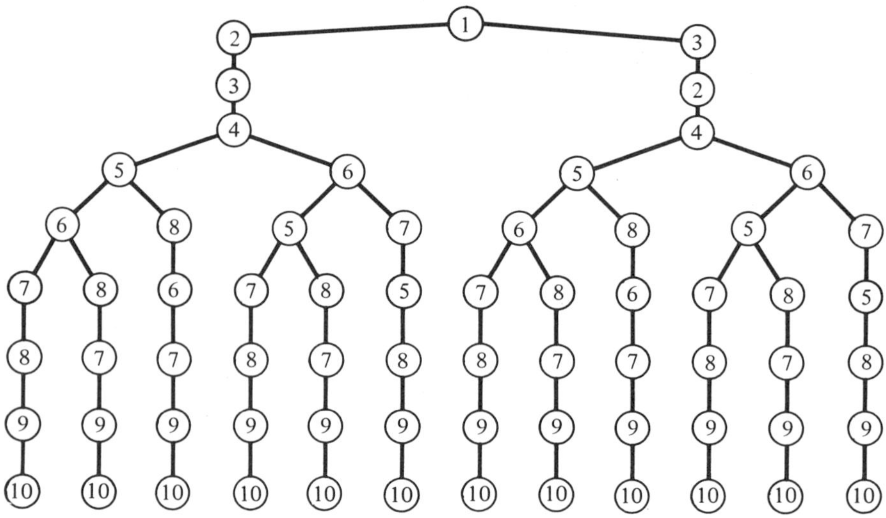

Figure 4.4 Tree Diagram of the possible orderings of the Elements in Figure 4.3

Consider, for example, the ordering {1, 3, 2, 4, 5, 6, 7, 8, 9, 10} with corresponding element times of 5, 5, 10, 2, 7, 5, 10, 2, 5, and 7 minutes, respectively. We can group together any number of adjacent elements in this set to form an operation, so long as the sum of the element times in the operation does not exceed the cycle time. Using a cycle time of 10 minutes, for example, we could group the elements from the above ordering into the operations shown in Table 4.5. The resulting idle time for each operation is also shown. Note that operation 4 is idle half the time, meaning that in an 8-hour day we would get only 4 hours of productive work from this work station.

The balance shown in Table 4.5 is for only 1 of 12 possible permutations of the 10-element problem. To find the absolute best line balance, we would have to perform a similar analysis on each of the other 11 permutations. For each balance we would compute the total idle time and then choose as our final solution that balance with the least total idle time.

Table 4.5 Balance for Elements in Fig. 4.3; Cycle Time Equals 10 Minutes

Operation	Elements	Σ Element Times	Idle Time
1	1, 3	10	0
2	2	10	0
3	4, 5	9	1
4	6	5	5
5	7	10	0
6	8, 9	7	3
7	10	7	3
		58	12

We could determine a balance for all 12 permutations without a great deal of effort. This approach is not practical, however, for the larger problems found in industry. For complex products there could be millions of permutations. It is not practical to examine each permutation, even with the assistance of a large digital computer.

Much research has been conducted into the assembly line balancing problem, and several rather good procedures are now available for determining good balances, though none of these procedures can assure an optimum balance. (See, for example, Helgeson and Birnie 1961; Ignall 1965; Kilbridge and Wester 1961; and Moodie and Young 1965.) One of these procedures is discussed in the following section. It was selected, not because it is better than the others, but rather because the rationale underlying it is very easy to understand.

*Ranked Positional Weight Technique.** The rationale of this particular line balancing method is: when assigning work elements to operations, attempt to assign first those elements separated from the final completion point by the greatest duration time. This assignment can be accomplished by the following procedure:

Step 1. Determine the positional weight of each element by adding together the duration time (t_i) of the elements itself and of all the elements which must follow it. The positional weights of the elements in Table 4.1 are shown in Table 4.6 (the precedence relationships are shown in Fig. 4.3).

Table 4.6 Positional Weights of Work Elements in Fig. 4.3

Element (i)	1	2	3	4	5	6	7	8	9	10
Duration Time (t_i), in minutes	5	10	5	2	7	5	10	2	5	7
Positional Weight (W_i)	53	48	43	38	21	27	22	14	12	7

*See Helgeson and Birnie (1961) for details.

Step 2. The elements are ranked and rearranged in decreasing order according to their positional weights. Table 4.7 shows the new arrangement. The last row in the table shows for each element the element(s) that must immediately precede it.

Table 4.7 Elements Rearranged According to their Positional Weights

Element (i)	1	2	3	4	6	7	5	8	9	10
Duration Time (t_i), in minutes	5	10	5	2	5	10	7	2	5	7
Positional Weight (W_i)	53	48	43	38	27	22	21	14	12	7
Preceding Element(s)	—	1	1	2, 3	4	6	4	5	7, 8	9

Step 3. Specify the cycle time. (The comments on cycle time in the preceding section apply here.) We use a cycle time of 10 minutes for this example.

Step 4. Assign elements to operations on the basis of the ranking shown in Table 4.7. The sum of the element times assigned to an operation cannot exceed the cycle time. Precedence restrictions must be observed. If an element violates either of the two constraints just mentioned, it is passed over and the next element (in decreasing order of positional weight) is tried. This process is continued until no further elements can be assigned to the operation. We then repeat the above process for the next operation, beginning with the first of any elements passed over on the previous iteration. This assignment process is continued until all elements have been assigned to an operation. For our example, the resulting line balance is shown in Table 4.8. This balance is simply a feasible one, not necessarily the best we can obtain. We attempt to improve the balance by manually manipulating elements between operations.

Table 4.8 Initial Balance for 10-Element Problem; Cycle Time Equals 10 Minutes

Operation	Elements	Σt_i	Idle Time, Minutes
1	1, 3	$5 + 5 = 10$	0
2	2	10	0
3	4, 6	$2 + 5 = 7$	3
4	7	10	0
5	5, 8	$7 + 2 = 9$	1
6	9	5	5
7	10	7	3
			12

Step 5. Compute maximum possible efficiency:

$$e_{\max} = \frac{\sum_i t_i}{nC}$$

where t_i = duration time of work element i
C = cycle time
n = minimum integer number of operations

$$= (\sum_i t_i)/C$$

In our example,

$$e_{\max} = \frac{58}{(6)(10)} = 0.966$$

Step 6. Compute efficiency of the feasible balance obtained in Step 4:

$$e = \frac{\sum_i t_i}{mC}$$

where m = number of operations obtained.
In our example,

$$e = \frac{58}{(7)(10)} = 0.83$$

Step 7. If $e < e_{\max}$, attempt to increase e by rearranging certain elements. In doing this, we look first to those operations with large idle times. We rearrange elements among operations if it will result in a better balance and if precedence restrictions permit. Looking at our initial balance in Table 4.8, we see that there are several ways in which we could rearrange the elements with a resulting balance delay of 12 minutes, but no other balance has a delay of less than 12 minutes. Thus, the initial balance obtained in Table 4.8 is as good as we can do with this particular problem. If we had evaluated all 12 permutations (see Fig. 4.4) in the preceding section, we would have arrived at the same conclusion.

The method just presented is but one of several line balancing procedures now available. Research into this problem is continuing, and we can expect better methods to be developed in the future. Most line balancing procedures require a digital computer for applications to large industrial problems.

An important output from the line balancing procedure that is used as input to the remainder of the operations planning function is the resulting production rate and the resources necessary to obtain that production rate. In the example just completed, the cycle time of 10 minutes assures a production rate of 12,000 units per shift in a work year of 250 days. If we assign one man to each operation, seven men are needed for this assembly line. This assignment amounts to 14,000 man-hours per year. However, only (12,000)(58)/(60) = 11,600 hours of this time is productive effort, the remaining 2,400 hours are wasted idle time. The important point here is that total production time must be considered in operations planning, not just total productive time.

TABULAR AND GRAPHICAL AIDS

A number of tabular and graphical procedures can assist us in developing an operations plan. Since some of these can be used as decision-making processes, they may be called transfer functions. Their main use, however, is in conjunction with other operations planning transfer functions. We discuss four tabular and graphical aids: bills of material, operations sheets, production flow charts, and requirement time profiles.

To illustrate the use of each of these aids, we consider an example in which we must plan for the production of two different all-metal bookcases. The small bookcase has three shelves, while the large bookcase has six shelves. The larger bookcase is made up of two small bookcases joined by four plastic connectors. Line drawings of the two bookcases are shown in Fig. 4.5. The legs are bent into a hollow square from sheet metal. The shelves are also made from sheet metal. Screws connect the legs to the shelves, and plastic inserts are placed in both ends of each leg.

Bills of Material. A bill of material is a listing of all components and materials of a product or subassembly. Each component is distinguished by part number

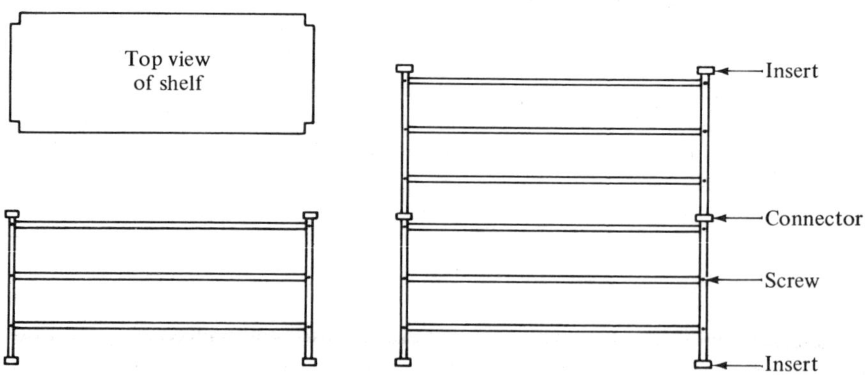

Figure 4.5 Three and Six-Shelf Bookcases

BILL OF MATERIAL			
Product Description: Bookcase, metal, 3 shelves. *Stock Number*: 1			
Component		Quantity (Amount) Required	Source
Stock No.	Description		
3	Shelf	3	Manufacturing
4	Leg	4	Manufacturing
5	Inserts	8	Purchasing
6	Screws	12	Purchasing

BILL OF MATERIAL			
Product Description: Bookcase, metal, 6 shelves. *Stock Number*: 2			
Component		Quantity (Amount) Required	Source
Stock No.	Description		
3	Shelf	6	Manufacturing
4	Leg	8	Manufacturing
5	Inserts	8	Purchasing
6	Screws	24	Purchasing
7	Connectors	4	Purchasing

BILL OF MATERIAL			
Product Description: Shelf *Stock Number*: 3			
Component		Quantity (Amount) Required	Source
Stock No.	Description		
8	Sheet metal	3 square feet	Purchasing

BILL OF MATERIAL			
Product Description: Leg *Stock Number*: 4			
Component		Quantity (Amount) Required	Source
Stock No.	Description		
8	Sheet metal	2 square feet	Purchasing

Figure 4.6 Bills of Material for two Bookcases having common components

and description. The number of units of each part needed in one unit of the product or subassembly is shown. The normal source (purchased or manufactured) is also indicated. Figure 4.6 shows example bills of material for the two metal bookcases.

Note in Fig. 4.6 that four bills of material are needed. Two describe the finished product, while one is needed for the shelves and one for the legs. In the quantity required column, the amount of each material or subassembly needed to make one unit of the subassembly or product is indicated.

It is often desirable to represent the bill of material structure as an "explosion chart" that shows the relationships of components, materials, and subassemblies. Figure 4.7 shows the explosion chart for each bookcase. The stock numbers from Fig. 4.6 are shown in the bubbles, and the number of units required for one unit of the next higher level item is shown just above each bubble.

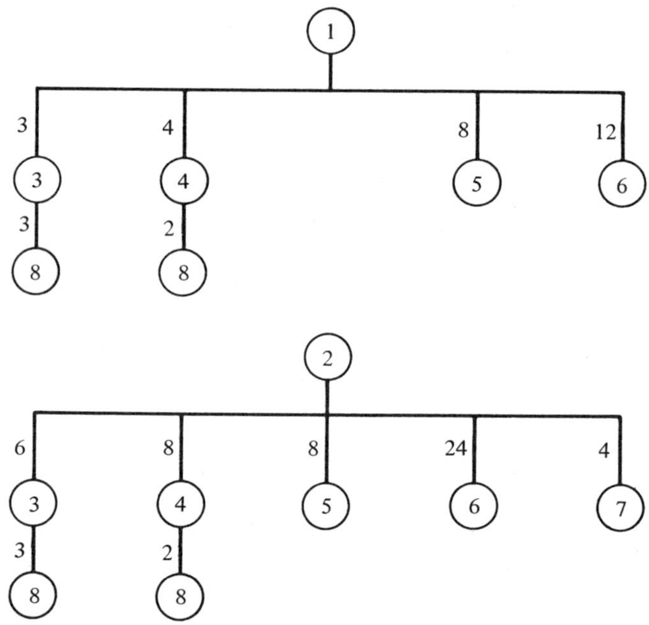

Figure 4.7 Explosion Charts for two sizes of Bookcases

We can determine material requirements from the explosion charts by successively multiplying the unit values for each material. In order to produce 10 units each of stock numbers 1 and 2, we would need $10 \times 3 \times 3 + 10 \times 4 \times 2 + 10 \times 6 \times 3 + 8 \times 2 = 510$ square feet of sheet metal. This process can become clumsy with many complex products having common components. We later discuss a better method for accomplishing this function in the section entitled "The Gozinto Procedure."

Operations Sheets. It is helpful in the planning function to have available for each finished product a listing of the operations sequence, showing the department number, the operation performed, process and set-up times, and material and tool requirements.

Three operations are required to produce a bookcase: stamping, bending, and assembling. Suppose a separate department (Departments 1, 2, and 3, respectively) performs each of these three operations. Figure 4.8 shows an operations sheet for the three-shelf bookcase. The operations sheet for the six-shelf bookcase would be similar. Much of the information on the operations sheet comes from Engineering.

OPERATIONS SHEET

Product Description: Bookcase, metal; 3 shelves
Stock Number: 1

Operation Number	Dep't	Operations Description	Materials	Amount	Tools	Process Time (min)	Set-up Time (min)
1	1	Stamp leg	sheet steel	2 sq. ft	A36	0.50	60
2	1	Stamp shelf	sheet steel	3 sq. ft.	A40	0.67	60
3	2	Bend leg	stamped leg	1	C17	1.00	120
4	2	Bend shelf	stamped shelf	1	C19	0.75	120
5	3	Assemble				1.00	60
		Stock Number 1	leg	4			
			shelf	3			
			screws	12			
			inserts	8			

Figure 4.8 Operations Sheet for Three-Shelf Bookcase

Production Flow Charts. In Fig. 4.1 and the accompanying descriptive text, it was mentioned several times that production and distribution lead times, as well as purchasing lead times, have to be considered in the planning function. These considerations are especially important when lead times are long or when we are producing to customer order with a given due date. It is helpful in such situations to show all factors of production on one time-scaled diagram called a production flow chart.

Figure 4.9 shows a typical production flow chart for a hypothetical product. The time scale provides a convenient way of showing all lead times and thus the latest dates on which each activity must commence if the due date is to be met.

Suppose we are producing bookcases to customer order rather than for inventory and that we receive an order for 1,000 three-shelf bookcases. To construct a production flow chart, we must first determine the amount of production time this order requires for each of the operations shown in Fig. 4.8. An extension of the information in Fig. 4.8 results in 33, 33, 67, 38, and 17 hours

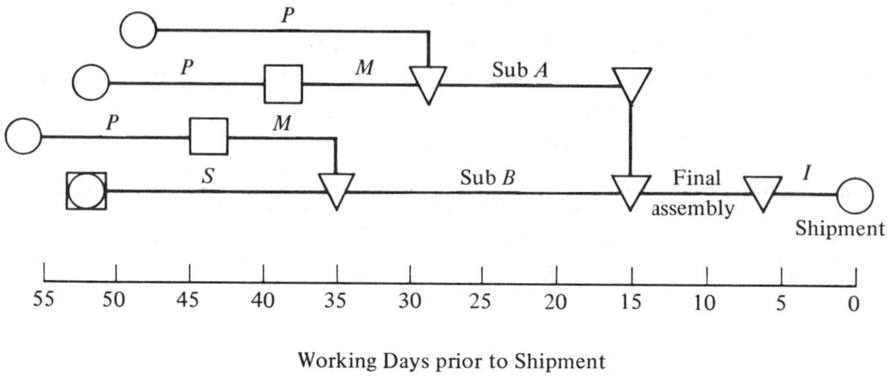

Purchased part: *P*
Manufactured part: *M*
Subcontract part: *S*
Subassembly: Sub
Inspection: *I*

Figure 4.9 Flow Chart for a hypothetical product

of work required for operations 1, 2, 3, 4, and 5, respectively, to satisfy the order for 1,000 bookcases. If we convert hours to days (8 working hours per day) and asssume that all legs are stamped before being moved to the bending operation and that all shelves are stamped before being moved to bending, we obtain the production flow chart shown in Fig. 4.10. Purchasing lead times of 6, 3, and 6 days are assumed for sheet steel, screws, and inserts, respectively.

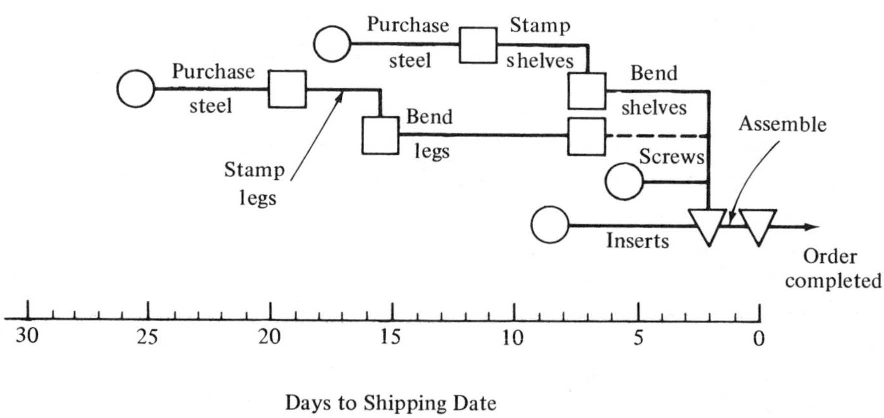

Figure 4.10 Production Flow Chart for 1000 units of Three-Shelf Bookcases

The production flow chart in Fig. 4.10 was drawn under the assumption that shelves and legs cannot be stamped or bent simultaneously. Note that Department 1, Stamping, is idle part of the time and that all legs are bent and stored for about 4 days while the shelves are being bent. It should be recognized that if we were producing for inventory rather than to customer order, we would probably set up a continuous operation and have more than one stamping and one bending machine available.

Requirement-Time Profiles. It is helpful in operations planning to represent various requirements as a function of time. Consider, for example, the demand trend shown in Fig. 3.9. When projected into the next 12 months, the transfer function yields the forecasted demand values shown in Table 3.10. These values, along with the cumulative demand values, are shown in Table 4.9. The cumulative demands are plotted in Fig. 4.11 as a solid line.

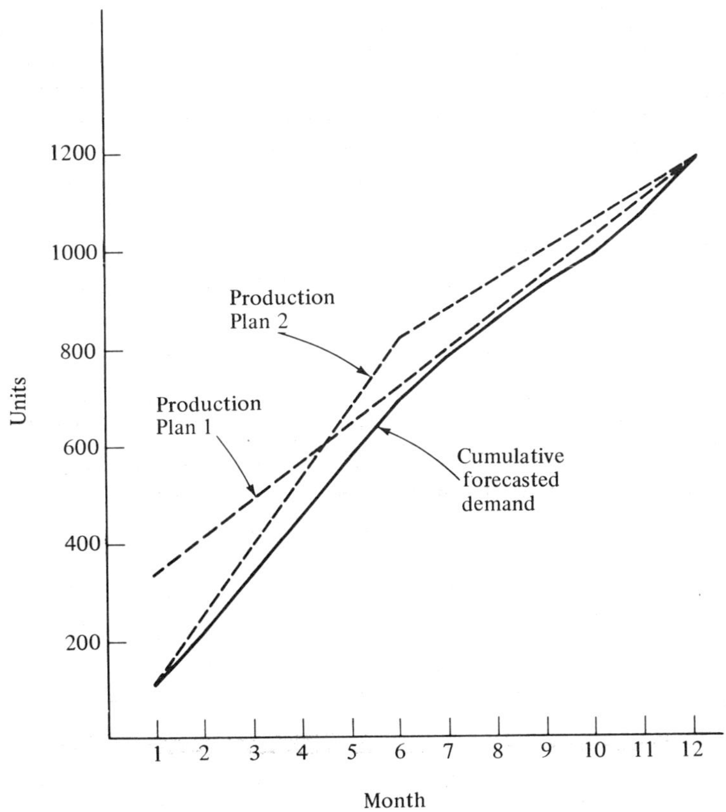

Figure 4.11 Cumulative Demand Forecast for Table 4.9

Table 4.9 Monthly and Cumulative Demand Forecasts, in Units, from Table 3.10

Month	1	2	3	4	5	6	7	8	9	10	11	12
Demand	103	117	115	121	123	109	89	74	71	73	81	98
Cumulative	103	220	335	456	579	688	777	851	922	995	1076	1174

The significance of the cumulative demand curve is that through a combination of inventory and production, we must have available at least the cumulative values by each of the corresponding time periods. At any one time our combined inventory and production can exceed the cumulative demand, but it cannot be less (if no backordering is permitted).

Two possible production plans (production plus inventory) are plotted as dashed lines. Plan 1 permits a level production rate of about 70 units per month, but assumes that we begin the year with about 340 units in inventory. Note also that we end the year with almost no inventory. This condition could be undesirable for next year's planning. Plan 2 employs a production rate of 120 units per month for the first 6 months and 60 units per month for the last 6 months. In Plan 2, we must start the year with only 100 units in inventory, but this plan also leaves us with almost no inventory at the end of the year. The two plans are summarized in Table 4.10.

The plans discussed here are but two of many different ones that could be used to satisfy the expected demand. It is the purpose of operations planning

Table 4.10 Production Plans 1 and 2 of Fig. 4.11

			Production Plan 1			Production Plan 2		
Month	Demand Forecast	Cumulative Forecast	Begin Inventory	Production	End Inventory	Begin Inventory	Production	End Inventory
1	103	103	340	70	307	100	120	117
2	117	220	307	70	260	117	120	120
3	115	335	260	70	215	120	120	125
4	121	456	215	70	164	125	120	124
5	123	579	164	70	111	124	120	121
6	109	688	111	70	72	121	120	132
7	89	777	72	70	53	132	60	103
8	74	851	53	70	49	103	60	89
9	71	922	49	70	48	89	60	78
10	73	995	48	70	45	78	60	65
11	81	1076	45	70	34	65	60	44
12	98	1174	34	70	6	44	60	6

to devise the best plan; i.e., the one that results in lowest cost to the company. The primary costs we must consider are: (1) the cost of carrying inventory and (2) the cost of production. If backordering is permitted, we would also have to consider the cost associated with it. We would attempt to devise a production plan that would minimize the total of these costs. The methods for devising least cost plans are discussed in subsequent sections.

We can convert units to man-hours and machine hours and tabulate these requirements as a function of time. The profile would be quite similar to the forecast demand curve in Fig. 4.11. The requirement-time profiles for man-hours and machine hours would need to be summarized for each department on the basis of total requirements for all products.

THE GOZINTO PROCEDURE*

We have seen previously that demand for finished goods is generated in two basic ways, sales forecast (when we produce for stock) and customer orders (when we produce to customer order). Regardless of how the demand for finished goods is generated, it in turn generates a requirement for subassemblies. The demand for subassemblies generates a requirement for fabricated parts and purchased parts. The demand for fabricated parts generates a demand for raw materials. We can see how this process works by reviewing the example involving the two bookcases. In the example, we determined the number of square feet of sheet metal required to produce 10 units of each of the two types of bookcases. We could determine other parts and material requirements in the same manner for this small problem, but this procedure would be too clumsy for larger problems involving more products and more parts.

Vazsonyi (1958) developed a systematic procedure based upon matrix algebra that provides a good framework for performing the planning function when large numbers of products and parts are involved.

Basic Concepts. Every manufacturing organization has (or should have) bills of material, such as those introduced in Fig. 4.6. Corresponding to the bills of material are explosion charts, such as those in Fig. 4.7.

Let us now consider just what bills of material and explosion charts tell us. The bill of material tells us the simple, direct relationship of each raw material and lower level of assembly to the assembly (or subassembly) which the bill of material describes. For this purpose, our bookcase example required four bills of material: (1) one for the three-shelf bookcase, (2) one for the six-shelf bookcase, (3) one for shelves (for either bookcase), and (4) one for legs (for either bookcase). No bill of material is required for inserts, screws, or connectors, since these are purchased parts.

*This section requires a knowledge of basic matrix algebra, a brief summary of which is presented in Appendix A.

The explosion charts (Fig. 4.7) attempt to relate in graphic form both the simple (direct) and compound (indirect) relationships among raw materials, purchased parts, subassemblies, and assemblies. However, in order to do computation, a better method is needed.

Let us first examine the bills of material, since they give simple relationships. Let us also establish some notation. Since we are interested in the number of raw materials, purchased parts, or subassemblies that constitute (or Gozinto) a particular assembly, let us denote this number by n. Then we can say that n_{ij} can be defined as the number of a given raw material, purchased part, or subassembly i used directly to make one unit of assembly j.

In our bookcase example, this notation would result in the following sets of direct relationships:

For the three-shelf bookcase, stock number 1:

$$\begin{aligned} n_{31} &= 3 \text{ (shelves used in bookcase 1)} \\ n_{41} &= 4 \text{ (legs use in bookcase 1)} \\ n_{51} &= 8 \text{ (inserts used in bookcase 1)} \\ n_{61} &= 12 \text{ (screws used in bookcase 1)} \end{aligned}$$

For the six-shelf bookcase, stock number 2:

$$\begin{aligned} n_{32} &= 6 \text{ (shelves used in bookcase 2)} \\ n_{42} &= 8 \text{ (legs used in bookcase 2)} \\ n_{52} &= 8 \text{ (inserts used in bookcase 2)} \\ n_{62} &= 24 \text{ (screws used in bookcase 2)} \\ n_{72} &= 4 \text{ (connectors used in bookcase 2)} \end{aligned}$$

For the shelves, stock number 3:

$$n_{83} = 3 \quad \text{(square feet of sheet metal used in one shelf)}$$

For the legs, stock number 5:

$$n_{84} = 2 \quad \text{(square feet of sheet metal used in one leg)}$$

We can easily see that our simple example is a two-level system. One level converts a common item, sheet metal, to shelves and legs. The second level converts shelves, legs, and purchased items into bookcases.

The notation used for the simple bill of material relations suggests matrix notation. To explore this, let us form a matrix of the upper level (bookcases) relations. Call this matrix $\mathbf{N}_1$.

$$\mathbf{N}_1 = \begin{array}{c|c|c|} & 1 & 2 \\ \hline 3 & 3 & 6 \\ \hline 4 & 4 & 8 \\ \hline 5 & 8 & 8 \\ \hline 6 & 12 & 24 \\ \hline 7 & 0 & 4 \\ \hline \end{array} \tag{4.1}$$

Let us also show lower-level relationships (sheet metal to legs, shelves, and other items used at the upper level) in matrix notation. Let us call this matrix $\mathbf{N}_2$.

$$\mathbf{N}_2 = \begin{array}{c|c|c|c|c|c|} & 3 & 4 & 5 & 6 & 7 \\ \hline 8 & 3 & 2 & 0 & 0 & 0 \\ \hline \end{array} \tag{4.2}$$

Note now that we could multiply $\mathbf{N}_2 \times \mathbf{N}_1$.

$$\mathbf{N}_2\mathbf{N}_1 = \begin{array}{c|c|c|} & 1 & 2 \\ \hline 8 & 17 & 34 \\ \hline \end{array} \tag{4.3}$$

This matrix indicates that 17 square feet of sheet metal are used for one three-shelf bookcase, and 34 square feet for a six-shelf bookcase. We can confirm that these figures are correct in this simple case by referring to the bills of material and explosion charts in Fig. 4.6 and 4.7. These are second-level (indirect) relationships.

Let us now show all of the bill of material relationships in a matrix $\mathbf{N}$:

$$\mathbf{N} = \begin{array}{c|cc|ccccc|c|} & 1 & 2 & 3 & 4 & 5 & 6 & 7 & 8 \\ \hline 1 & 0 & 0 & 0 & 0 & 0 & 0 & 0 & 0 \\ 2 & 0 & 0 & 0 & 0 & 0 & 0 & 0 & 0 \\ \hline 3 & 3 & 6 & 0 & 0 & 0 & 0 & 0 & 0 \\ 4 & 4 & 8 & 0 & 0 & 0 & 0 & 0 & 0 \\ 5 & 8 & 8 & 0 & 0 & 0 & 0 & 0 & 0 \\ 6 & 12 & 24 & 0 & 0 & 0 & 0 & 0 & 0 \\ 7 & 0 & 4 & 0 & 0 & 0 & 0 & 0 & 0 \\ \hline 8 & 0 & 0 & 3 & 2 & 0 & 0 & 0 & 0 \\ \hline \end{array} \tag{4.4}$$

Note that we have partitioned $\mathbf{N}$ in such a manner that $\mathbf{N}_1$ and $\mathbf{N}_2$ are preserved and all other submatrices are null matrices (all zero). Denoting our new matrix by $\mathbf{N}_p$, we have:

$$\mathbf{N}_p = \begin{bmatrix} 0 & 0 & 0 \\ \mathbf{N}_1 & 0 & 0 \\ 0 & \mathbf{N}_2 & 0 \end{bmatrix} \tag{4.5}$$

Let us now multiply $\mathbf{N}_p$ by itself (the reader should confirm that this is possible because of conformability of the $\mathbf{N}_p$ matrix and all submatrices).

$$\mathbf{N}_p^2 = \begin{bmatrix} 0 & 0 & 0 \\ 0 & 0 & 0 \\ \mathbf{N}_2\mathbf{N}_1 & 0 & 0 \end{bmatrix} \tag{4.6}$$

This multiplication generates the second-level relationship that we obtained from direct multiplication of $\mathbf{N}_2\mathbf{N}_1$. We now note that if we multiply $\mathbf{N}_p^2$ by $\mathbf{N}_p$, we would obtain a null matrix. (The reader should confirm this statement.)

It now appears that if we add $\mathbf{N}_p$, $\mathbf{N}_p^2$, and $\mathbf{N}_p^3$ (the latter trivially), we would obtain a consumption matrix $\mathbf{C}$ representing both simple (direct) and compound (indirect) requirements:

$$\mathbf{C} = \mathbf{N}_p + \mathbf{N}_p^2 + \mathbf{N}_p^3 \tag{4.7}$$

This matrix would be:

$$\mathbf{C} = \begin{bmatrix} 0 & 0 & 0 \\ \mathbf{N}_1 & 0 & 0 \\ \mathbf{N}_2\mathbf{N}_1 & \mathbf{N}_2 & 0 \end{bmatrix} \tag{4.8}$$

By the rules of partitioning, we can drop the subscript $_p$ and obtain:

$$\mathbf{C} = \mathbf{N} + \mathbf{N}^2 + \mathbf{N}^3 \tag{4.9}$$

Using this rule on our original $\mathbf{N}$ matrix, we obtain:

$\mathbf{C} =$ (4.10)

	1	2	3	4	5	6	7	8
1	0	0	0	0	0	0	0	0
2	0	0	0	0	0	0	0	0
3	3	6	0	0	0	0	0	0
4	4	8	0	0	0	0	0	0
5	8	8	0	0	0	0	0	0
6	12	24	0	0	0	0	0	0
7	0	4	0	0	0	0	0	0
8	17	34	3	2	0	0	0	0

Note that the cells c_{ij} of this matrix can be defined as the amount of the ith item (raw material, purchased part, or subassembly) used in the jth higher-level subassembly or assembly. Let us also now define a total requirements matrix **T** where:

$$t_{ij} = \begin{bmatrix} c_{ij}, & \text{if } i \neq j \\ 1 + c_{ij}, & \text{if } i = j \end{bmatrix} \tag{4.11}$$

Note that the entries for **T** are the same as those for **C**, except on the principal diagonal, and that the principal diagonal will consist of 1's. Thus the principal diagonal comprises the identity matrix **I**.

Continuing our example, our matrix **T** would be:

$$\mathbf{T} = \begin{array}{c|cccccccc|} & 1 & 2 & 3 & 4 & 5 & 6 & 7 & 8 \\ \hline 1 & 1 & 0 & 0 & 0 & 0 & 0 & 0 & 0 \\ 2 & 0 & 1 & 0 & 0 & 0 & 0 & 0 & 0 \\ 3 & 3 & 6 & 1 & 0 & 0 & 0 & 0 & 0 \\ 4 & 4 & 8 & 0 & 1 & 0 & 0 & 0 & 0 \\ 5 & 8 & 8 & 0 & 0 & 1 & 0 & 0 & 0 \\ 6 & 12 & 24 & 0 & 0 & 0 & 1 & 0 & 0 \\ 7 & 0 & 4 & 0 & 0 & 0 & 0 & 1 & 0 \\ 8 & 17 & 34 & 3 & 2 & 0 & 0 & 0 & 1 \\ \hline \end{array} \tag{4.12}$$

This introduction of the identity matrix is somewhat arbitrary, but the diagonal 1's do represent the inherent self-identity of a given part. Furthermore, the **T** matrix is more useful than the **C** matrix in later sections.

We now observe that:

$$\mathbf{T} = \mathbf{I} + \mathbf{C}. \tag{4.13}$$

We can now generalize the relationship

$$\mathbf{C} = \mathbf{N} + \mathbf{N}^2 + \mathbf{N}^3 \tag{4.14}$$

to

$$\mathbf{C} = \mathbf{N} + \mathbf{N}^2 + \mathbf{N}^3 + \cdots + \mathbf{N}^m \tag{4.15}$$

where m is the number of levels in the system. (Note that $\mathbf{N}^{m+1}$ is the null matrix.)

From Eqs. 4.13 and 4.15 we get:

$$\mathbf{T} = \mathbf{I} + \mathbf{N} + \mathbf{N}^2 + \cdots + \mathbf{N}^m \tag{4.16}$$

$$(\mathbf{I} - \mathbf{N})\mathbf{T} = (\mathbf{I} - \mathbf{N})(\mathbf{I} + \mathbf{N} + \mathbf{N}^2 + \cdots + \mathbf{N}^m) \tag{4.17}$$

$$(\mathbf{I} - \mathbf{N})\mathbf{T} = \mathbf{I} \tag{4.18}$$

and finally,

$$\mathbf{T} = (\mathbf{I} - \mathbf{N})^{-1} \tag{4.19}$$

Computation of the Total Requirements Matrix. One way to compute the total requirements matrix **T** would be to use Eq. 4.16. This computation would require repeated matrix multiplication and would, for large systems, be very time consuming.

A better way is to form the $(\mathbf{I} - \mathbf{N})$ matrix and use a standard matrix inversion technique. This method is reasonably efficient, but the job can be done still faster by noting that the matrix **N** is strictly triangular and that both $(\mathbf{I} - \mathbf{N})$ and **T** are triangular. Matrices which are triangular have some special properties which make them easy to invert.

First, it should be noted that the matrices in our example are triangular because of the way our bills of material are numbered. In a real situation, the bills of material may have to be renumbered, or indexed, to force this triangularity. This indexing can be done by following the rule that the part number (index) of a given subassembly or assembly must be smaller than the part number (index) of every subassembly, part, or raw material required in its assembly and by then assigning both row and column indices in the **N** matrix in the same order, starting at the upper left corner.*

If we examine our **N** and **T** matrices, we note that the diagonal of **T** consists of cells each with a value of 1, placed there by definition of **T** and operation of Eq. 4.11. Also, several columns have the diagonal 1 as the only non-zero entry in the column. These columns represent raw materials or purchased parts.

We can also note that every column of **T**, except for the diagonal 1, is a linear combination of the columns to the right of it, with the coefficients of the combination being specified by the corresponding columns of the **N** matrix. We also see that a given cell of the **N** matrix need be used only once in the computation. Also, the order of multiplication and addition of columns to form the linear combination does not matter.

These facts suggest that we could start at the lower right of the matrices and

*This renumbering, or re-indexing, can be done very quickly, with punch card or computer equipment.

compute **T** by reading n_{ij} values from the bottom of the columns.* Based on these observations, we can compute the jth column of **T**:

$$\mathbf{t}_j = \sum_{k=j+1}^{n} n_{kj}\mathbf{t}_k \qquad \text{if } j < i \qquad (4.20)^\dagger$$

$$t_{ij} = 1 \qquad \text{if } j = i \qquad (4.21)$$

$$t_{ij} = 0 \qquad \text{if } j > 1 \qquad (4.22)$$

In these expressions, $\mathbf{t}_j$ is the jth column vector of **T** for cells below the diagonal; $\mathbf{t}_k$ is the kth column vector of **T** for $k > j$; and n is the total number of elements (raw materials, purchased parts, subassemblies, and assemblies) in the system. Computation starts with the right-most column of **T** and proceeds column by column to the left.

The $\mathbf{T} = (\mathbf{I} - \mathbf{N})^{-1}$ *Algorithm.* The basic algorithm is:

1. Form the matrix **N** according to the rules:
   - Assign a row and a column for each raw material, purchased part, subassembly, and assembly. The sequence of rows and columns must be such that each assembly, etc., precedes all subassemblies, parts, etc., used in its fabrication. (If reindexing, discussed in the previous section, has been done, then row and column assignment will be in numerical order.) The order of rows and columns must be identical.
   - Insert in each cell n_{ij} the quantity (if any) of the ith part (raw material or subassembly) used in the jth assembly (subassembly).
   - Define all other cells as zero. The matrix so defined should be strictly lower triangular.
2. Prepare a format for the matrix **T**, with the same row and column designations in the same order as in the matrix **N**.
3. Place 1's on the diagonal of the **T** matrix.
4. Start computation at the lower right corner cell. The computation will proceed column by column, from right to left, until the left-most column has been completed.
5. Establish a column vector **v**, initially zeros, outside the **T** matrix. This vector is a working vector and should contain as many cells as there are cells below plus the diagonal cell in the column of **T** currently being

*In programming this computation for the computer, only **T** needs be stored, and the n_{ij}'s can be read from input, one at a time.

†The letter n is used in two ways in Eq. 4.20: (1) the n over the summation sign Σ refers to the total number of elements in the system and (2) n_{kj} is the element in the matrix **N** in the intersection of the kth row and jth column.

computed. The top cell of **v** corresponds to the diagonal element of the column being computed, while the bottom cell of **v** corresponds to the bottom cell of the column being computed. (Note: In the first iteration there is but one cell, the diagonal itself. Although computation of all columns representing raw materials and purchased parts is trivial, it is necessary in computer operation from a conceptual point of view.)

6. Look at the corresponding column of the **N** matrix (start at the bottom of the column). If this cell is non-zero, determine the row designation of the cell in question, and find the column with the same designation. If the cell is zero, repeat Step 6 for the cell next above.

7. Multiply the column so obtained by the value of the non-zero cell, and add the result to **v**. Repeat Steps 6 and 7, working up the column, until the diagonal is reached. When the diagonal is reached, add the vector **v** to the column of the **T** matrix being computed. (The only non-zero entry in this column up to this point will have been the 1 on the diagonal.) Choose the column of the **T** matrix immediately to the left, and repeat from Step 5. When all columns have been computed, a triangular matrix will have been generated, with 1's on the principal diagonal and zeros above this diagonal.

An Illustrative Example. We can now use our bookshelf problem to illustrate our algorithm. Using the data from the **N** matrix in Eq. 4.4, we can illustrate the computation of the **T** matrix in Eq. 4.12.

Perhaps the best way to understand the algorithm is to compute one column of the **T** matrix in detail. Since the first several columns are trivial or easy to compute, let us look at column 2. Assume that all columns to the right—that is, columns 3 through 8—have been computed according to the rules of the algorithm, and that columns 2 and 1 have 1's on the diagonal, because of Step 3 of the algorithm. We are ready to start on column 2, reading cell values from the bottom up in column 2 of the **N** matrix. We are on Step 5 of the algorithm.

First, a column vector **v** of 7 cells, initially zeros, is established. Then, in Step 6, we see that the bottom cell of column 2 of the **N** matrix is zero. Therefore, we repeat Step 6, and the second from the bottom cell is found to be 4. The row designation of this cell is 7. The multiplication of the correspondingly labeled column of the **T** matrix (column 7) by the cell value and the addition to the **v** vector is shown in Table 4.11.

Step 6 is then repeated. We then find 24 in the next cell above, with a row index of 6. The multiplication of the correspondingly labeled column of **T** by the cell value and addition to the previous **v** vector is shown in Table 4.12.

Table 4.11 Computation of Column 2—First Step

Column 7 (From **T**)		Cell Value (From **N**)		Product	Vector **v**
$\begin{bmatrix} 1 \\ 0 \end{bmatrix}$	$\times$	[4]	$=$	$\begin{bmatrix} 4 \\ 0 \end{bmatrix}$	$\begin{bmatrix} 0 \\ 0 \\ 0 \\ 0 \\ 0 \\ 4 \\ 0 \end{bmatrix}$

Table 4.12 Computation of Column 2—Second Step

Column 6 (From **T**)		Cell Value (From **N**)		Product	Vector **v**
$\begin{bmatrix} 1 \\ 0 \\ 0 \end{bmatrix}$	$\times$	[24]	$=$	$\begin{bmatrix} 24 \\ 0 \\ 0 \end{bmatrix}$	$\begin{bmatrix} 0 \\ 0 \\ 0 \\ 0 \\ 24 \\ 4 \\ 0 \end{bmatrix}$

In another repetition of Step 6, an 8 is encountered with row index 5, and the subsequent multiplication and addition results in the vector:

$$\mathbf{v} = \begin{bmatrix} 0 \\ 0 \\ 0 \\ 8 \\ 24 \\ 4 \\ 0 \end{bmatrix} \tag{4.23}$$

The next repetition of Step 6 is more interesting. A value of 8 is encountered, with row index 4. The subsequent choice of column 4 of **T**, the multiplication, and the addition to **v** is shown in Table 4.13. Similarly, the next repetition of Step 6 finds a 6 with row index 3. Table 4.14 shows the computation.

Table 4.13 Computation of Column 2—Fourth Step

Column 4 (From **T**)		Cell Value (From **N**)		Product	Vector **v**
$\begin{bmatrix}1\\0\\0\\0\\2\end{bmatrix}$	×	[8]	=	$\begin{bmatrix}8\\0\\0\\0\\16\end{bmatrix}$	$\begin{bmatrix}0\\0\\8\\8\\24\\4\\16\end{bmatrix}$

Table 4.14 Computation of Column 2—Fifth Step

Column 3 (From **T**)		Cell Value (From **N**)		Product	Vector **v**
$\begin{bmatrix}1\\0\\0\\0\\0\\3\end{bmatrix}$	×	[6]	=	$\begin{bmatrix}6\\0\\0\\0\\0\\18\end{bmatrix}$	$\begin{bmatrix}0\\6\\8\\8\\24\\4\\34\end{bmatrix}$

The next iteration of Steps 6 and 7 encounters the diagonal cell, causing the addition of vector **v** (Table 4.14) to column 2 of **T**, with the result shown in column 2 of Eq. 4.12.

The algorithm would then proceed to calculate column 1 of the **T** matrix in the same fashion that column 2 has just been determined. This process would complete the calculation of the **T** matrix.

The method presented in this section of computing the **T** matrix is but one way of achieving this and is especially convenient for manual calculations. If a canned computer program is available for inverting matrices, it might be easier to form the (**I** − **N**) matrix and obtain its inverse, as required by Eq. 4.19.

The use of the **T** matrix is discussed in the following section. The manner in which **T** is used is independent of the particular method used for determining it.

Use of the **T** *Matrix.* Total requirements matrix **T** can now be used directly in obtaining parts requirements for customer orders or for forecasted demand. Suppose, for example, that we receive an order for 200 three-shelf bookcases and 100 six-shelf bookcases. Forming a vector **b** of this forecast and multiplying by matrix **T** we obtain:

$$\begin{bmatrix} 1 & & & & & & & \\ 0 & 1 & & & & & & \\ 3 & 6 & 1 & & & & & \\ 4 & 8 & 0 & 1 & & & & \\ 8 & 8 & 0 & 0 & 1 & & & \\ 12 & 24 & 0 & 0 & 0 & 1 & & \\ 0 & 4 & 0 & 0 & 0 & 0 & 1 & \\ 17 & 34 & 3 & 2 & 0 & 0 & 0 & 1 \end{bmatrix} \begin{bmatrix} 200 \\ 100 \\ 0 \\ 0 \\ 0 \\ 0 \\ 0 \\ 0 \end{bmatrix} = \begin{bmatrix} 200 \\ 100 \\ 1200 \\ 1600 \\ 2400 \\ 4800 \\ 400 \\ 6800 \end{bmatrix} \tag{4.24}$$

Note that we can now express our problem in terms of the vector **x** for any demand vector **b**:

$$\mathbf{x} = \mathbf{Tb} \tag{4.25}$$

In order to develop the several concepts discussed above we have taken many steps that are unnecessary in the application of the Gozinto procedure. Let us now summarize the steps actually needed in an application.

Step 1. Construct the **N** matrix, as in Eq. 4.4.

Step 2. Compute the **T** matrix, as in Eq. 4.12 (use the algorithm in the previous section, or simply determine $(\mathbf{I} - \mathbf{N})^{-1}$.)

Step 3. Determine parts requirements for customer orders or forecasted demand, as in Eq. 4.24.

Large applications are normally performed on a computer. The solution could be obtained using a linear programming software package or by other matrix inversion schemes, but the algorithm presented is much faster. In all computer applications of the Gozinto procedure, efficient schemes for packing matrices must be employed because of the large number of empty elements in the matrices.

Gozinto Decisions. The procedure for determining parts requirements outlined in the previous section is most useful for processing customer orders for a single time period. When we want to determine requirements for forecasted demand over several time periods, an extended procedure is needed. Such a procedure, developed in this section, will also generate a great deal more information needed in making several other planning decisions.

The decisions to be rendered are:

1. Time-phased parts requirements (gross).
2. Time-phased parts requirements (netted against stock on hand and outstanding production and purchase orders).
3. From decision 2, we can obtain these decisions:
   - Whether to order raw materials; how much to order.
   - Whether the requirements can be met with present capacity, considering machine time and man-hours available. Possibility of extra shifts and overtime is considered.

In order to arrive at the above decisions, we need the following information as input to the Gozinto transfer function:

1. **T:** the total requirements matrix, as in Eq. 4.12.
2. **D:** time-phased demand matrix.
3. **e:** vector showing desired ending inventory (represents safety stock).
4. **s:** vector showing current stock on hand for each stock number.
5. **P:** matrix showing outstanding production and purchase orders and the time period in which they are scheduled to be available.
6. **r:** vector showing reorder point for each purchased item.
7. **B:** matrix showing production rate of each work center in terms of added minutes per unit for each stock number.
8. **L:** matrix showing current load on each work center.

Note that some of the above inputs would come from the Parameter File, while others would come from the Variable Status File. The reader should relate each of these inputs to Fig. 4.1 and to the list of inputs presented on pages 101 through 102.

We illustrate the determination of each of the Gozinto decisions by once again considering the bookcase example. We assume that demand for the two bookcases has been forecasted for a 3-month period: January, February, and March. We have some stock on hand for the bookcases as well as for the parts. We have certain production and purchase orders scheduled for arrival. We also wish to have a certain amount of each stock number on hand as safety stock after all demands for the 3-month period have been satisfied.

We now list each of the inputs for our example:

1. **T:** the **T** matrix shown in Eq. 4.12.

2. **D:** time-phased demand, from forecast or customer orders:

$$\mathbf{D} = \begin{array}{c} \text{Jan.} \quad \text{Feb.} \quad \text{Mar.} \\ \begin{bmatrix} 500 & 500 & 500 \\ 400 & 450 & 500 \\ 0 & 0 & 0 \\ 0 & 0 & 0 \\ 0 & 0 & 0 \\ 0 & 0 & 0 \\ 0 & 0 & 0 \\ 0 & 0 & 0 \end{bmatrix} \end{array}$$

3. **e:** desired ending inventory (amount we want on hand at end of March, after satisfying demands):

$$\mathbf{e} = \begin{bmatrix} 300 \\ 300 \\ 1{,}000 \\ 1{,}500 \\ 2{,}000 \\ 10{,}000 \\ 1{,}200 \\ 3{,}000 \end{bmatrix}$$

4. **s:** stock on hand (current inventory status of all stock numbers).

$$\mathbf{s} = \begin{bmatrix} 250 \\ 100 \\ 8{,}000 \\ 1{,}200 \\ 2{,}500 \\ 20{,}000 \\ 1{,}000 \\ 42{,}000 \end{bmatrix}$$

5. **P:** outstanding production and purchase orders, scheduled for period j.

$$\mathbf{P} = \begin{array}{c} \text{Jan.} \quad \text{Feb.} \quad \text{Mar.} \\ \begin{bmatrix} 250 & 100 & 0 \\ 100 & 0 & 0 \\ 0 & 0 & 0 \\ 400 & 0 & 0 \\ 1{,}000 & 500 & 0 \\ 10{,}000 & 10{,}000 & 0 \\ 2{,}000 & 0 & 0 \\ 8{,}000 & 4{,}000 & 0 \end{bmatrix} \end{array}$$

P′: add vector **s** to column 1 of **P** to obtain time-phased matrix showing availability of scheduled inventory. Scheduled inventory for each stock number is that inventory already on hand plus any due in on previously placed orders.

$$\mathbf{P}' = \begin{array}{c} \text{Jan.} \quad \text{Feb.} \quad \text{Mar.} \\ \begin{bmatrix} 500 & 100 & 0 \\ 200 & 0 & 0 \\ 8{,}000 & 0 & 0 \\ 1{,}600 & 0 & 0 \\ 3{,}500 & 500 & 0 \\ 30{,}000 & 10{,}000 & 0 \\ 3{,}000 & 0 & 0 \\ 50{,}000 & 4{,}000 & 0 \end{bmatrix} \end{array} \begin{array}{c} \text{Total} \\ \left.\begin{array}{r} 600 \\ 200 \\ 8{,}000 \\ 1{,}600 \\ 4{,}000 \\ 40{,}000 \\ 3{,}000 \\ 54{,}000 \end{array}\right\} \mathbf{P}'_T \end{array}$$

Sum each row to obtain total scheduled inventory vector $\mathbf{P}'_T$.

6. **r**: reorder point for each purchased item:

$$\mathbf{r} = \begin{bmatrix} - \\ - \\ - \\ - \\ 8{,}000 \\ 20{,}000 \\ 2{,}000 \\ 20{,}000 \end{bmatrix}$$

7. **B**: work center production rate, in terms of added minutes per unit for each stock number.

$$\mathbf{B} = \begin{array}{c} \\ \text{Dep't 1} \\ \text{Dep't 2} \\ \text{Dep't 3} \end{array} \begin{array}{c} \begin{array}{cccccccc} 1 & 2 & 3 & 4 & 5 & 6 & 7 & 8 \end{array} \\ \begin{bmatrix} 0 & 0 & 0.67 & 0.5 & 0 & 0 & 0 & 0 \\ 0 & 0 & 0.75 & 1.0 & 0 & 0 & 0 & 0 \\ 1.0 & 1.5 & 0 & 0 & 0 & 0 & 0 & 0 \end{bmatrix} \end{array}$$

8. **L:** current load on each work center, in terms of number of hours per time period.

$$\mathbf{L} = \begin{array}{c} \text{Dep't 1} \\ \text{Dep't 2} \\ \text{Dep't 3} \end{array} \overset{\begin{array}{ccc} \text{Jan.} & \text{Feb.} & \text{Mar.} \end{array}}{\begin{bmatrix} 100 & 140 & 80 \\ 160 & 120 & 100 \\ 80 & 80 & 80 \end{bmatrix}} \overset{\text{Total}}{\left.\begin{matrix} 320 \\ 380 \\ 240 \end{matrix}\right\}} \mathbf{L}'$$

The above inputs to the Gozinto procedure will yield the several decisions listed earlier. From **T**, **P′**, **D**, and **e**, we can make the first two decisions:

Decision 1: **X**—time-phased parts requirements (gross),

$$\mathbf{X} = \mathbf{TD}:$$

$$\mathbf{X} = \begin{bmatrix} 1 & & & & & & & \\ 0 & 1 & & & & & & \\ 3 & 6 & 1 & & & & & \\ 4 & 8 & 0 & 1 & & & & \\ 8 & 8 & 0 & 0 & 1 & & & \\ 12 & 24 & 0 & 0 & 0 & 1 & & \\ 0 & 4 & 0 & 0 & 0 & 0 & 1 & \\ 17 & 34 & 3 & 2 & 0 & 0 & 0 & 1 \end{bmatrix} \overset{\overbrace{\text{Jan.} \quad \text{Feb.} \quad \text{Mar.}}^{\text{Forecast}} \quad \text{Inventory}}{\begin{bmatrix} 500 & 500 & 500 & 300 \\ 400 & 450 & 500 & 300 \\ 0 & 0 & 0 & 1{,}000 \\ 0 & 0 & 0 & 1{,}500 \\ 0 & 0 & 0 & 2{,}000 \\ 0 & 0 & 0 & 10{,}000 \\ 0 & 0 & 0 & 1{,}200 \\ 0 & 0 & 0 & 3{,}000 \end{bmatrix}}$$

$$= \overset{\begin{array}{cccc} \text{Jan.} & \text{Feb.} & \text{Mar.} & \text{Inventory} \end{array}}{\begin{bmatrix} 500 & 500 & 500 & 300 \\ 400 & 450 & 500 & 300 \\ 3{,}900 & 4{,}200 & 4{,}500 & 3{,}700 \\ 5{,}200 & 5{,}600 & 6{,}000 & 5{,}100 \\ 7{,}200 & 7{,}600 & 8{,}000 & 6{,}800 \\ 15{,}600 & 16{,}800 & 18{,}000 & 20{,}800 \\ 1{,}600 & 1{,}800 & 2{,}000 & 2{,}400 \\ 22{,}100 & 23{,}800 & 25{,}500 & 24{,}300 \end{bmatrix}} \overset{\text{Total}}{\left.\begin{matrix} 1{,}800 \\ 1{,}650 \\ 16{,}300 \\ 21{,}900 \\ 29{,}600 \\ 71{,}200 \\ 7{,}800 \\ 95{,}700 \end{matrix}\right\}} = \mathbf{X}_T$$

Each row is summed to obtain the total gross requirement vector, $\mathbf{X}_T$. Note that $\mathbf{X}$ and $\mathbf{X}_T$ do not consider parts already on hand or on order. These parts are considered in the following decision.

Decision 2: $\mathbf{X}_N$—the netted requirements matrix is determined in three steps:

Step 1. Subtract the scheduled inventory matrix $\mathbf{P}'$ from the demand matrix $\mathbf{D}$; call this $\mathbf{Y}$.

$$\mathbf{Y} = \begin{array}{c} \begin{matrix} \text{Jan.} & \text{Feb.} & \text{Mar.} \end{matrix} \\ \begin{bmatrix} 500 & 500 & 500 \\ 400 & 450 & 500 \\ 0 & 0 & 0 \\ 0 & 0 & 0 \\ 0 & 0 & 0 \\ 0 & 0 & 0 \\ 0 & 0 & 0 \\ 0 & 0 & 0 \end{bmatrix} \end{array} - \begin{array}{c} \begin{matrix} \text{Jan.} & \text{Feb.} & \text{Mar.} \end{matrix} \\ \begin{bmatrix} 500 & 100 & 0 \\ 200 & 0 & 0 \\ 8{,}000 & 0 & 0 \\ 1{,}600 & 0 & 0 \\ 3{,}500 & 500 & 0 \\ 30{,}000 & 10{,}000 & 0 \\ 3{,}000 & 0 & 0 \\ 50{,}000 & 4{,}000 & 0 \end{bmatrix} \end{array}$$

$$= \begin{array}{c} \begin{matrix} \text{Jan.} & \text{Feb.} & \text{Mar.} \end{matrix} \\ \begin{bmatrix} 0 & 400 & 500 \\ 200 & 450 & 500 \\ -8{,}000 & 0 & 0 \\ -1{,}600 & 0 & 0 \\ -3{,}500 & -500 & 0 \\ -30{,}000 & -10{,}000 & 0 \\ -3{,}000 & 0 & 0 \\ -50{,}000 & -4{,}000 & 0 \end{bmatrix} \end{array}$$

Note that the values in $\mathbf{Y}$ for stock numbers 1 and 2 (the two types of bookcases) represent the *additional* number of units that must be manufactured for each of the 3 months. The negative values for the other stock numbers represent the number of units that will not have to be manufactured or purchased. These are already available or are scheduled to be available.

Step 2. Add the desired ending inventory vector $\mathbf{e}$ as the last column of $\mathbf{Y}$. Call this $\mathbf{Y}'$. Now, $\mathbf{W} = \mathbf{TY}'$.

$$
\mathbf{W} = \begin{bmatrix}
1 & & & & & & & \\
0 & 1 & & & & & & \\
3 & 6 & 1 & & & & & \\
4 & 8 & 0 & 1 & & & & \\
8 & 8 & 0 & 0 & 1 & & & \\
12 & 24 & 0 & 0 & 0 & 1 & & \\
0 & 4 & 0 & 0 & 0 & 0 & 1 & \\
17 & 34 & 3 & 2 & 0 & 0 & 0 & 1
\end{bmatrix}
\begin{array}{c}
\begin{array}{rrrr} \text{Jan.} & \text{Feb.} & \text{Mar.} & \text{Inventory} \end{array} \\
\begin{bmatrix}
0 & 400 & 500 & 300 \\
200 & 450 & 500 & 300 \\
-8{,}000 & 0 & 0 & 1{,}000 \\
-1{,}600 & 0 & 0 & 1{,}500 \\
-3{,}500 & -500 & 0 & 2{,}000 \\
-30{,}000 & -10{,}000 & 0 & 10{,}000 \\
-3{,}000 & 0 & 0 & 1{,}200 \\
-50{,}000 & -4{,}000 & 0 & 3{,}000
\end{bmatrix}
\end{array}
$$

$$
= \begin{array}{c}
\begin{array}{rrrr} \text{Jan.} & \text{Feb.} & \text{Mar.} & \text{Inventory} \end{array} \\
\begin{bmatrix}
0 & 400 & 500 & 300 \\
200 & 450 & 500 & 300 \\
-6{,}800 & 3{,}900 & 4{,}500 & 3{,}700 \\
0 & 5{,}200 & 6{,}000 & 5{,}100 \\
-1{,}900 & 6{,}300 & 8{,}000 & 6{,}800 \\
-25{,}200 & 5{,}600 & 18{,}000 & 20{,}800 \\
-2{,}200 & 1{,}800 & 2{,}000 & 2{,}400 \\
-70{,}400 & 18{,}100 & 25{,}500 & 24{,}300
\end{bmatrix}
\end{array}
$$

Step 3. Negative values in **W** indicate excessive inventory above needed requirements for that month. We distribute the excess inventory over following months, thereby giving us our final netted requirements matrix, $\mathbf{X}_N$.

$$
\mathbf{X}_N = \begin{array}{c}
\begin{array}{rrrr} \text{Jan.} & \text{Feb.} & \text{Mar.} & \text{Inventory} \end{array} \\
\begin{bmatrix}
0 & 400 & 500 & 300 \\
200 & 450 & 500 & 300 \\
0 & 0 & 1{,}600 & 3{,}700 \\
0 & 5{,}200 & 6{,}000 & 5{,}100 \\
0 & 4{,}400 & 8{,}000 & 6{,}800 \\
0 & 0 & 0 & 19{,}200 \\
0 & 0 & 1{,}600 & 2{,}400 \\
0 & 0 & 0 & 0
\end{bmatrix}
\end{array}
\quad
\begin{array}{c}
\text{Total} \\
\left.\begin{array}{r}
1{,}200 \\ 1{,}450 \\ 5{,}300 \\ 16{,}300 \\ 19{,}200 \\ 19{,}200 \\ 4{,}000 \\ -2{,}500
\end{array}\right\} \mathbf{X}'_N
\end{array}
$$

The netted requirements for each stock number are summed to give us the total netted requirements vector $\mathbf{X}'_N$. The values in $\mathbf{X}'_N$ are the minimum total number of additional parts that must be manufactured or purchased sometime during the 3-month planning period if the expected demand and desired ending inventories are to be satisfied. The values in $\mathbf{X}_N$ are the minimum number of additional parts that must be manufactured or purchased *for that month's requirements*. For example, we must manufacture at least 200 additional units of stock number 2 (the six-shelf bookcase) for January's requirements. We must produce at least 450 additional units for February's requirements. Thus, by the end of February, we must have completed at least 650 units of stock number 2. By the end of March, we must have completed at least 1450 units in order to have satisfied all demands plus the desired ending inventory.

Note that no additional stock number 8 (sheet steel) is required. In fact, even after completing all requirements, we will still have 2,500 square feet of sheet metal in inventory.

The decision on exactly how to provide the netted requirements in $\mathbf{X}_N$ depends on the loads on the work centers, economic manufacturing quantities, economic ordering quantities, and reorder levels. Some of these decisions are discussed below.

Decision 3a: whether to order raw materials. This decision would have to be made periodically, for example, at the beginning of each month. The current stock on hand, vector **s**, would be compared to the reorder level, vector **r**, for this decision. We see for January that stock numbers 5, 6, and 7 should be ordered, since current stock on hand for these items is either at or below the reorder level. A similar comparison would have to be made at the beginning of February. The amount to order would depend on the economic order quantity for each item. This decision is discussed in much greater detail in Chap. 5.

Decision 3b: comparison of requirements to available capacity. We now determine the total amount of work generated for each work center.

$$\mathbf{M}_T = \mathbf{B}\mathbf{X}_N'$$

$$\mathbf{M}_T = \begin{bmatrix} 0 & 0 & 0.67 & 0.5 & 0 & 0 & 0 & 0 \\ 0 & 0 & 0.75 & 1.0 & 0 & 0 & 0 & 0 \\ 1.0 & 1.5 & 0 & 0 & 0 & 0 & 0 & 0 \end{bmatrix} \begin{bmatrix} 1{,}200 \\ 1{,}450 \\ 5{,}300 \\ 16{,}300 \\ 19{,}200 \\ 19{,}200 \\ 4{,}000 \\ 0 \end{bmatrix}$$

$$= \begin{bmatrix} 11{,}701 \\ 20{,}275 \\ 3{,}375 \end{bmatrix} \text{minutes} = \begin{bmatrix} 195 \\ 338 \\ 56 \end{bmatrix} \text{hours}$$

We see then that the netted requirements for the 3 months generate 195, 338, and 56 additional hours of work for departments 1, 2, and 3, respectively. Note that these are total requirements for the 3 months' demand plus desired ending inventory. If we wanted to express these requirements for each month, we would use $\mathbf{X}_N$ rather than $\mathbf{X}'_N$.

$$\mathbf{M} = \mathbf{B}\mathbf{X}_N$$

$$\mathbf{M} = \begin{bmatrix} 0 & 0 & 0.67 & 0.5 & 0 & 0 & 0 & 0 \\ 0 & 0 & 0.75 & 1.0 & 0 & 0 & 0 & 0 \\ 1.0 & 1.5 & 0 & 0 & 0 & 0 & 0 & 0 \end{bmatrix} \begin{array}{c} \begin{array}{cccc} \text{Jan.} & \text{Feb.} & \text{Mar.} & \text{Inventory} \end{array} \\ \begin{bmatrix} 0 & 400 & 500 & 300 \\ 200 & 450 & 500 & 300 \\ 0 & 0 & 1{,}600 & 3{,}700 \\ 0 & 5{,}200 & 6{,}000 & 5{,}100 \\ 0 & 4{,}400 & 8{,}000 & 6{,}800 \\ 0 & 0 & 0 & 19{,}200 \\ 0 & 0 & 1{,}600 & 2{,}400 \\ 0 & 0 & 0 & 0 \end{bmatrix} \end{array}$$

$$= \begin{array}{c} \begin{array}{cccc} \text{Jan.} & \text{Feb.} & \text{Mar.} & \text{Inventory} \end{array} \\ \begin{bmatrix} 0 & 2{,}600 & 4{,}072 & 5{,}029 \\ 0 & 5{,}200 & 7{,}200 & 7{,}875 \\ 300 & 1{,}075 & 1{,}250 & 750 \end{bmatrix} \end{array} \text{minutes} = \begin{array}{c} \begin{array}{cccc} \text{Jan.} & \text{Feb.} & \text{Mar.} & \text{Inventory} \end{array} \\ \begin{bmatrix} 0 & 43 & 68 & 84 \\ 0 & 87 & 120 & 131 \\ 5 & 18 & 21 & 12 \end{bmatrix} \end{array} \text{hours}$$

Note that the time requirements for ending inventory must be assigned in some manner to the three monthly periods. This assignment depends on other demands on the departments.

We now consider the current work load on each department. Assume that for each month there are 160 hours per shift available at each work station. At the beginning of January, the departments are committed to the following work loads, not including the new requirements we are attempting to assign:

$$\mathbf{L} = \begin{array}{c} \begin{array}{ccc} \text{Jan.} & \text{Feb.} & \text{Mar.} \end{array} \\ \begin{bmatrix} 100 & 140 & 80 \\ 160 & 120 & 100 \\ 80 & 80 & 80 \end{bmatrix} \end{array} \quad \begin{array}{c} \text{Total} \\ \left.\begin{array}{c} 320 \\ 380 \\ 240 \end{array}\right\} \mathbf{L}' \end{array}$$

Uncommitted capacity for each department is obtained by subtracting each value in $\mathbf{L}'$ from 480, the total 3-month capacity. This subtraction results in uncommitted capacities of 160, 100, and 240 hours for departments 1, 2, and 3, respectively. Comparing these values to those in $\mathbf{M}_T$, we see that department 3 can easily handle the additional 56 hours, but departments 1 and 2 cannot handle the added requirements on a one-shift basis. Perhaps department 1 could accommodate the additional requirements by working 35 hours overtime, but department 2 probably should work two shifts (perhaps for 2 months) or add additional capacity. We would also want to consider the particular month during which the new requirements are needed. The monthly requirements are shown in matrix **M**.

Let us recall the step in the procedure in which we computed $\mathbf{X}_N$, the netted requirements matrix. We could have included at that point consideration of the quality factor. Since our manufactured or purchased items are rarely of perfect quality, we usually must plan to produce and buy more than the final number that we need. Suppose, for example, that only 95 percent of the shelves we fabricate are of acceptable quality. We must then plan to produce $1/0.95 = 1.05$ times the number of shelves we actually need. We can have a quality factor for each stock number and can represent these factors in the following matrix:

$$\mathbf{Q} = \begin{bmatrix} 1.04 &&&&&&& \\ & 1.04 &&&&&& \\ && 1.05 &&&&& \\ &&& 1.05 &&&& \\ &&&& 1.11 &&& \\ &&&&& 1.02 && \\ &&&&&& 1.11 & \\ &&&&&&& 1.03 \end{bmatrix}$$

To include the quality factor in our planning, we simply multiply $\mathbf{X}_N$ by $\mathbf{Q}$; $\mathbf{X}_{NQ} = \mathbf{Q}\mathbf{X}_N$.

$$\mathbf{X}_{NQ} = \begin{bmatrix} 1.04 &&&&&&& \\ & 1.04 &&&&&& \\ && 1.05 &&&&& \\ &&& 1.05 &&&& \\ &&&& 1.11 &&& \\ &&&&& 1.02 && \\ &&&&&& 1.11 & \\ &&&&&&& 1.03 \end{bmatrix} \begin{array}{c} \begin{array}{rrrr} \text{Jan.} & \text{Feb.} & \text{Mar.} & \text{Inventory} \end{array} \\ \begin{bmatrix} 0 & 400 & 500 & 300 \\ 200 & 450 & 500 & 300 \\ 0 & 0 & 1{,}600 & 3{,}700 \\ 0 & 5{,}200 & 6{,}000 & 5{,}100 \\ 0 & 4{,}400 & 8{,}000 & 6{,}800 \\ 0 & 0 & 0 & 19{,}200 \\ 0 & 0 & 1{,}600 & 2{,}400 \\ 0 & 0 & 0 & 0 \end{bmatrix} \end{array}$$

$$= \begin{array}{c} \begin{array}{cccc} \text{Jan.} & \text{Feb.} & \text{Mar.} & \text{Inventory} \end{array} \\ \begin{bmatrix} 0 & 416 & 520 & 312 \\ 208 & 468 & 520 & 312 \\ 0 & 0 & 1{,}680 & 3{,}885 \\ 0 & 5{,}460 & 6{,}300 & 5{,}355 \\ 0 & 4{,}884 & 8{,}880 & 7{,}548 \\ 0 & 0 & 0 & 19{,}584 \\ 0 & 0 & 1{,}776 & 2{,}664 \\ 0 & 0 & 0 & 0 \end{bmatrix} \end{array}$$

$\mathbf{X}_{NQ}$ would then be used instead of $\mathbf{X}_N$ in all other Gozinto decisions.

We have illustrated some of the ways in which the Gozinto procedure can be used in making decisions for the planning function. These examples are merely to illustrate the versatility of the method and by no means exhaust the possible applications.

The Gozinto procedure is basically a planning tool and is not intended to generate detailed schedules or departmental loads. It provides the overall framework within which the scheduling function may assign specific work activities to specific work centers.

PRODUCTION SMOOTHING

In the preceding section, we used the Gozinto procedure to determine the time-phased requirements for finished goods, subassemblies, fabricated parts, and purchased parts. We can apply the Gozinto procedure to the forecasted demand pattern of each of our products for an entire planning period, thereby obtaining the number of each stock number needed for each month's expected demand. We can now convert unit requirements to machine hours, or simply to processing time in each department, and then summarize each department's time requirements for each operating period (usually 1 month) in the planning period (usually 1 year).

If the demand patterns for our products are nearly level throughout the year, it is very easy to plan production. We simply hire the right number of men and maintain the right amount of equipment needed to manufacture our products at a constant rate. In such a situation, we need very little inventory—just enough to take care of random fluctuations in demand. We have described a very stable system, one that is rather easy to control.

If the demand for our products fluctuates significantly, such as in a seasonal pattern or in an erratic pattern, then our system is no longer stable and becomes difficult to control. Planning for such a system is difficult. One alternative is to manufacture at a constant rate and carry enough inventory (of raw material as well as of finished goods) to accommodate the fluctuations. Another alternative is to vary the size of the work force from month to month, depending on the demand level for that month. This alternative reduces the amount of inventory we must carry but results in a large amount of hiring and firing. There are many disadvantages to the second alternative:

1. It is difficult to maintain a skilled work force. Skilled workers usually do not want to work for companies which lay off people as a matter of policy.
2. We usually must pay higher wages to compensate for the lack of security.
3. This alternative results in poor community and union relations. Labor legislation provides so much protection to the employee that this alternative is rarely a feasible policy.
4. There are very high costs associated with hiring, firing, and training employees. It is usually better to try to minimize these costs.
5. Even if we could hire and fire employees at will, we cannot buy and sell equipment at will. Therefore, equipment would be idle during slack periods.

There are certain industries, such as many food-processing operations, in which employment levels vary due either to the seasonality of the demand or to the seasonality of raw material availability. In these industries, nonpermanent employment is accepted as an inherent characteristic. In most manufacturing firms, however, steady employment practices are much more desirable.

Many companies which market seasonal products attempt to offset a particular seasonal pattern with an opposing pattern. One product group might have generally high sales during the winter months and low sales during the summer months. If another product group with an opposite demand trend could be added, level employment could be achieved without excessive inventory.

In this section we discuss a procedure for smoothing production over the planning period. The procedure attempts to balance the costs of inventory and overtime for one department.

Let us consider again the bookcase problem.* Suppose that our forecasting transfer function has estimated demand for each bookcase for the next 12 months. By using the Gozinto method, or any other appropriate summarizing procedure, demand in units for each bookcase can be converted to processing time, and these times can be summarized for each department. Suppose these

*This example is independent of previous examples involving the two bookcases in terms of demand and number of machines in each department.

steps have been performed, with the results shown in the last three columns of Table 4.15. The first three columns indicate the month, number of working days,* and number of regular time hours available per machine per shift. We assume that time requirements at each department are netted values, as discussed in the Gozinto section.

Table 4.15 Regular Hours Available and Production Hours Needed for Three Departments

Month	Working Days	Regular Hours/ Machine/ Shift	Hours Needed Dep't 1	Hours Needed Dep't 2	Hours Needed Dep't 3
January	22	176	4,200	3,330	420
February	20	160	4,000	3,150	400
March	21	168	4,600	3,640	460
April	22	176	4,800	3,800	480
May	21	168	4,500	3,570	450
June	21	168	4,000	3,150	400
July	18†	144	4,000	3,150	400
August	15†	120	3,600	2,850	360
September	21	168	4,200	3,330	420
October	23	184	4,500	3,570	450
November	18	144	4,800	3,800	480
December	21	168	5,000	3,960	500
Total	243	1,944	52,200	41,300	5,220

†Entire plant closes during last week in July and first week in August for vacation.

We demonstrate the production smoothing procedure by considering only department 1 and leave as an exercise for the reader the application of the procedure to departments 2 and 3.

We see from Table 4.15 that 52,200 machine hours are required in department 1 for the total year's production. This is equivalent to $52{,}200/1{,}944 = 27$ machines on a one-shift basis. Other possibilities are 14 machines on a two-shift basis or 13 machines plus some overtime on a two-shift basis. Still another possibility is 9 machines on a three-shift basis, which would consume our total capacity and leave no flexibility for emergencies.

Suppose we decide to use 10 machines. We will work 10 machines for two shifts and 7 machines on the third shift. We assume that the first shift and the second shift each may work up to 4 hours overtime per day on the three machines normally idle during the third shift. The regular labor rate is $3.00 per hour and overtime is $4.50 per hour. Thus, the overtime premium is $1.50 per hour.

*Based on 1969 calendar.

It is possible to produce more than is needed during a particular month for sale during some following month. We will assume a carrying charge of five cents per month for each hour's production carried over to following months. For example, if we work 100 hours in January making items needed in March, the carrying charge would be $(\$0.05) \times (100) \times (2) = \$10.00$, since the items are carried for two months.

We want to smooth production over the entire year in such a way that the sum of inventory carrying cost and overtime premium cost is minimized. To facilitate computations, we will employ a production plan spread sheet, such as that shown in Fig. 4.12.

In the first column of Fig. 4.12 are listed the month numbers in which the production time shown in the second column is required. The time requirements are those for department 1 in Table 4.15. Across the top are listed the month numbers during which the production will occur. Within each month across the top are two columns:

R: number of regular time hours available; obtained by multiplying column 3 of Table 4.15 by 27, the number of machine-shifts available. For example, regular time for January $= 176 \times 27 = 4752$ hours.

O: number of overtime hours available; obtained by multiplying column 3 of Table 4.15 by 3, the number of machines available for overtime on the third shift. For example, overtime for January $= 176 \times 3 = 528$ hours.

Within each month down the left side of Fig. 4.12 are three rows:

A: number of machine-hours available, both regular time and overtime, for that month. Note that, along the diagonal, these are the same values as those under *R* and *O*, but as these available hours are assigned to a particular month's production, the amount available for any following month is reduced.

C: incremental cost of an additional machine-hour. All overtime hours incur an incremental cost of \$1.50. Production in one month for a following month also carries \$0.05 per month carrying charge for each machine-hour used.

P: planned production, both regular and overtime, for each month.

The last two rows summarize the planned production, both regular and overtime, for each month.

In filling out the body of the table, the objective is to arrive at the most economical production plan. To accomplish this, we satisfy each month's requirements by using the least expensive machine time available. Detailed study of Fig. 4.12 will reveal the mechanics of how this is done.

We see in Fig. 4.12 that we can meet the year's requirements with no overtime. It is interesting that we must use 264 machine-hours in January to meet the demand for December. Let us now determine the incremental cost of this plan. Since no overtime is used, the only incremental costs are carrying costs, summarized in Table 4.16.

Table 4.16 Total Incremental Costs for 27 Regular-Shift Machines

January:	(264) × (0.55) =	$ 145.20
February:	(64) × (0.05) + (48) × (0.10) + (8) (0.45) + (200) × (0.50) =	111.60
May:	(36) × (0.30) =	10.80
June:	(112) × (0.05) + (360) × (0.10) + (64) (0.25) =	57.60
September:	(336) × (0.10) =	33.60
October:	(468) × (0.05) =	23.40
	Total Inventory Carrying Costs	382.20
	288 idle machine-hours @$3.00/hr	864.00
	Total Incremental Costs, Idle Time Permitted	$1,246.20
	Total Incremental Costs, Idle Time Not Permitted	555.00

Note that 4752 − 4464 = 288 regular machine-hours in January are not used in this plan. In reality we would probably use all these hours and build up a small amount of inventory. If we did this and carried it for 12 months, our incremental costs would increase by (288) × ($0.60) = $172.80, rather than $864.00 if idle time is permitted. Thus, it is much better to use the available time in January to produce extra material.

We now ask ourselves whether the production plan we have developed is the best such plan. The only way to answer this question is to go through the same steps for other possible plans and compare the resulting incremental costs with the incremental cost obtained in Table 4.16.

If we had a computer program available for computing the production plan spread sheet, we could enter any number of alternative plans we wished and obtain the resulting incremental costs. We could consider these plans in conjunction with any other factors and select the best plan. Relate this concept to the last step in Fig. 4.1.

We could, for example, operate 10 machines during the first and second shifts and only six machines during the third shift. We would then allow the first shift and the second shift each to work up to 4 hours overtime per day on the four machines normally idle during the third shift. Using the same cost values as before, we get the production plan shown in Fig. 4.13. The incremental costs for this plan are shown in Table 4.17.

Month in which to be produced

Month required	Time required		1		2		3		4		5	
			R	*O*	*R*	*O*	*R*	*O*	*R*	*O*	*R*	*O*
			4752	528	4320	480	4536	504	4752	528	4536	504
1	4200	*A*	4752	528								
		C	0	1.50								
		P	4200									
2	4000	*A*	552	528	4320	480						
		C	0.05	1.55	0	1.50						
		P			4000							
3	4600	*A*	552	528	320	480	4536	504				
		C	0.10	1.60	0.05	1.55	0	1.50				
		P			64		4536					
4	4800	*A*	552	528	256	480	0	504	4752	528		
		C	0.15	1.65	0.10	1.60		1.55	0	1.50		
		P			48				4752			
5	4500	*A*	552	528	208	480		504	0	528	4536	504
		C	0.20	1.70	0.15	1.65		1.60		1.55	0	1.50
		P									4500	
6	4000	*A*	552	528	208	480		504		528	36	504
		C	0.25	1.75	0.20	1.70		1.65		1.60	0.05	1.55
		P										
7	4000	*A*	552	528	208	480		504		528	36	504
		C	0.30	1.80	0.25	1.75		1.70		1.65	0.10	1.60
		P										
8	3600	*A*	552	528	208	480		504		528	36	504
		C	0.35	1.85	0.30	1.80		1.75		1.70	0.15	1.65
		P										
9	4200	*A*	552	528	208	480		504		528	36	504
		C	0.40	1.90	0.35	1.85		1.80		1.75	0.20	1.70
		P										
10	4500	*A*	552	528	208	480		504		528	36	504
		C	0.45	1.95	0.40	1.90		1.85		1.80	0.25	1.75
		P										
11	4800	*A*	552	528	208	480		504		528	36	504
		C	0.50	2.00	0.45	1.95		1.90		1.85	0.30	1.80
		P			8						36	
12	5000	*A*	552	528	200	480		504		528	0	504
		C	0.55	2.05	0.50	2.00		1.95		1.90		1.85
		P	264		200							
Total production planned		*RT*	4464		4320		4536		4752		4536	
		OT		0		0		0		0		0

Figure 4.12 Production Plan Spread Sheet for Department 1, using 10 machines for 2 shifts and 7 machines on the third shift

Month in which to be produced

6		7		8		9		10		11		12	
R	*O*	*R*	*O*	*R*	*O*	*R*	*O*	*R*	*O*	*R*	*O*	*R*	*O*
4536	504	3888	432	3240	360	4536	504	4968	552	3888	432	4536	504
4536 0 4000	504 1.50												
536 0.05 112	504 1.55	3888 0 3888	432 1.50										
424 0.10 360	504 1.60	0	432 1.55	3240 0 3240	360 1.50								
64 0.15	504 1.65		432 1.60	0	360 1.55	4536 0 4200	504 1.50						
64 0.20	504 1.70		432 1.65		360 1.60	336 0.05	504 1.55	4968 0 4500	552 1.50				
64 0.25 64	504 1.75		432 1.70		360 1.65	336 0.10 336	504 1.60	468 0.05 468	552 1.55	3888 0 3888	432 1.50		
0	504 1.80		432 1.75		360 1.70	0	504 1.65	0	552 1.60	0	432 1.55	4536 0 4536	504 1.50
4536		3888		3240		4536		4968		3888		4536	
	0		0		0		0		0		0		0

Figure 4.12 (Continued)

Month in which to be produced

Month required	Time required		1 R	1 O	2 R	2 O	3 R	3 O	4 R	4 O	5 R	5 O
			4576	704	4160	640	4368	672	4576	704	4368	672
1	4200	*A*	4576	704								
		C	0.00	1.50								
		P	4200									
2	4000	*A*	376	704	4160	640						
		C	0.05	1.55	0.00	1.50						
		P			4000							
3	4600	*A*	376	704	160	640	4368	672				
		C	0.10	1.60	0.05	1.55	0.00	1.50				
		P	72		160		4368					
4	4800	*A*	304	704	0	640	0	672	4576	704		
		C	0.15	1.65		1.60		1.55	0.00	1.50		
		P	224						4576			
5	4500	*A*	80	704		640		672	0	704	4368	672
		C	0.20	1.70		1.65		1.60		1.55	0.00	1.50
		P	80								4368	52
6	4000	*A*	0	704		640		672		704	0	620
		C		1.75		1.70		1.65		1.60		1.55
		P										
7	4000	*A*		704		640		672		704		620
		C		1.80		1.75		1.70		1.65		1.60
		P										
8	3600	*A*		704		640		672		704		620
		C		1.85		1.80		1.75		1.70		1.65
		P										
9	4200	*A*		704		640		672		704		620
		C		1.90		1.85		1.80		1.75		1.70
		P										
10	4500	*A*		704		640		672		704		620
		C		1.95		1.90		1.85		1.80		1.75
		P										
11	4800	*A*		704		640		672		704		620
		C		2.00		1.95		1.90		1.85		1.80
		P										
12	5000	*A*		704		640		672		704		620
		C		2.05		2.00		1.95		1.90		1.85
		P										
Total production planned		*RT*	4576		4160		4368		4576		4368	
		OT		0		0		0		0		52

Figure 4.13 Production Plan Spread Sheet for Department 1, using 10 machines for 2 shifts and 6 machines on the third shift

Month in which to be produced

6		7		8		9		10		11		12	
R	*O*	*R*	*O*	*R*	*O*	*R*	*O*	*R*	*O*	*R*	*O*	*R*	*O*
4368	672	3744	576	3120	480	4368	672	4784	736	3744	575	4368	672
4368 0.00 4000	672 1.50												
368 0.05 256	672 1.55	3744 0.00 3744	576 1.50										
112 0.10 112	672 1.60	0	576 1.55	3120 0.00 3120	480 1.50 368								
0	672 1.65		576 1.60	0	112 1.55	4368 0.00 4200	672 1.50						
	672 1.70		576 1.65		112 1.60	168 0.05	672 1.55	4784 0.00 4500	736 1.50				
	672 1.75		576 1.70		112 1.65	168 0.10 168	672 1.60	284 0.05 284	736 1.55 28	3744 0.00 3744	576 1.50 576		
	672 1.80		576 1.75		112 1.70	0	672 1.65	0	708 1.60	0	0	4368 0.00 4368	672 1.50 632
4368		3744		3120		4368		4784		3744		4368	
	0		0		368		0		28		576		632

Figure 4.13 (Continued)

Table 4.17 Total Incremental Costs for 26 Regular-Shift Machines

Month	Computation		Cost
January:	(72) × (0.1) + (224) × (0.15) + (80) × (0.2)	=	$ 56.80
February:	(160) × (0.05)	=	8.00
May:	(52) × (1.50)	=	78.00
June:	(256) × (0.05) + (112) × (0.1)	=	24.00
August:	(368) × (1.50)	=	552.00
September:	(168) × (0.1)	=	16.80
October:	(284) × (0.05) + (28) × (1.55)	=	57.60
November:	(576) × (1.50)	=	864.00
December:	(632) × (1.50)	=	948.00
	Total		$ 2,605.20

It is clear that the plan using 27 regular-shift machines is superior to the one using 26 regular-shift machines. It is possible, however, that some other plan would be superior to the one using 27 regular-shift machines.

We must remember that we have considered only department 1. A similar analysis for departments 2 and 3 would be necessary. The resulting production plans for the three departments would have to be mutually compatible and would have to satisfy the cumulative demand requirements for the two bookcases. This planning can be done by identifying the most critical department (i.e., the one with the least flexibility) and adjusting the production plans of the other two departments accordingly.

After our production plans for all departments are firm, we must plan the acquisition of purchased parts to conform to our plan. Lead times for purchased items must be considered. This aspect is covered in the next chapter.

Feedback and Corrective Action

In the previous section we discussed several transfer functions available for the execution of the operations planning function. Some of the transfer functions are capable of making more than one of the required planning decisions, but none will make all of the decisions. We must therefore select several transfer functions which, when integrated properly, will yield all the decisions required of the operations planning function.

No matter which set of transfer functions we select, they cannot be expected to generate an operations plan that will anticipate precisely the actual requirements over the entire planning period. An operations plan is only a framework within which detailed activities are to be accomplished. As we progress through the planning period, the following variables can be expected to deviate from their "planned" values: demand, production, material availability, labor availability, equipment performance, and most costs associated with production.

Furthermore, these variables interact in complex ways. It is imperative, therefore, that we continuously monitor actual results, compare actual occurrences with planned occurrences, and take appropriate corrective action as required.

SOURCES OF FEEDBACK

Accurate and timely information feedback is essential for successful monitoring of the variables pertinent to the planning function. Several sources of feedback are discussed briefly in the following paragraphs. This feedback is used to update the two information files for the planning function, P.II (the Parameter File) and VS.II (the Variable Status File).

Forecasting. The Demand Forecasting function informs operations planning whenever a significant change in the cause system (i.e., the demand trend) is detected (please refer to Fig. 3.1).

Operations. From Operations we receive feedback on actual production, frequency of equipment malfunctions, actual manufacturing lead times, and actual quality of manufactured items.

Inventory Control. The Inventory Planning function furnishes feedback regarding changes in safety stock, lead times, scrap material factor, carrying costs, set-up costs, order costs, out-of-stock costs, and material costs (see Fig. 5.4).

Engineering. The Engineering function informs operations planning of all changes in manufacturing methods, materials, components, routings, standard times, set-up times, facility layouts, and facility capacities.

Financial Control. From Financial Control we receive any changes in budgetary considerations.

Marketing. All new, cancelled, or altered marketing programs are sent to Operations Planning.

Personnel. Personnel provides information regarding the availability of the number and types of skills planned for.

External Perturbations. Any significantly unusual condition outside the control of the company is reported. These would include large rush orders, new competition, labor problems, material availability and general economic fluctuations.

SHORT-TERM CORRECTIVE ACTION

Corrective action usually must be taken whenever actual occurrences deviate significantly from those we had planned. It is meaningful to distinguish between short-term corrective action and long-term corrective action for the operations planning function.

In the planning function, short term refers to within the planning period, which is usually 1 year. In the operations plan, rarely would changes be made more often than at monthly intervals. In most other functions, however, short term implies a much shorter review period than it does in operations planning. In Operations Scheduling, for example, schedules are often revised daily.

Short-term corrective action for the planning function is instigated when it has been determined that a significant deviation from the plan has occurred. The action consists of performing the series of transfer functions selected to make the decisions depicted in Fig. 4.1. The transfer functions executed make use of updated information in the information files, and a revised plan is generated.

Very often a revised operations plan involves judgmental decisions and management approval. It is not unusual, therefore, for human judgment to supplement, and even replace, one or more of the formal transfer functions.

LONG-TERM CORRECTIVE ACTION

Long-term corrective action for the operations planning function pertains to revisions in the methods by which planning is performed. Such revisions normally would not be made during a planning period. Long term, therefore, implies at least the length of a planning period.

Long-term corrective action for this function involves keeping abreast of new planning transfer functions that are published in the professional literature and adapting these improved decision processes to the particular control system used by the company.

It is difficult to determine whether a new technique should be incorporated into the control system. Operations planning generates a considerable amount of information needed in other functions of our control system. Any new planning transfer function must be compatible with the information requirements of other transfer functions. Another consideration is the cost of installing and maintaining a new technique compared to the potential savings that could be expected from its use.

Interfaces

The operations planning function interacts with many other functions of the enterprise. These interactions were discussed in detail in previous sections of this chapter. Figure 4.14 is included to help us consolidate the previous discussions of the interfaces with the planning function. The dashed line boundary separates elements of the operations planning and control system from those outside the system. In Fig. 4.14, one time period normally means 1 month. Short-term adjustments and long-term adjustments have the same meanings as were discussed in the previous section.

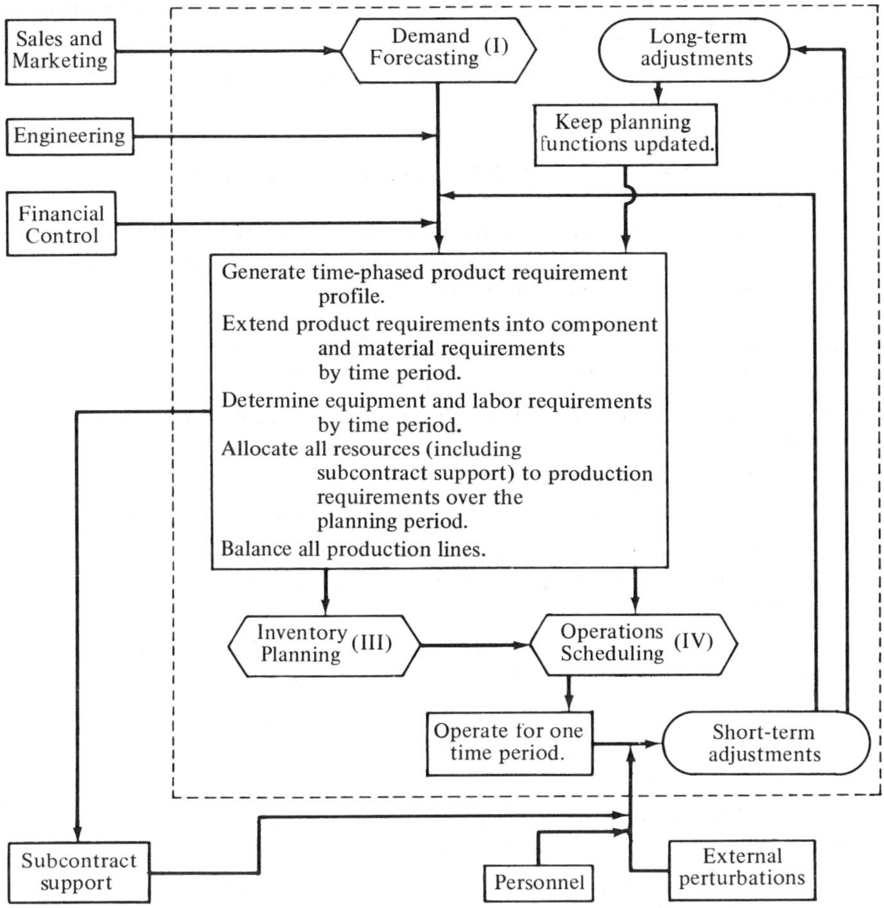

Figure 4.14 Interfaces for Planning Function

REFERENCES

Buffa, E. S. 1968. *Production-Inventory Systems Planning and Control*. Homewood, Illinois: Richard D. Irwin, Inc.

Helgeson, W. B.; and Birnie, D. P. 1961. Assembly Line Balancing Using the Ranked Positional Weight Technique. *Journal of Industrial Engineering* 12: 394-98.

Ignall, E. J. 1965. A Review of Assembly Line Balancing. *Journal of Industrial Engineering* 16: 244-54.

Kilbridge, M. D.; and Wester, L. 1961. A Hueristic Method of Assembly Line Balancing. *Journal of Industrial Engineering* 12: 292-98.

Moodie, C. L.; and Young, H. H. 1965. A Hueristic Method of Assembly Line

Balancing for Assumptions of Constant or Variable Work Element Times. *Journal of Industrial Engineering* 16: 23-29.

Vazsonyi, A. 1958. *Scientific Programming in Business and Industry.* New York: John Wiley & Sons, Inc.

SOURCES OF ADDITIONAL TRANSFER FUNCTIONS

Listed below are several additional sources of planning transfer functions. The list, not intended as an exhaustive bibliography on the planning function, is a representative sample of the planning techniques available.

Church, F. L. 1963. Requirements Generation, Explosions, and Bills of Material. *IBM Systems Journal* 2: 268.

Galbraith, J. R. 1969. Solving Production Smoothing Problems. *Management Science* 15: B-665.

Hubbell, J. P.; and Ekey, D. C. 1963. The Application of Input-Output Theory to Industrial Planning and Forecasting. *Journal of Industrial Engineering* 14: 49.

Lee, E. S.; and Shah, P. D. 1969. Optimization of Production Planning by the Generalized Newton-Raphsen Method. *Proceedings of the 1969 AIIE National Conference* (Houston), p. 359.

Lee, E. S.; and Shaikh, M. A. 1969. Optimal Production Planning by a Gradient Technique I: First Variations. *Management Science* 16: 109.

Lippmann, S. A.; Rolfe, A. J.; Wagner, H. M.; and Yuan, J. S. C. 1967. Optimal Production Scheduling and Employment Smoothing with Deterministic Demands. *Management Science* 14: 127.

Mansoor, E. M.; and Ben-Tuvia, S. 1966. Optimizing Balanced Assembly Lines. *Journal of Industrial Engineering* 17: 126.

Moodie, C. L. 1968. New Directions in Assembly Line Balancing. Paper presented at the 1968 AIIE National Conference (Tampa, Florida).

Nuttle, H. L. W. 1969. The Application of Dynamic Programming to an Employment Planning Problem. *Proceedings of the 1969 AIIE National Conference* (Houston), p. 461.

Smith, S. B. 1965. An Input-Output Model for Production and Inventory Planning. *Journal of Industrial Engineering* 16: 64.

Sabel, M. J. 1969. Production Smoothing with Stochastic Demand I: Finite Horizon Case. *Management Science* 16: 195.

Thelwell, R. R. 1967. An Evaluation of Linear Programming and Multiple Regression for Estimating Manpower Requirements. *Journal of Industrial Engineering* 18: 227.

5 Inventory Planning and Control

Concepts

The Demand Forecasting function (Chap. 3) provides us with an estimate of demand for each of our products. Forecasts are generated for each operating period, and these values are summed to determine cumulative forecasts for the entire planning period. The Operations Planning function (Chap. 4) converts the forecasts into a time-phased production program. Units of finished product are translated into material requirements of raw materials, purchased parts, fabricated parts, and assemblies. Inventory planning and control must now provide procedures that, while assuring that materials and products will be available when they are needed and in the quantities required, will also guard against the costs of excessive inventories.

PURPOSE OF INVENTORY PLANNING AND CONTROL

Inventories serve a vital function in the production and distribution of goods in our economy. As consumers, we depend on merchants to maintain inventories of the goods we wish to purchase. Merchants depend on their suppliers to maintain inventories of the goods that all their customers wish to purchase. A production process depends upon the stock room to maintain inventories of raw materials and other components of production. Innumerable other examples could be cited which would illustrate our great dependency upon inventories. A modern, dynamic economy simply could not function without them; neither, in fact, could a primitive economy.

Despite their obvious necessity, inventories are often regarded as generating unwanted costs. Many managers regard inventories as necessary evils, to be tolerated only until management science finds a way to eliminate them altogether. Inventory costs are really no different from other costs, such as labor, machines, and overhead. They are all unwanted but necessary. The purpose of the inventory planning and control function is to determine appropriate inventory policies and to keep all associated costs at a minimum.

In general, inventories are necessary because of the lack of synchronization between the following pairs of factors:

- The supplier and our production process.
- Successive stages within the production process.
- The production process and demand for the product.

These factors are represented diagramatically in Fig. 5.1 as input/output processes. The three pairs of factors give rise to the most common classification of *types of inventory*:

- Raw materials and purchased parts.
- In-process.
- Finished goods.

It would be instructive at this point to relate these concepts to Fig. 1.1.

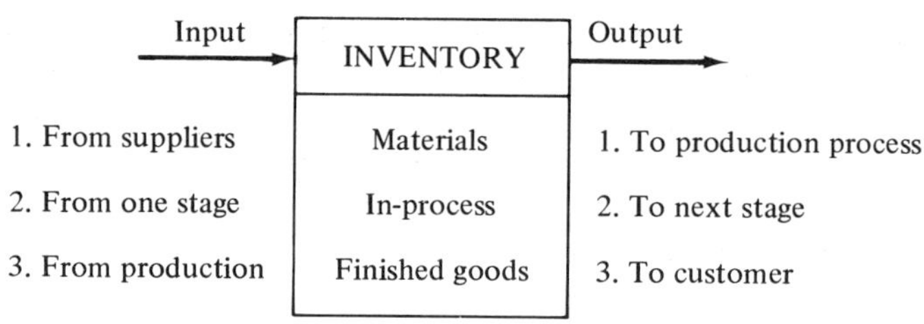

Figure 5.1 Input/Output representation of the three types of Inventory

The inventory level of each type fluctuates because of the varying rates of the associated input and output. During a particular month, for example, customer demand might be less than expected, while the production rate was greater than expected. This discrepancy would result in the finished goods inventory level being greater than expected for that month. The opposite result could occur next month. Analogous examples could be given for the other two types of inventory.

Note that the levels of the three types of inventory are not independent. A significantly excessive inventory of finished goods could trigger a reduction of production rates, which in turn could cause raw material inventory to become larger than planned. Very low inventories of raw materials could reduce production runs, thus causing finished goods inventory to fall below planned levels. Inventory policy must be formulated to cope with such occurrences.

We have discussed several of the reasons that make it necessary to maintain inventories. There are also reasons that make it economically desirable to maintain inventories. One such reason is the desirability to order or manufacture certain components in relatively large quantities, rather than in the exact number needed on a short-term basis. This problem is the classical one of economic lot size, to which we devote major emphasis in this chapter. Another reason to maintain inventories has to do with the production smoothing problem, discussed in detail in the previous chapter. The production plan shown in Fig. 4.12 requires that production in certain months be carried as inventory for use in later months. It is quite common to compensate for the seasonality of demand through the use of inventories. Still another reason is that economic benefits may often be gained by buying in large quantities to take advantage of quantity discounts or other favorable market conditions. If we suspect that a particular raw material will become more expensive, we might choose to purchase a large amount prior to the price increase.

BASIC STRUCTURE OF INVENTORY PROBLEMS

Most inventory problems* are concerned with answering two fundamental questions:

- *How much* to order (either from a supplier or from our own production facilities) at one time.
- *When* (or *how often*) to place an order.

There are opposing costs in each of these questions. In the first question, there is a cost associated with ordering too much at one time and another associated with ordering too little at one time. In the second question, there is a cost associated with placing orders too frequently and another associated with not placing orders frequently enough. Our general objective is to determine the course of action that minimizes the sum of all such costs.

The two fundamental questions are not independent. The answer to one question greatly affects, and in many cases completely determines, the answer to the other. Consider a particular purchased part. Suppose, based upon the forecasts for all finished goods in which this part is used, the expected annual usage of this part is estimated at 10,000 units. Let us consider the different ways that we could obtain 10,000 units during a 1-year period:

*Particularly those concerned with raw materials and finished goods. In-process inventories, somewhat different, are treated in a later section.

- We could order the entire 10,000 units at one time and use them as needed throughout the year.
- We could order one unit at a time, as it is needed.
- We could order some fraction of the total requirement, resulting in several orders per year, each of the same size.

Which of these methods is best? The answer to this question depends entirely on our objective. If our objective is to maintain the lowest possible inventory at all times, then method 2 would be best. Such a policy, however, would result in very high ordering costs, since 10,000 separate orders would have to be placed. If the objective is to maintain the lowest number of ordering clerks in the purchasing department, then method 1 would be best. This method, however, would result in the expense of maintaining a very high average inventory for the year.

Usually, our objective, to minimize the total costs associated with acquiring and maintaining inventory, results in following method 3, in which some fraction of the total requirement is ordered several times during the year. This decision leaves us with the problem of determining the proper amount to order at one time so that the sum of the costs is minimized. We can express these costs as a total cost equation and then use mathematical procedures for finding the optimum amount to order at one time.

The expression for total costs depends on the particular situation. Several models are developed in the section entitled Transfer Functions. It would be instructive at this point, however, to identify the major classes of inventory costs.

Procurement Cost. The procurement cost of a purchased item is the order cost and consists of the clerical cost of making up and processing a purchase requisition. In general, order cost includes any cost whose size or amount is affected by the number of orders processed during a given time period. The procurement cost of a manufactured item is called the set-up cost, which is the cost of altering the production process to produce the new item plus any clerical and administrative costs involved in preparing and sending a production order to Operations.

Carrying Cost. Carrying cost is the cost of holding inventory and includes several component costs:

- The cost of money invested in inventory which could be used in other ways.
- The cost of storage space—warehousing costs, utilities such as heat and light, etc.

- The cost of obsolescence, spoilage, and pilferage.
- The cost of insurance and taxes on the items in inventory.

Out-of-Stock Cost. This cost is associated with running out of a finished good or a material item. Either of two situations can occur when we run out of a finished good:

- The order is filled later; called a "backorder." In this case, the sale is not lost, but there is an extra cost resulting from expediting, special handling, extra transportation costs, etc.

- The sale is lost. In this case, we not only lose the profit from this particular sale, but we might also lose future sales and loss of customer goodwill.

When we run out of a raw material, purchased part, or assembly, the out-of-stock cost depends upon the severity of the effect of the stock-out. Some possible situations are:

- We could be in the middle of a production run when we run out of a needed material. It may be possible to change the production process over to another product until the needed material arrives and then change back. In this case, the out-of-stock cost would be simply the cost of the two additional set-ups.
- We might be able to substitute another material for the out-of-stock material. In this case, the cost would be the cost of the more expensive material* plus the cost of any adjustments in the product or process necessitated by the substitution.
- If we run out of a critical material, without which our production process must cease, then the out-of-stock cost would include the cost of lost production time, the cost of any lost sales, the cost of overtime and extra shifts needed to get back on schedule, and any costs of inferior quality resulting from the process interruption.

The total cost equation for an inventory problem will include some, but not necessarily all, of the above cost categories. Again, the objective is always to determine that course of action (how much to order and when to order) which minimizes the sum of the cost factors included in the model.

We should mention that the above costs are difficult to measure, particularly such costs as loss of goodwill, etc. Nevertheless, it is useful to structure inventory problems analytically and to use the best possible approximations for the various costs. Fortunately, good results can be obtained even with reasonable

*If it were a less expensive material, then we would have been using it in the first place.

errors in the measurement of costs. A good discussion of cost sensitivity analysis can be found on pages 83–84 and 176–181 of Starr and Miller (1962).

The Classical Inventory Model. To complete our treatment of the basic structure of inventory problems, we now discuss the most elementary form of an inventory model. This model has been with us for over 50 years. Even though based upon highly idealistic conditions, it serves to demonstrate the general approach to solving inventory problems quantitatively. More realistic models are discussed in the Transfer Function section of this chapter.

The classical inventory model determines the optimal number of units to order on each order placed. This number is derived under the following assumptions:

- The usage rate of the item is linear and known with certainty.
- The order is received instantaneously when placed; i.e., lead time is zero.
- The first two assumptions imply a third: that no stock-out can occur, therefore out-of-stock cost is omitted from the model.

An inventory system operating under these idealized conditions would behave as shown in Fig. 5.2. Under the conditions we have assumed, we can wait until the inventory level reaches zero, then place an order for the desired amount. This order is received immediately. Our objective is to determine the number that should be ordered at one time.

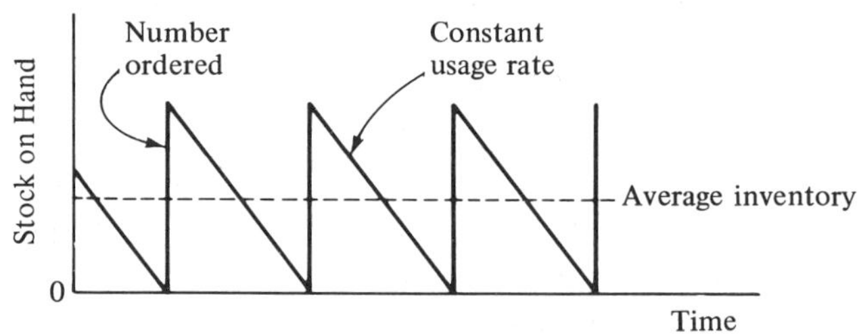

Figure 5.2 An idealized Inventory System

Since stock-out costs are omitted from this model, we can express the total cost as the sum of procurement cost and carrying cost:

$$\text{Total Cost} = \text{Procurement Cost} + \text{Carrying Cost}$$

We use the following symbols in the development of the model:

TC = total cost

PC = procurement cost for each order

CC = carrying cost per unit per year

D = annual demand (or usage) of the item

Q = quantity ordered on each order placed (lot size)

Q_0 = optimal lot size

To determine the particular lot size that will result in the lowest value for total cost, we develop a total cost expression in which the two opposing costs, procurement cost and carrying cost, are in terms of Q, the quantity ordered.

We can express the procurement cost in terms of Q by recognizing that the total annual procurement is equal to PC (the procurement cost for each order) multiplied by the number of orders placed during the year. (The number of orders placed during the year is just the total annual demand divided by the number of units ordered on each order.) Symbolically, therefore, we have

$$\text{Annual Procurement Cost} = (PC)(D/Q)$$

We can express the carrying cost in terms of Q by recognizing that the total annual carrying cost is equal to CC (carrying cost per unit per year) multiplied by the average number of units held in inventory. It is clear from Fig. 5.2 that the average number of units held in inventory is just $Q/2$. Therefore,

$$\text{Annual Carrying Cost} = (CC)(Q/2)$$

Our total cost equation can now be written as:

$$TC = (PC)(D/Q) + (CC)(Q/2) \tag{5.1}$$

For a particular inventory problem we will know values for PC, CC, and D. With these values fixed, total cost becomes a function only of Q, the lot size. Suppose, for example, $PC = \$10$ per order, $CC = \$0.20$ per unit per year, and $D = 10{,}000$ units. If we order 2,000 units at a time, then from Eq. 5.1 we obtain

$$\begin{aligned} TC &= (\$10)(10{,}000/2{,}000) + (\$0.20)(2000/2) \\ &= \$50 + \$200 = \$250 \end{aligned}$$

Other total cost values would be obtained using different lot sizes. In fact, it is instructive to plot several values of Q, as has been done in Fig. 5.3, to show its effect on each cost component in Eq. 5.1 as well as on TC itself.

Our objective is to determine the particular value of Q that will minimize our total cost equation. In Fig. 5.3, we see that the minimum value on the TC

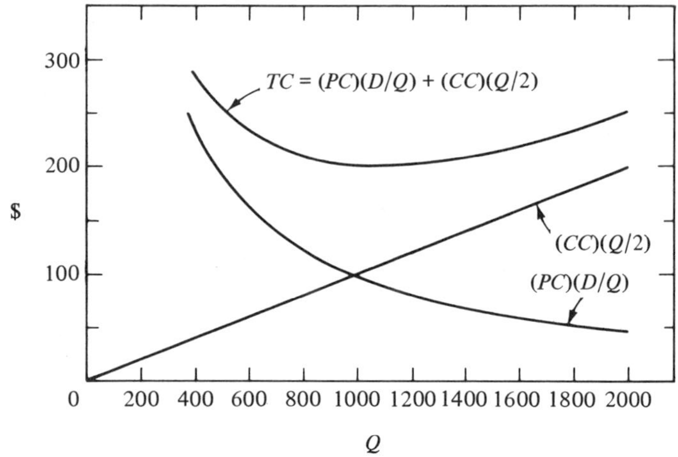

Figure 5.3 Total Cost Function for classical Inventory Model. D = 10,000 units per year, CC = \$.20 per unit per year, and PC = \$10.00 per order

curve occurs at a point where the curve has zero slope. We know from differential calculus that the first derivative of a mathematical function gives the slope (instantaneous rate of change) of the function at a particular point. In our case, we are hunting the point where the slope is zero. Our approach, then, is to take the first derivative of the total cost equation, Eq. 5.1, set the derivative equal to zero, and then solve for Q_0, the optimal lot size:

$$\frac{d(TC)}{dQ} = -(PC)(D/Q^2) + \frac{CC}{2} \tag{5.2}$$

$$-(PC)(D/Q^2) + \frac{CC}{2} = 0 \tag{5.3}$$

So

$$Q_0 = \sqrt{\frac{2(PC)(D)}{CC}} \tag{5.4}$$

By substituting our example values into Eq. 5.4, we obtain

$$Q_0 = \sqrt{\frac{2(\$10)(10{,}000)}{\$0.20}} = 1{,}000 \text{ units}$$

We see from Fig. 5.3 that a lot size of 1,000 units does indeed result in lowest total cost of \$200.

The classical inventory model can be used as a transfer function for situations in which the stated assumptions are valid. Other transfer functions are developed in a later section in which we relax many of the restrictions of the highly idealized model just presented. The purpose of introducing the classical model at this point is to illustrate the manner by which we attempt to balance the opposing costs that are in effect in a particular inventory problem.

Considerable attention, both theoretical and practical, has been directed toward the inventory function. Much of the work in the so-called scientific management movement during the early part of this century was concerned with inventory. Since World War II, the fields of management science and operations research have brought forth a large and growing body of knowledge which might properly be called inventory theory. Much of this theory is contained in such books as Brown (1967), Buchen and Koenigsberg (1963), and Starr and Miller (1962).

It is beyond the scope of this text to present a complete treatment of inventory theory. Instead, we will concentrate on certain basic inventory models and emphasize the place of inventory planning and control in our total operations planning and control system.

We now turn our attention to the specific decisions that must be made in the inventory planning and control function. Two levels of decisions are made in this function. One level of decisions, inventory planning, has to do with the specification of inventory system procedures, such as the type of ordering system, economic lot sizes, safety stock, etc. The second level, inventory control, deals with the actual operating decisions, such as the placement of production orders and purchase orders.

INVENTORY PLANNING DECISIONS

An inventory system consists of several procedures and decision rules that govern the operation of the system. It is convenient to discuss these according to the three types of inventory: purchased items, manufactured items, and in-process inventory.

Purchased Items. Whether to purchase or manufacture a certain item is not decided by the inventory planning function. All such decisions are made by Engineering and are indicated on the bills of materials. We are presently concerned, therefore, with inventory planning decisions for all purchased items.

The major inventory planning decision for purchased items is the type of inventory monitoring and ordering system to be used. These systems must contain rules that lead to economical order sizes and order intervals. That is, they must tell us how much to order and when to order. We must also make decisions on how much safety stock to provide, what lead time to anticipate, and whether discount order quantities should override our economical batch sizes.

We will often wish to group our purchased parts by vendor so that we can order several items on the same purchase order, thereby reducing the cost of placing orders. Our system for ordering purchased items must also accommodate the time-phased operations plan, as discussed in Chap. 4.

Manufactured Items. We are now concerned with finished goods, subassemblies, and fabricated components. The first decision that must be made for each such item is whether it should be stocked at all. The frequency of demand for many items is so low that we cannot justify maintaining inventories of them. Whenever an order does come in, we can manufacture the necessary number of items at that time.

Our inventory system for manufactured items must adhere closely to the time-phased operations plan. This plan provides a broad framework regarding how many of the several manufactured items to produce during the planning period. The operations plan does not provide detailed instructions, however, as to exactly how many of each item to produce at one time and when each production run should be made. These decisions are made by the inventory function and are released to Operations Scheduling as production orders.

Just as vendor lead time must be considered for purchased items, manufacturing lead time must be considered for manufactured items. This delay necessitates safety stock for manufactured items.

Whenever several manufactured items must be processed on the same facilities, it is often not possible nor desirable to manufacture the items in the quantities determined when each item is considered individually. A decision must be made as to the best lot sizes for items competing for the same facilities.

In-Process Buffer Sizes. For reasons explained earlier, we often use inventory to "uncouple" successive stages in our production process. Thus each stage may operate relatively independently of other stages, thereby permitting high operating efficiency in each stage. There are practical limits, however, to the amount of inventory we can permit between stages. The decision that must be made is how large the buffer size should be between successive stages of production.

Adjustments for Constraints. The two primary constraints that limit our actions in inventory planning and control are physical space limitations and monetary limitations. We must be able to adjust our inventory policy to conform to these two constraints, even if it means foregoing an "optimal" inventory policy.

INVENTORY CONTROL DECISIONS

The decisions discussed in the preceding section are concerned with setting up the systems and procedures by which the inventory control function may be performed. These procedures determine for us the safety stock, reorder level, reorder quantity, etc., for each item over which we must maintain inventory control.

We now turn our attention to operating decisions. Conceptually, we can say that in the previous section we were concerned with designing our inventory system, whereas in this section we are concerned with running the system.

Whether to Place an Order. The most fundamental operating decision that must be made in an inventory system is whether to place an order, whether it be a purchase or a production order. This decision must be made for each item that we carry in inventory. The decision is made according to the particular set of rules and procedures set up in the previous section. The manner in which this decision is made is discussed in a later section.

How Much to Order. Once it has been decided that a purchase or production order should be placed, we must then determine how large the order should be. This decision is highly dependent upon the previous one.

Whether an Order Should be Expedited. We occasionally will be faced with a dangerously low inventory level for either a purchased or manufactured item. This situation can result from any combination of the following factors:

1. Unusually long lead time from a vendor or from our production process.
2. Unusually heavy demand or usage of the item during an order period.
3. Unusually high rejection rate on an order by quality control inspection.

When faced with this situation, we must decide what type of expediting, if any, we should perform. If we are dealing with a purchased item, we can put a tracer on the order to determine when it can be expected. We can sometimes speed up delivery by using other means of transportation. If we are dealing with a manufactured item, we can expedite this order in our own plant by requesting a higher priority from Scheduling.

Unusual Market or Production Conditions. We will be confronted occasionally with some unusual market or production condition that affects the operating decisions that must be made in our inventory control system. We may be offered a bargain price on a certain raw material if we will buy in a quantity larger than the regular economic order quantity. Our production facilities may encounter a slack period, which causes us to produce a certain amount of a manufactured item before we normally would do so. We may be aware of an approaching labor strike or other condition that would cause a raw material to become scarce. In such a case, we might like to build up a large reserve of the particular raw material. We may suspect that the price of a raw material will soon be increased. Again, we may wish to purchase more of this item than usual prior to the price increase.

The Purchasing function often has access to information such as that discussed above. The necessity for close cooperation and effective communication between Purchasing and the inventory function is obvious.

Figure 5.4 represents the manner in which the above decisions are made in the context of the total control system. The Inventory Planning Decisions, which determine in detail the procedures and decision rules that govern the operating decisions, can be expected to remain in effect for a relatively long period of time. Only when a major change occurs in a parameter, such as a price change or a change in the operations plan, would the inventory system be modified. The Inventory Control Decisions, however, are made on an essentially continuous basis. Continuous monitoring of inventory levels is usually necessary.

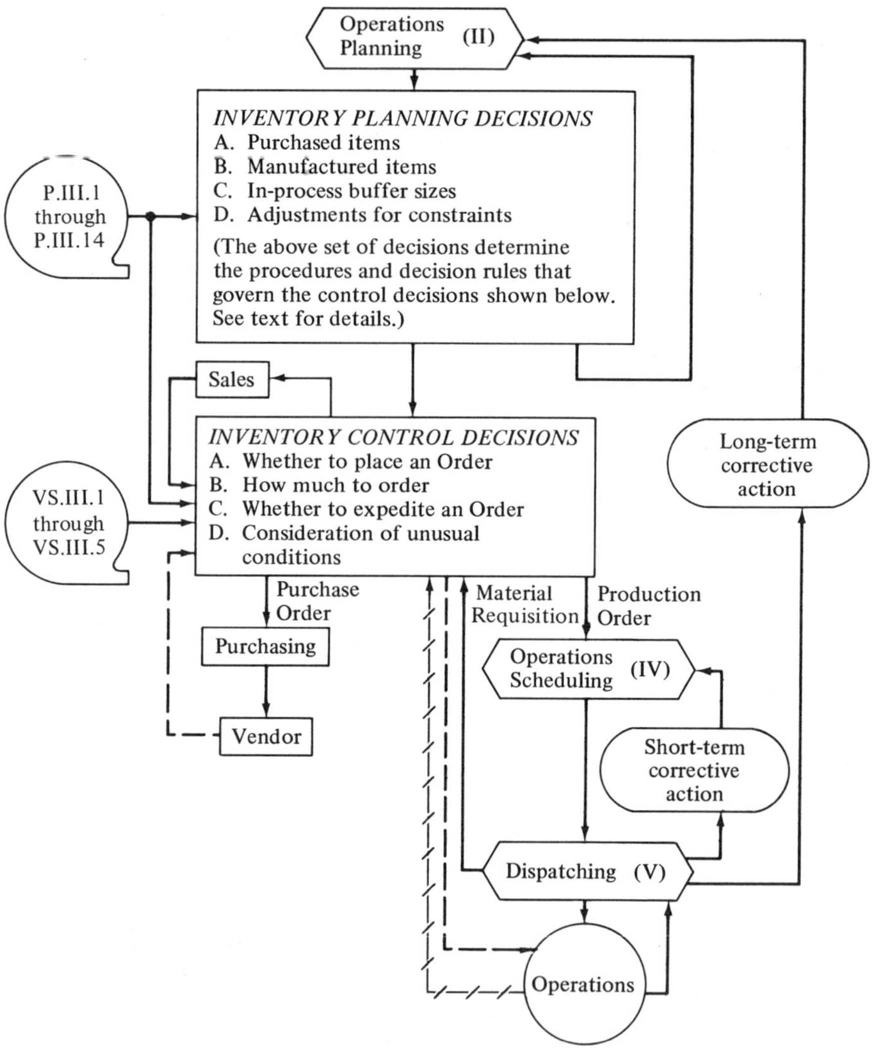

Figure 5.4 Flow Diagram of General Inventory Planning and Control Procedure

Note that the decisions from Inventory Planning are returned to the Operations Planning function. These decisions are the parameters (P. III), such as safety stock, lead times, scrap factor, etc., needed to develop an operations plan. This information is shown as input to the Operations Planning function in Fig. 4.1. This again illustrates the circular nature of the functions within the total control system.

Inputs

The inventory planning and control function requires information from several sources. In general, the Parameter File (P.III) is needed for Inventory Planning Decisions, and the Variable Status File (VS.III) is needed for Inventory Control Decisions.

PARAMETER FILE FOR INVENTORY (P.III)

P.III.1 Time-phased requirements for finished products, components, raw materials, assemblies, and supplies. Source: Operations Planning.

P.III.2 Customer service policy. Source: Management.

P.III.3 Ordering costs. Source: Purchasing.

P.III.4 Carrying costs. Source: Engineering.

P.III.5 Material prices with discount quantities. Source: Purchasing.

P.III.6 Out-of-stock costs. Source: Engineering.

P.III.7 Set-up costs. Source: Engineering.

P.III.8 Scrap factors for purchased and manufactured items. Source: Engineering.

P.III.9 Lead times. Source: Purchasing (for purchased items) and Engineering (for manufactured items).

P.III.10 Production rates. Source: Engineering.

P.III.11 Constraints. Source: Financial Control and Engineering.

P.III.12 *A*, *B*, *C* designation (usage/value classification).

P.III.13 Economic lot size.

P.III.14 Reorder level or reorder period.

VARIABLE STATUS FILE FOR INVENTORY (VS.III)

VS.III.1 Demand for all items. Source: Sales (for finished goods) and Dispatching (for components, raw materials, assemblies, and supplies).

VS.III.2 Receipts of each stock number.

VS.III.3 Amount on order and anticipated arrival date.

VS.III.4 Amount backordered.

VS.III.5 Current stock on hand.

UPDATING THE FILES

It is important to keep both the parameter file and the variable status file updated. If any significant changes occur in the operations plan, the resulting effect on the inventory system must be determined and taken into account. All costs must be kept current. The accounting function is usually responsible for keeping many costs current, such as holding and out-of-stock costs. Purchasing also is responsible for certain costs and for material prices. Records must be kept on each vendor to determine his lead time distribution and quality performance.

All information in the variable status file must be kept current on a real-time basis. As soon as any transaction or event occurs, it should be entered into the system immediately.

Transfer Functions

In Chap. 2, we defined a transfer function as "any process by which information is converted to a decision." It follows, therefore, that a transfer function for inventory control is a process by which the pertinent information is converted to one or more of the required operating decisions:

1. Whether an order should be placed.
2. How large the order should be.
3. Whether an order should be expedited.
4. Whether to deviate from normal practice because of unusual market or production conditions.

In this section, we examine some of the processes (usually called inventory models) available for making the above decisions. We again find it convenient to group the transfer functions according to the type of inventory involved, e.g., purchased items, manufactured items, and in-process inventory.

PURCHASED ITEMS

In an earlier section, we developed the classical inventory model, in which several idealistic conditions were assumed. We now develop inventory models in which we relax these assumptions. We discuss two basic types of inventory systems for purchased items.

Fixed Order Size System. In this system, the same number of units is always ordered. The time interval between orders varies according to fluctuations in usage of the item. Figure 5.5 illustrates the behavior of stock on hand (SOH) over time under such a system. Note we have relaxed the first assumption of the classical inventory model: our usage rate no longer is known with certainty, nor is it linear. We still assume, however, that lead time is zero.

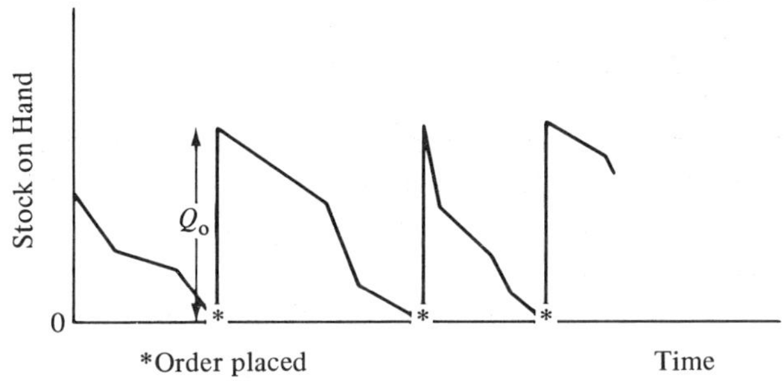

Figure 5.5 Fixed Order Size System, Fluctuating Usage Rate, Zero Lead Time

The transfer function for this situation is the same as that for the classical model; except in this case the expected annual demand is used as D in Eq. 5.4. Using the same values that we used in the previous example (PC = \$10.00 per order, CC = \$0.20 per unit per year, and expected demand D = 10,000 units per year), we again obtain

$$Q_0 = \sqrt{\frac{2(PC)(D)}{CC}} = \sqrt{\frac{2(\$10)(10{,}000)}{\$0.20}} = 1{,}000 \text{ units}$$

The complete transfer function for our required operating decisions (when to order and how much to order) consists of this simple rule: *whenever* $SOH =$ 0, *order 1000 units.*

We might also be interested in the expected number of orders placed during the year and the average time between orders in months.*

$$N = D/Q_0 = \sqrt{\frac{(CC)(D)}{2(PC)}} \tag{5.5}$$

$$T = 12/N = \sqrt{\frac{288(PC)}{(CC)(D)}} \tag{5.6}$$

where

*Other time units, such as weeks or days, can be found easily by appropriate adjustment of all cost and demand units.

$N =$ expected number of orders placed during the year for which D is the expected demand.

$T =$ average time (months) between orders.

For our example,

$N = 10{,}000/1{,}000 = 10$ orders.

$T = 12/10 = 1.2$ months between orders.

Let us now eliminate the second (and perhaps most unrealistic) assumption of the classical inventory model, that lead time is zero. The presence of a lead time implies that we must place our order in advance of when it will be needed, so that the order will arrive before SOH actually reaches zero. Our problem now is to determine RL, the reorder level (the particular value of SOH), at which an order of size Q_0 should be placed. Figure 5.6 demonstrates the behavior of an inventory system with lead time (denoted LT) and a constant usage rate.

The complete transfer function for this situation is: *whenever* $SOH = RL$, *order* Q_0 *units.*

The numerical value of RL is obtained by determining the demand that will occur during the lead time period.

$$RL = (D/12)(LT) \tag{5.7}$$

where LT is expressed in months. In the present example, if lead time is one half of a month, then the reorder level would be computed as follows:

$$RL = (10{,}000/12)(0.5) = 416 \text{ units}$$

The value $RL = 416$ would be used in the transfer function. Notice that Q_0 is computed from Eq. 5.4, as before.

It is now in order to relax simultaneously the first and second assumptions of the classical inventory model. We are now concerned with inventory systems in which the usage rate is variable and the lead time is greater than zero. Our problem here, as in the previous case, is to determine the reorder level, RL. The problem is more complex in the present case, however, since the demand during lead time is unknown. One approach to this problem is to consider the linear usage rate line in Fig. 5.6 as the expected (or average) usage rate, and then to estimate a reasonable maximum usage rate value from historical records (Magee and Boodman 1967, p. 119). RL is determined to be the amount of SOH just sufficient to satisfy this reasonable maximum usage rate during the lead time.

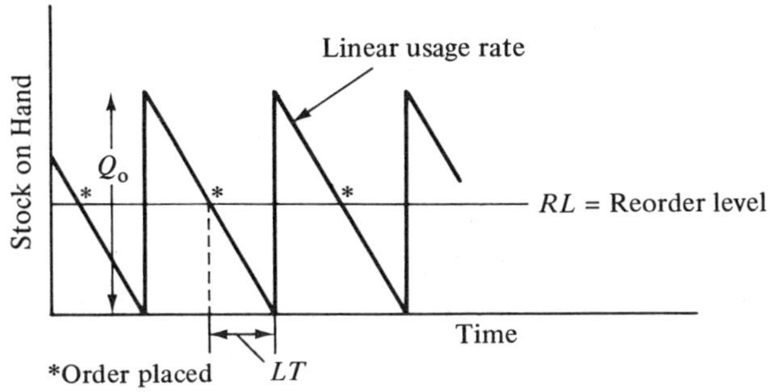

Figure 5.6 Fixed Order Size System, Constant Usage Rate, with Lead Time

To develop the transfer function for this case, we need to examine the problem in a very fundamental way. At some point in time, we must place an order for Q_0 items. This order will arrive at some time in the future, after a lead time of LT. We do not know what the usage rate will be during the lead time period. However, if proper records have been kept, we can determine the distribution of the random variable, usage rate during lead time. Figure 5.7 shows a distribution that might describe this random variable. The reasonable maximum usage rate x_{max} is determined by specifying a proportion of the time we are willing to run out of stock. This proportion is denoted as α in Fig. 5.7. For example, if $\alpha = 0.005$, we are willing to run out of stock in only 5 of each 1000 times we place an order.

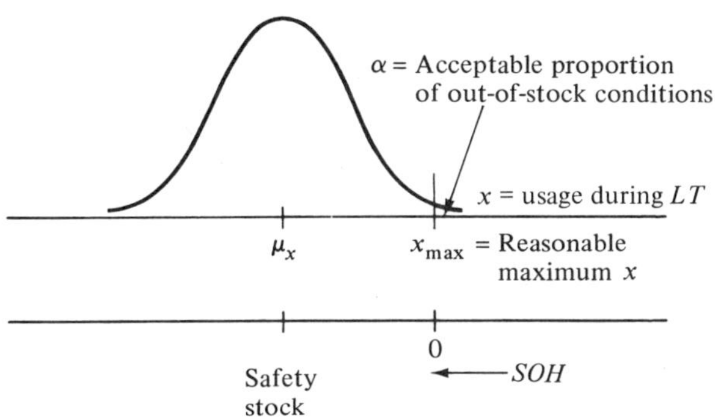

Figure 5.7 Distribution of Usage Rate during Lead Time with corresponding Stock on Hand (SOH) Scale

The average usage rate during the demand period is indicated as μ_x. If the distribution of x is symmetrical, as is the case with the normal distribution, for example, then during approximately one half of the lead times the usage rate will be less than μ_x, and during the other half it will be more than μ_x.

Also shown in Fig. 5.7 is a stock on hand (*SOH*) scale on which values increase in the leftward direction. On this scale we have our first encounter with the concept of safety stock (denoted *SS*). On the average, we want to have μ_x stock left on hand when an order arrives. We need this much safety stock to protect us during those lead times when the usage rate is x_{max}.

We now translate the information in Fig. 5.7 to our familiar *SOH* versus time graph. The results for one order period are shown in Fig. 5.8. The distribution shown is for the random variable, *SOH* when an order arrives. Note that this distribution is identical to that in Fig. 5.7. This similarity is explained by the fact that *SOH* when an order arrives is a direct function of usage during lead time.

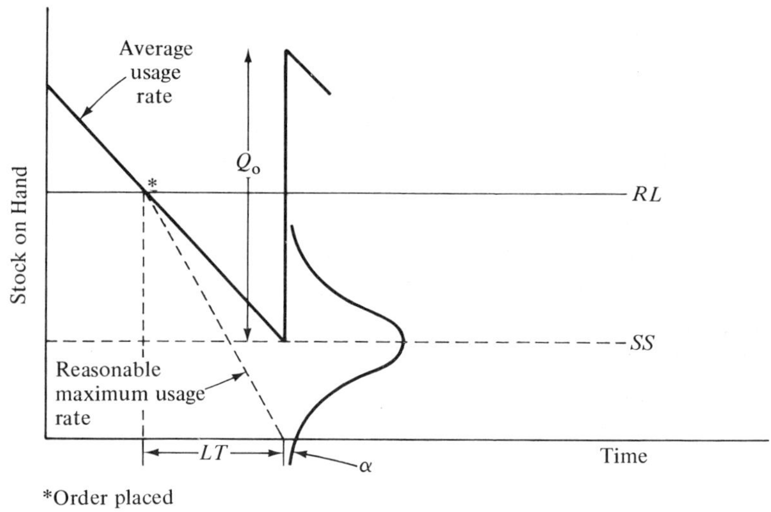

Figure 5.8 Distribution of SOH at the time an Order arrives

The average usage rate is shown as a solid line in Fig. 5.8, while the maximum reasonable usage rate during lead time is shown as a dashed line. Note that it does not matter to us what the usage rate is prior to placing the order; we have a fixed reorder level, and whenever *SOH* falls to this level we place an order. Thus, variation in demand prior to placing an order does not bother us; we simply order sooner or later, depending on the direction of variation. We are very much concerned, however, with variation of demand during the lead time period.

Our basic problem in this inventory situation is to determine *RL* such that the probability of running out of stock prior to the arrival of the order is no greater than an acceptable level α. As can be seen from Fig. 5.8, *RL* is the sum of average usage during lead time and safety stock.

$$RL = (D/12)(LT) + SS \tag{5.8}$$

If we assume that annual demand is normally distributed with mean $= D$ and standard deviation $= \sigma_D$, SS can be determined from the following expression:*

$$SS = (K_\alpha)(\sigma_M)(\sqrt{LT}) \tag{5.9}$$

where

K_α = standard normal deviate for a specified α value

σ_M = standard deviation of monthly demand

$= (\sigma_D)(\sqrt{1/12})$

LT = lead time, in months.

In the example we have been discussing, suppose annual demand is a normally distributed random variable with mean = 10,000 units and standard deviation = 1,000 units. Suppose we specify α to be 0.05. Reference to a table of normal curve values gives us $K_\alpha = 1.645$. $\sigma_M = (1{,}000)\sqrt{1/12} = 288$. Using a lead time of one half month,

$SS = (1.645)(288)(0.707) = 335$ units

$RL = 416 + 335 = 751$ units

The complete transfer function for this situation is as follows: *whenever* $SOH = 751$ *units, order* 1,000 *units*. Remember that Q_0 had been determined as 1,000 units. The behavior of a typical system of this type is illustrated in Fig. 5.9.

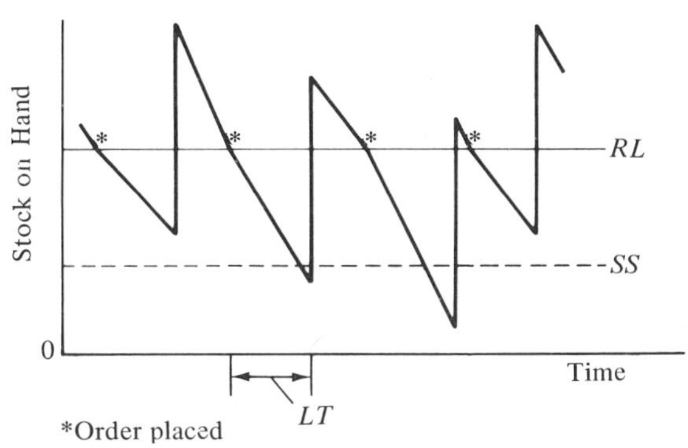

Figure 5.9 Fixed Order Size System, Variable Usage Rate, with Lead Time

*The derivation of this expression can be found in several inventory texts, such as those listed in the references at the end of this chapter.

Note that in all the transfer functions discussed in this section, the same number of units Q_0 is always ordered. The time interval between orders varies. In the next section, we discuss an inventory system having opposite characteristics.

Fixed Order Interval System. In this system, orders are placed at equally spaced, predetermined points in time. The size of the order varies according to fluctuations in usage of the item between orders.

Generally, we follow the same procedure as in the previous section. We begin by relaxing the first assumption of the classical inventory model: we allow the usage rate to vary, but we consider lead time as zero. Figure 5.10 illustrates the behavior of *SOH* versus time under such a system. Orders are placed at fixed reorder points, denoted RP_1, RP_2, etc. The size of each order is the difference between *SOH* and the desired maximum inventory *IMAX*.

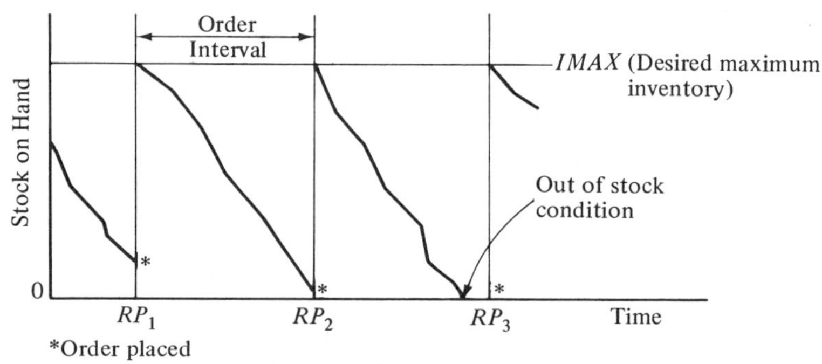

Figure 5.10 Fixed Order Interval System, Fluctuating Usage Rate, Zero Lead Time

Note in Fig. 5.10 that we ran out of stock before reaching RP_3, even though we assumed a lead time of zero. This occurrence points up an essential difference between the fixed order interval system and the fixed order size system. Because we are rigidly tied to predetermined reorder times in the fixed order interval system, we must provide safety stock, even when lead time is zero. (No safety stock was needed in the fixed order size system when lead time was zero, as seen in Fig. 5.5.)

In determining the amount of safety stock to provide, we must estimate the reasonable maximum usage rate for an entire order interval (denoted *OI*). From inventory records we determine the distribution of the random variable, usage during an order interval. Figure 5.11 shows a distribution that might describe this random variable.

In Fig. 5.11, α is defined exactly as it was in Fig. 5.6. By specifying α, we can determine x_{max} from normal curve tables (assuming that x is normally distributed). The average amount of *SOH* we will have when an order is placed

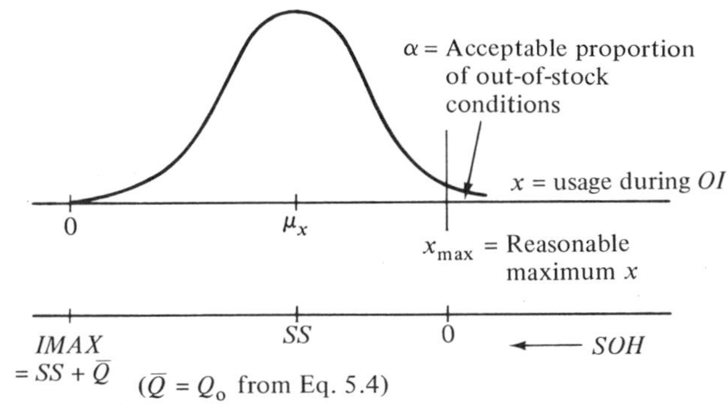

Figure 5.11 Distribution of Usage during Order Interval with corresponding SOH Scale; Lead Time is Zero

is $\mu_x = SS$. We need this much safety stock to protect us during those order intervals when the usage rate is x_{max}.

Note that if usage rate was always its expected value, μ_x, then we would once again have the classical inventory model. We recall that, in the classical model, we always ordered an amount Q_o, determined by Eq. 5.4. In the present case, however, the order size varies but the average order size $\bar{Q}$ is equal to Q_o. We must begin each order interval with enough stock on hand to last the entire order interval at maximum usage rate. We see from Fig. 5.11 that this rate is x_{max} on the x scale and *IMAX* on the *SOH* scale. Our ordering rule, therefore, is to order enough units Q to bring us back up to the *IMAX* level. (Remember that we receive the order at the same time we place it, since lead time is zero.)

We now translate the information in Fig. 5.11 to our familiar *SOH* versus time graph. The results for one order period are shown in Fig. 5.12. The distribution shown is for the random variable, *SOH* when an order is placed. In general, a greater amount of safety stock is required for the fixed order interval system than for the fixed order size system. This fact becomes even more evident when we include lead time in the fixed order interval system.

Our basic problem in this system is to determine the optimum order interval OI_o and the desired maximum inventory level *IMAX*. The optimum order interval is determined directly from Eq. 5.6. *IMAX* is determined as follows:

$$IMAX = \bar{Q} + SS \tag{5.10}$$

$\bar{Q} = Q_o$ and is determined from Eq. 5.4. If we assume that annual demand is normally distributed with mean $= D$ and standard deviation $= \sigma_D$, SS can be determined from:

$$SS = (K_\alpha)(\sigma_M)(\sqrt{OI_o}) \tag{5.11}$$

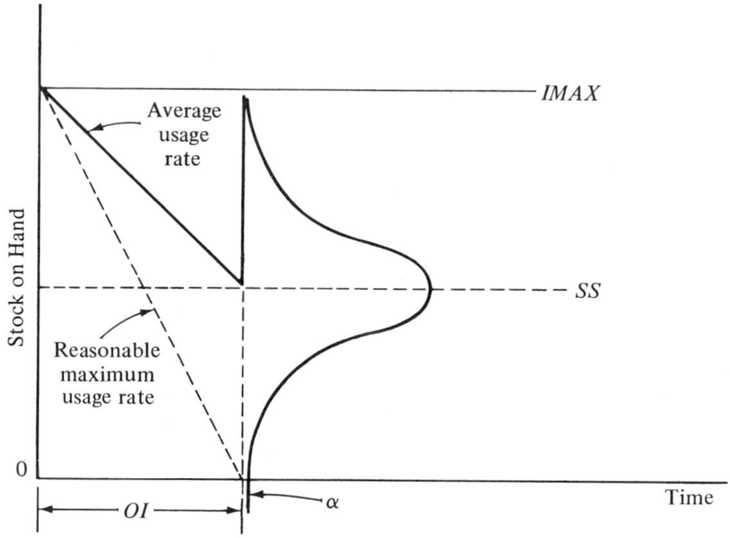

Figure 5.12 Distribution of SOH at the time an Order is placed; Fixed Order Interval System

where

OI_0 = order interval, in months; K_α and σ_M are the same as in Eq. 5.9.

Our complete transfer function for this case is: *every OI_o time units, order an amount $Q = IMAX - SOH$.*

In the example we have been discussing throughout this chapter, we would obtain the following parameters for our transfer function:

$$OI_o = \sqrt{\frac{288(10)}{(0.20)(10{,}000)}} = 1.2 \text{ months, from Eq. (5.6)}$$

$$SS = (1.645)(288)(\sqrt{1.2}) = 520 \text{ units, from Eq. (5.11)}$$

$$IMAX = 1{,}000 + 520 = 1{,}520 \text{ units, from Eq. (5.10)}$$

For our example, the transfer function would be: *every 1.2 months, order an amount $Q = 1520 - SOH$.* Note that safety stock for this case is considerably greater than for the fixed order size case, even with zero lead time for this system and a lead time of one half month for the former system. The difference is even greater when we consider lead time for the fixed order interval system.

We now relax simultaneously the first and second assumptions of the classical inventory model.* We are now concerned with a fixed order interval system

*We omit the development of a transfer function in which only the second assumption (lead time and constant usage rate) is relaxed. This is almost identical to the situation represented in Fig. 5.6.

in which the usage rate is variable and the lead time is greater than zero. The behavior of such a system is illustrated in Fig. 5.13. A very important point to observe in Fig. 5.13 is that we must maintain enough safety stock to allow for the reasonable maximum usage rate during lead time plus an entire order interval. We see that the fixed order interval system commits us for a longer time period than did the fixed order size system.

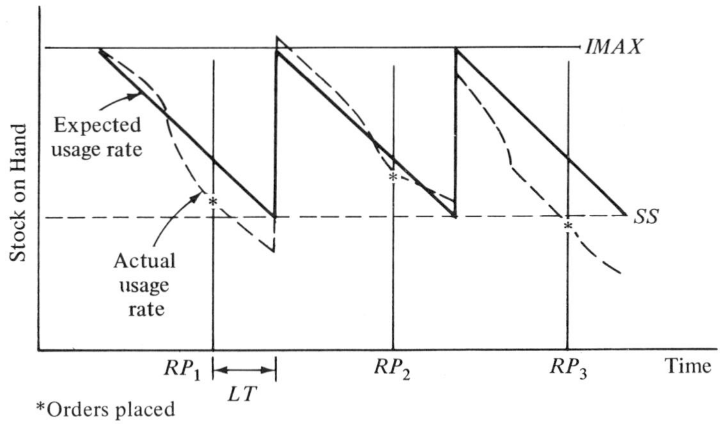

Figure 5.13 Fixed Order Interval System, Variable Usage Rate, with Lead Time

Our transfer function for this case is identical to that for the preceding case, except that we must provide additional safety stock to cover us during the lead time period. Safety stock for this case may be determined from:*

$$SS = (K_\alpha)(\sigma_M)(\sqrt{OI_o + LT}) \tag{5.12}$$

Let us again consider lead time to be one half of a month in our continuing example.

$$SS = (1.645)(288)(\sqrt{1.2 + 0.5}) = 617 \text{ units}$$

Our only remaining problem is to determine *IMAX*, the desired maximum inventory level. On the average, we want *SS* units on hand when our order arrives, and on the average, we order $\bar{Q}$ units on each order. Therefore, *IMAX* will be equal to $\bar{Q}$ plus *SS* plus an amount sufficient to carry us through lead time at the average usage rate. Symbolically,

*This expression is derived in several inventory texts, such as those listed in the references at the end of this chapter.

$$IMAX = (D/12)(OI_o + LT) + SS \tag{5.13}$$

For our example,

$$\begin{aligned} IMAX &= (10{,}000/12)(1.2 + 0.5) + 617 \\ &= 2{,}035 \text{ units} \end{aligned}$$

These values, together with those found earlier, give us our transfer function: *every 1.2 months, order an amount* $Q = 2{,}035 - SOH$.

It would be instructive at this point to cast the transfer function just derived into the context of the general transfer function shown in Fig. 2.4. This is done in Fig. 5.14. The transfer function shown in Fig. 5.15 is for a fixed order size inventory system. The essential difference between the two systems can be seen by comparing the two diagrams.

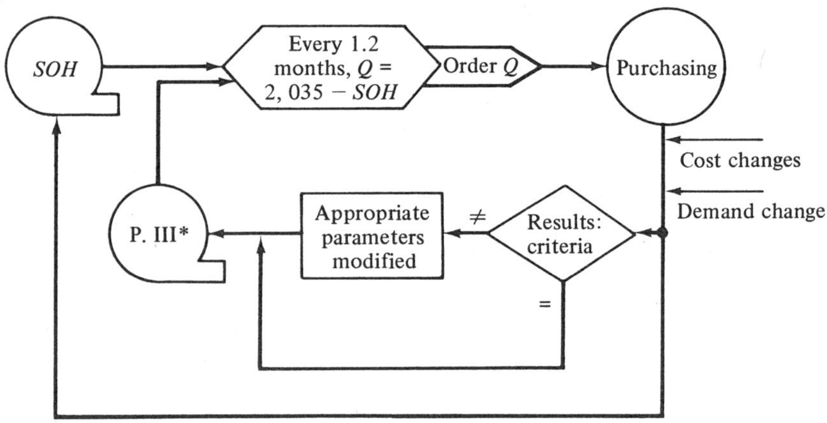

*Parameter file includes demand distribution, lead time, all costs, OI_o, SS, and $IMAX$

Figure 5.14 Transfer Function representation for Fixed Order Interval Inventory System

Comparison of the Two Systems. We have already observed that a greater amount of safety stock is required for the fixed order interval system than for the fixed order size system. This greater safety stock results in a higher carrying cost over a year. The fixed order interval system has an advantage, however, that at least partially offsets the higher carrying cost. In general, less clerical activity is involved in a fixed order interval system because inventory levels are monitored only once during each order interval. In the fixed order size system, inventory levels must be monitored continuously. The costs of operating the inventory system should be considered when choosing between the two systems. Starr and Miller (1962, Chap. 9) call these "systemic costs" and provide an

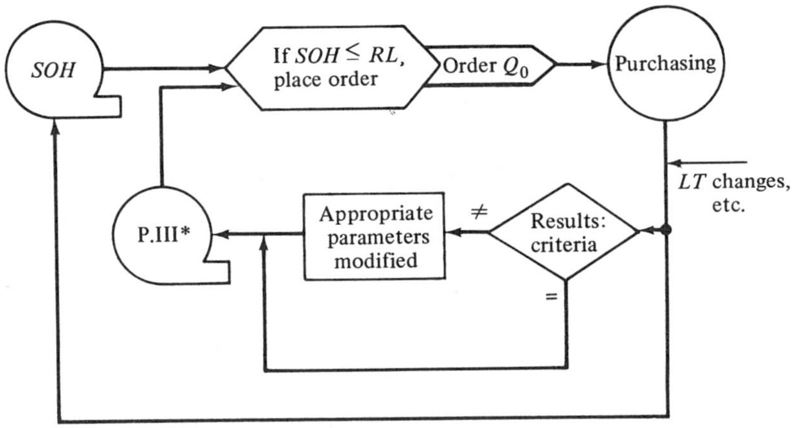

*Parameter file includes demand distribution, lead time, all costs, Q_o, RL, and SS

Figure 5.15 Transfer Function representation for Fixed Order Size Inventory System

excellent development of how these costs are included in the analytical framework of inventory problems.

It is often desirable to group orders for several individual items from the same supplier, not only to reduce ordering costs but also to reduce transportation costs. The fixed order interval system is more adaptable to this kind of ordering system.

It can be argued that the fixed order size system is more compatible with the Operations Planning function because the operations plan (considering all finished goods) usually calls for uneven production of the various finished goods over time. This plan results in an uneven usage rate of the components that must be ordered. The fixed order size system can react to fluctuating usage rates better than can the fixed order interval system. Reference to Fig. 5.9 shows that during heavy usage periods, we would simply order more often than usual, while during light usage periods, we would order less often than usual. The fixed order interval system, by definition, cannot react as readily. The problem of tying the ordering policy to the production plan is discussed further in a later section.

ABC Classification. Often, a small proportion of the items held in inventory account for a rather large proportion of total inventory investment. A rather large proportion of inventory items account for a very small proportion of total inventory investment. We can also usually identify a middle group of items that account for a moderate proportion of total inventory investment.

When inventory items are grouped according to the three categories mentioned, we have what is known as the *ABC* classification. While no general rules can be stated for all companies, a typical *ABC* classification might be as follows:

A *items:* The high-value items that account for 75 percent of the total inventory investment but only 10 percent of the total number of items carried in inventory.

B *items:* The medium-value items that account for 20 percent of total inventory investment and 30 percent of the inventory items carried.

C *items:* The low-value items that account for only 5 percent of total inventory investment but 60 percent of the total number of items carried in inventory.

Particular companies may choose to group their inventory items into more than three classifications, but the principle is the same. The high value items receive the most careful attention in the inventory system. The class *A* items are monitored frequently, perhaps perpetually. Class *C* items are not controlled as closely, with review occurring once a month or even less frequently. Whenever the reorder level of a class *C* item is reached, a large order is placed since relatively little money is tied up in these low value items. The degree of attention given to class *B* items is somewhere between that given to class *A* and to class *C*.

Quantity Discounts. The price of a purchased item often depends upon the quantity ordered at one time. A supplier will often offer a quantity discount if a certain size order is placed. Whether we should accept this offer depends upon whether the amount saved because of the discount is greater than the additional cost of varying from our economic order quantity. Figure 5.3 shows that any deviation from Q_o results in an increase in the *TC* value.

Let us consider again the example we have been using throughout this chapter. Our economic order quantity is 1,000 units. Suppose the supplier offers a \$0.03 per unit discount if we will purchase in quantities of 2,500 instead of 1,000. This order would result in a savings of \$75.00 for the 2,500 items. The expression for total cost for any order quantity Q is given by Eq. 5.1. We now use this expression to determine total costs using the two values of Q in question:

For $Q = 1{,}000$:

$$\begin{aligned} TC &= (PC)(D/Q) + (CC)(Q/2) \\ &= (\$10)(10{,}000/1{,}000) + (\$0.20)(1{,}000/2) \\ &= \$100 + \$100 = \$200 \end{aligned}$$

For $Q = 2{,}500$:

$$\begin{aligned} TC &= (\$10)(10{,}000/2{,}500) + (\$0.20)(2{,}500/2) \\ &= \$40 + \$250 = \$290 \end{aligned}$$

We see that, in this example, we should not accept the discount since there is more lost than gained. Note that if the supplier would offer a $0.04 per unit discount for purchase of 2,500 units, we should accept this offer.

There can be several price breaks at various quantities. A good discussion on how to handle these is found in Buffa (1968, p. 60).

Rounding Off the Order Interval. Under operating conditions, it is often more practical to round OI_o values off to the nearest week or month, rather than having highly irregular order points. Starr and Miller (1962, p. 82) present the following table as a guide to rounding off OI_o to divisors of 12:

If OI_o is between	round OI_o off to
0 —1.414	1
1.415—2.449	2
2.450—3.464	3
3.465—4.899	4
4.900—8.485	6
8.486—	12

The last row indicates that we will order no less frequently than once a year. The table can be extended to longer policy periods.

Variable Lead Times. In the transfer functions developed previously, we assumed that lead time was known with certainty. However, lead time is often a random variable. In such situations the determination of safety stock is more complex than in the inventory models we discussed. The usage rate during lead time is now a random variable having a joint probability distribution. Since the assumed statistical background of the readers of this text does not include joint probability distributions, we can only indicate here an intuitive approach to this problem. Probably the simplest approach is to determine the safety stock required for a reasonable maximum usage rate over a reasonable maximum lead time. The SS value so determined is then used in one of the transfer functions developed earlier.

Another approach to determining safety stock when lead time is a random variable is to experiment with different SS values in a computer simulation model. If simulated over enough time periods, a very good safety stock value can be arrived at.

Lead Time Greater than Order Interval. We often encounter a situation in which the lead time for acquiring purchased items is longer than the average or fixed time between orders. In this case, there is always at least one order on the way. Such an order is called an outstanding order. The total number of units included in outstanding orders is called stock on order and is denoted SOO. We can adjust for this condition in our transfer functions by replacing SOH in each case by $SOH + SOO$. For example, in a fixed order size inventory system, the transfer function would be: *when* $SOH + SOO \leq RL$, *order* Q_o *units.*

MANUFACTURED ITEMS

We now turn our attention to those items we manufacture on our own facilities—finished goods, subassemblies, and fabricated parts. The transfer functions we develop in this section are applicable to all three categories of items just mentioned, since they are alike insofar as the determination of how many items to make at one time is concerned.

A point of distinction is needed here. Many industrial processes are structured such that production is essentially continuous. For example, an electrical appliance may be produced continuously on an assembly line used for no other purpose. There is no lot size involved in this case.

Other industrial processes are structured such that production is intermittent. This is the case when the production rate of an item is significantly greater than its consumption rate or when one process produces a variety of products. As a matter of fact, many of the components making up the continuously produced electrical applicance discussed above may be produced intermittently. Metal screws, for example, are produced in large lots, or batches; enough may be made on one production run to last 6 months. We also recognize that the same metal screw may be used in many different products and assemblies.

Our basic problem in this section is to develop transfer functions that will determine the most economical quantity of an item to manufacture on one production run. We do this by first considering items individually, disregarding possible interactions between several items. We then consider several items simultaneously, hoping to find an inventory policy that will minimize the costs associated with all the items.

Economic Manufactured Quantity (*Instantaneous Replenishment*). The rate of production of some items, such as bolts, screws, etc., is so much greater than the rate of consumption that we can consider replenishment as occurring instantaneously. This inventory situation is identical in almost every respect to that of purchased items. The only real difference is that the procurement cost (*PC*) consists of set-up cost rather than of order cost. Set-up cost was defined earlier as those costs associated with changing over the production process to produce the new item plus any clerical and administrative costs involved in preparing and sending a production order to Operations.

The transfer functions developed in the previous section for purchased items are applicable directly to comparable inventory situations involving manufactured items. The lead time (*LT*) is manufacturing lead time for manufactured items. Records have to be kept on manufacturing lead time for each manufactured product, just as for purchased items. The meaning of safety stock (*SS*) for manufactured items is identical to that for purchased items. The two types of systems, fixed order size and fixed order interval, are also available for the present case. The discussions concerning *ABC* classification, quantity discounts, rounding off OI_o, variable lead time, and lead time greater than order interval, are also all applicable to inventory decisions involving manufactured items.

Economic Manufactured Quantity (*Noninstantaneous Replenishment*). The more realistic inventory problem involving manufactured items is that in which replenishment occurs over some period of time. Figure 5.16 illustrates the behavior of stock on hand over time under an idealized inventory system. Note that manufacturing lead time is zero and that both the production rate and the usage rate are linear. The inventory level increases during the time period T_1, the time during which the item is manufactured.

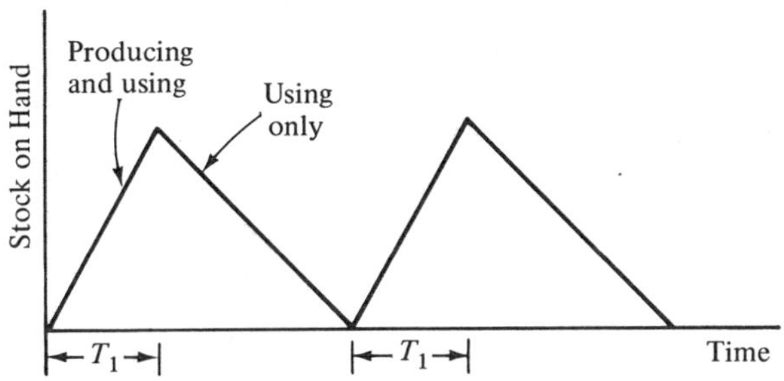

Figure 5.16 An idealized Inventory System; Replenishment occurs over Time

The development of an expression for the optimum manufactured lot size for this case is very similar to the classical inventory model. In addition to the notation defined earlier, we need three additional symbols:

PR = annual production rate if the product were manufactured continuously over an entire year.

MQ_o = optimal manufactured quantity

T_1 = number of days to produce MQ_o

The total cost equation is again the sum of set-up cost and carrying cost:

$$\text{Total Cost} = \text{Procurement Cost} + \text{Carrying Cost}$$

As in all inventory models, our approach is to express each of the two cost components in terms of the variable whose optimal value we are seeking. Thus, we want to express both procurement cost and carrying cost in terms of MQ.

Annual procurement cost for this case is identical to that for the classical inventory model:

$$\text{Annual Procurement Cost} = (PC)(D/MQ)$$

The annual carrying cost is:

$$\text{Annual Carrying Cost} = (CC)(\text{Average } SOH)$$

Our problem is to determine the average stock on hand value for this case. If we assume that there are 250 working days in one year,* then during T_1, SOH is increasing at the daily rate of $(PR - D)/250$ units. This simply reflects the fact that during T_1, we are producing items at a rate greater than the rate at which they are being used. If SOH increases by $(PR - D)/250$ each day during T_1, then it will increase by a total of $T_1(PR - D)/250$ units during the production run. This value is the maximum SOH value that will occur. Therefore,

$$\begin{aligned}\text{Average } SOH &= \tfrac{1}{2}\,(\text{maximum } SOH) \\ &= (T_1/2)(PR - D)/250\end{aligned}$$

Note, however, that this expression is not in terms of MQ. This expression is accomplished through the following relations:

$$\begin{aligned}MQ &= (\text{Daily production rate})\ (T_1) \\ T_1 &= MQ/(\text{Daily production rate}) \\ &= (MQ)/(PR/250) = 250\ MQ/PR\end{aligned}$$

We now substitute this expression for T_1 into the expression for average SOH:

$$\begin{aligned}\text{Average } SOH &= \left(\frac{250\ MQ}{2\ PR}\right)\left(\frac{PR - D}{250}\right) \\ &= MQ(1 - D/PR)/2\end{aligned}$$

Our total cost equation can now be written as:

$$TC = (PC)(D/MQ) + (CC)(MQ)(1 - D/PR)/2 \tag{5.14}$$

By differentiating TC with respect to MQ, setting the derivative to zero, and finally solving for MQ, we obtain:

$$MQ_0 = \sqrt{\frac{2(PC)(D)}{CC(1 - D/PR)}} \tag{5.15}$$

The optimal length of a production run is given by

$$T_1 = MQ_o/(PR/250) = 250\ MQ_o/PR \text{ days} \tag{5.16}$$

*This value later cancels out of the expression. Therefore, this assumption is not restrictive.

The transfer function for this case would be: *whenever $SOH = O$, begin a production cycle of T_1 days, producing at a rate of $PR/250$ units per day.*

The assumptions of zero manufacturing lead time and linear production usage rates are often unrealistic. To protect ourselves during lead times and against uncertain production and usage rates, we again resort to the use of safety stock. Safety stock values and reorder levels for our transfer functions are determined precisely as was done for purchased items. Figure 5.17 illustrates the behavior of stock on hand over time when there exists a manufacturing lead time and nonlinear production and usage rates.

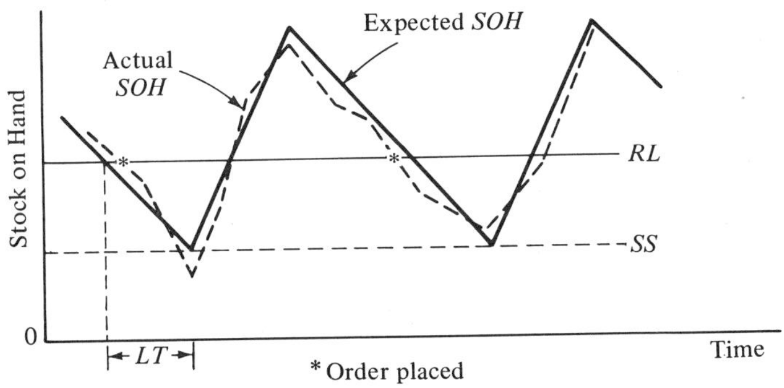

Figure 5.17 Inventory System for manufactured item, with Lead Time and Non-Linear Usage and Production Rates

Multi-Item Inventory Models. The inventory models discussed in the previous section are applicable to individual items manufactured on our production facilities. When several products are to be made in sequence on a single unit of equipment, or on one production line, the single-product model for determining lot size is usually inadequate. When applied to each product individually, the single-product model often results in the assignment of more than one job to the same machine during the same time period. Since the machine can perform only one job at a time, a decision has to be made as to which job is done first, which second, etc. In such cases, we usually cannot conform to the MQ_o values determined for each item individually, since we would run completely out of one item while the machine is producing MQ_o units of another.

One approach to the solution of the multiproduct situation is to establish repetitive production cycles during which each item is produced once. The objective is to determine the optimal fraction of annual usage to produce each product during each production cycle. The optimal fraction of annual usage is that amount that minimizes the sum of all set-up and holding costs for all products for the year.

Suppose that we have three products, A, B, and C, all of which are fabricated on the same equipment. We could produce these three items in, say, three different production cycles over the course of a year. If we did this, the loading chart would appear as in Fig. 5.18. A_1 indicates the production time during cycle 1 devoted to product A, and likewise for the other products and other cycles.

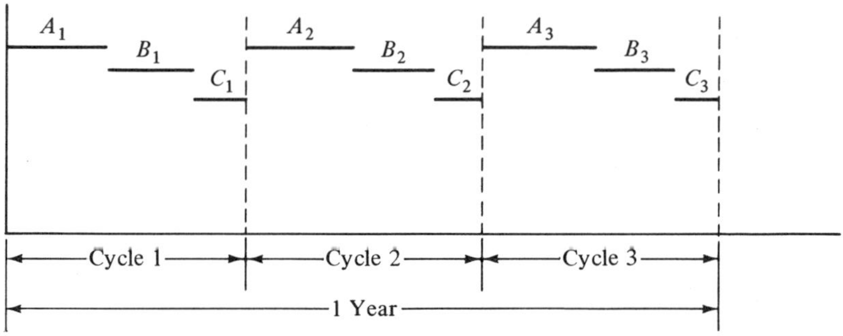

Figure 5.18 Machine Load Chart for three products in three Production Cycles over the Year

The following relationships are noted:

$$A_1 + A_2 + A_3 = A_T = \text{total usage of } A \text{ for one year}$$
$$B_1 + B_2 + B_3 = B_T = \text{total usage of } B \text{ for one year}$$
$$C_1 + C_2 + C_3 = C_T = \text{total usage of } C \text{ for one year}$$

If we impose the restriction that the same number of each product must be produced during each cycle, then the following relationships hold:

$$A_1 = A_2 = A_3 = A$$
$$B_1 = B_2 = B_3 = B$$
$$C_1 = C_2 = C_3 = C$$

The *fraction* of annual usage produced during each cycle for each item is $A/A_T, B/B_T, C/C_T$. Note that, since each item is produced each cycle, $A/A_T = B/B_T = C/C_T$. We call this common ratio F. Our objective is to determine that value of F that will result in lowest total cost for all the items involved. We need the following notation:

k = index used to denote a particular item

T_k = length of production runs in days for item k

F = function of annual usage produced each cycle, same for all items

PR_k = annual production rate for item k (amount of k the process could produce in one year if it did nothing else)

D_k = annual usage of item k

CC_k = annual cost of storing 1 unit of item k

PC_k = cost per set-up for item k

TC_k = total annual variable cost for item k

We assume 250 working days per year.

We must develop a total cost equation in which the cost components are expressed in terms of the variable which we are trying to optimize, in this case F. We first develop a total cost equation for a general item k and then convert this to an over-all total cost equation by summing the resulting expression over k.

$$TC_k = (\text{Annual set-up cost})_k + (\text{Annual holding cost})_k$$

$$(\text{Annual set-up cost})_k = PC_k\ (\text{Number of set-ups})_k$$

$$(\text{Number of set-ups})_k = D_k/(\text{Number of units produced on each set-up})_k$$

$$= D_k/(F)(D_k) = 1/F$$

therefore,

$$(\text{Annual set-up cost})_k = PC_k/F$$

This expression, in terms of F, is the annual set-up cost for product k. We must now develop a similar expression, also in terms of F, for the annual holding cost.

$$(\text{Annual holding cost})_k = (CC_k)\ (\text{Average } SOH)_k$$

$$(\text{Average } SOH)_k = (1/2)\ (\text{Maximum } SOH)_k$$

$$(\text{Maximum } SOH)_k = [(\text{Daily production rate})_k - (\text{Daily usage rate})_k]T_k$$

$$= [(PR_k - D_k)/250]T_k$$

$$(\text{Average } SOH)_k = (1/2)[(PR_k - D_k)/250]T_k$$

The expression is not yet in terms of F. To convert the expression, we must recognize a fundamental relationship between T_k and F.

$$T_k = \text{Number of units produced in } T_k/\text{Daily production rate}$$

$$= (F)(D_k)/(PR_k/250) = 250(F)(D_k)/PR_k$$

$$\begin{aligned}(\text{Average } SOH)_k &= (1/2)[(PR_k - D_k)/250]250(F)(D_k)/PR_k \\ &= (1/2)(F)(D_k)(1 - D_k/PR_k) \\ (\text{Annual holding cost})_k &= (1/2)(F)(D_k)(1 - D_k/PR_k)CC_k\end{aligned}$$

We now can add the two expressions to obtain a total cost equation for product k:

$$TC_k = PC_k/F + (1/2)(F)(D_k)(1 - D_k/PR_k)CC_k \qquad (5.17)$$

Since our objective is to minimize the total inventory costs of all items, we must sum both sides of Eq. 5.17 to obtain the final total cost equation for all items.

$$\sum_k TC_k = \sum_k (PC_k)/F + \sum_k [(1/2)(F)(D_k)(1 - D_k/PR_k)CC_k] \qquad (5.18)$$

By taking the first derivative of Eq. 5.18 with respect to F, setting the derivative equal to zero, and then solving for F, we obtain the following expression:

$$F = \sqrt{\frac{2\sum_k PC_k}{\sum_k [CC_k D_k(1 - D_k/PR_k)]}} \qquad (5.19)$$

Equation 5.19 determines the optimal fraction of annual demand for each item that should be produced during one production cycle, or one production run. To convert from fraction of annual demand to the actual quantity of each item to produce each cycle, we have

$$MQ_k = D_k F \qquad (5.20)$$

The number of cycles (NC) in one year is determined by:

$$NC = 1/F \qquad (5.21)$$

The length of each cycle (CL) in months is:

$$CL = 12/NC \qquad (5.22)$$

The transfer function for this case is: *every CL months, begin a new production cycle, and produce an amount* $MQ_k = D_k F$ *for each item k.*

To illustrate the use of this transfer function, let us consider the data for three products shown in Table 5.1. When used in Eq. 5.19, this data gives us $F = 0.153$.

Table 5.1 Data for Multi-Item Inventory Analysis

Item Number	Annual Demand	Set-Up Cost	Carrying Cost	Production Rate
k	D_k	PC_k	CC_k	PR_k
1	10,000	$10.00	$0.20	30,000
2	2,000	8.00	0.10	8,000
3	6,000	12.00	0.30	15,000

The number of cycles in one year would be 6.53. Each cycle length would be 1.84 months or 38.3 days. The number of units of each item produced each cycle and the number of days in each cycle devoted to the production of each item are shown in Table 5.2. The latter value is obtained by dividing MQ_k by the daily production rate. Note that the total days of production in Table 5.2 is 37.62, which is 98.3% of the 38.3 day production cycle. This is precisely the percentage of total production capacity used by annual demand of the three items.

Table 5.2 Production Allocation for Each Cycle

Item Number	Unit per Cycle	Days of Production of Item k
k	MQ_k	
1	1,530	12.75
2	306	9.57
3	918	15.30
		37.62

Once we have determined the number of each item to produce in each production cycle, there is still the decision as to the sequence in which the items are to be manufactured. In general, there are $n!$ possible sequences for n items produced in a cycle. In the above example involving three items, there are $3! = 6$ possible sequences. In some operations it makes no difference at all in what sequence the work is performed. In other cases, however, technological or economic considerations make one sequence better than another. For example, in fabric dyeing operations, it is generally desirable to dye lighter colors during the first part of the cycle, with progressively darker colors processed toward the end of the cycle.

Let us now consider a different approach to the problem discussed above; i.e., schedule conflicts resulting from the application of the lot size formula, Eq. 5.15, to individual items. Even though a MQ_o value can always be determined for each item, it is usually not possible to obtain a satisfactory production schedule consisting only of MQ_o values. It is often necessary, therefore, to deviate from the MQ_o values for at least some of the items.

The method presented below (see Eilon 1962, pp. 243–45) determines a range of MQ values for each item such that the resulting total inventory cost is not increased more than a previously stated allowable amount. Once the range of MQ values is established for each item, a feasible production schedule is generated by adjusting individual MQ values within the determined range.

The total cost function of a manufactured item is illustrated in Fig. 5.19. Any deviation from the optimal batch size MQ_o results in an increase in total cost. Since the cost function is relatively flat at its lowest point, we usually can deviate from MQ_o quite a lot without greatly increasing TC. Note, however, that we can deviate farther in the direction of larger lot sizes than we can in the direction of smaller lot sizes, for the same increase in total cost.

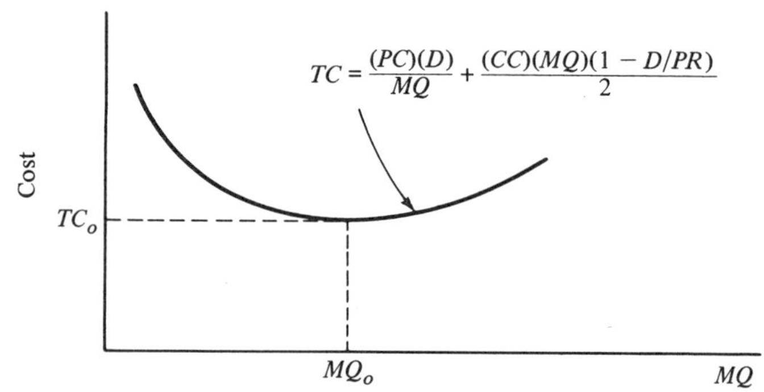

Figure 5.19 Total Cost Function of a manufactured item

Our approach is to specify an allowable increase in TC. The amount can be stated either as a percentage or as an upper limit on the numerical value of TC. If stated as an allowable percentage increase, the increase is denoted as $p = 1 +$ % increase. If stated as an upper limit, TC', on the numerical value of TC, the increase is determined as $p = TC'/TC_o$. The effects of the allowable increase are illustrated graphically in Fig. 5.20. Any lot size between the values MQ_L and MQ_U does not cause TC to be greater than the allowable maximum TC'. Our problem is to find the numerical values of the two limits, MQ_L and MQ_U for a given allowable total cost TC' or a given percentage increase p.

Eilon (1962, pp. 243–45) derives the following expressions:

$$\begin{aligned} MQ_L &= MQ_o(p - \sqrt{p^2 - 1}) \\ MQ_U &= MQ_o(p + \sqrt{p^2 - 1}) \end{aligned} \tag{5.23}$$

Suppose that for a particular item we determine $MQ_o = 1{,}000$ units. If we

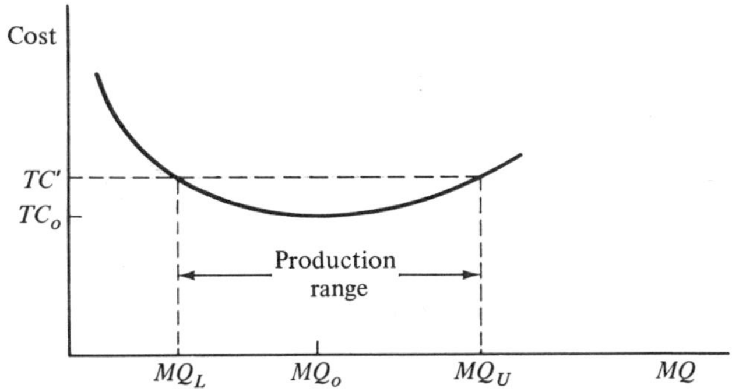

Figure 5.20 Production Range Concept

are willing to permit a 1 percent increase in costs (i.e., $p = 1 + 0.01 = 1.01$), then Eqs. 5.23 give the following limits on lot size:

$$MQ_L = 1{,}000(1.01 - \sqrt{(1.01)^2 - 1}) = 1000(0.868) = 868 \text{ units}$$
$$MQ_U = 1{,}000(1.01 + \sqrt{(1.01)^2 - 1}) = 1000(1.152) = 1{,}152 \text{ units}$$

Thus, when we determine our production schedule, we can produce any quantity between 868 and 1,152 units with the assurance that total inventory cost for this particular item will not increase by more than 1 percent. A similar range would be determined for other items as a guideline in adjusting individual MQ_o values for production scheduling.

IN-PROCESS BUFFER SIZES*

In the introductory section of this chapter, we identified the three basic types of inventory as raw materials and purchased parts, in-process, and finished goods. The inventory transfer functions developed in the preceding two sections are concerned with raw materials and purchased parts and finished goods inventories. We now discuss in-process inventory.

In-process inventory can occur between successive stages of multistage production processes. Only in those production processes in which the product is processed at a fixed speed or flow rate through all its stages does in-process inventory fail to occur. In all other cases, in-process inventory can be used for the uncoupling of successive production stages. Only when in-process inventory is provided can successive stages operate relatively independent of one another.

Relatively little attention has been given to in-process inventory in texts and

*The discussion in this section involves a few elementary concepts of queueing theory. While a prior knowledge of these concepts would certainly enhance the reader's understanding of the discussion, the mechanics of the methods discussed can be followed without such prior knowledge.

other professional literature. Consequently, practical procedures are not well developed. We restrict our attention here to a discussion of the basic problem and to two approaches to its solution.

Processing Rates Fixed and Equal. Consider two stages of a production process, as illustrated in Fig. 5.21. The product flows in the direction indicated by the arrow. Any units of product that accumulate between the two stages are stored in the buffer. The inventory accumulation will fluctuate up and down as a function of the output rates of the two stages, which are both random variables. Let us denote as y_{AB} the stock in-process between Stages A and B, or simply the number of units in the buffer. If the maximum buffer capacity is J units, then y_{AB} can range from 0 to J.

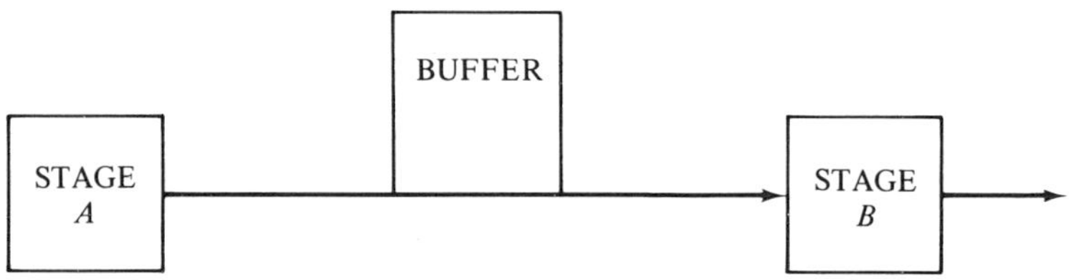

Figure 5.21 Two Stages of a Production Process, connected by a Buffer

Note that if $y_{AB} = 0$ when Stage B completes processing a unit, then Stage B must go idle until Stage A releases the next unit. If $y_{AB} = J$ when Stage A completes processing a unit, then Stage A must go idle until Stage B completes the unit it is now working on. Machine idle time is expensive and creates a cost we usually want to avoid.

The amount of machine idle time (for both stages) that results over several hours of operation is highly dependent upon J, the maximum buffer size. If $J = 0$, then a large amount of idle time results. If J is very large, then natural inventory accumulation in the buffer tends to decrease the amount of machine idle time.

We must recognize, however, that low machine idle time is achieved only at the expense of providing a large in-process buffer for inventory accumulation. Also, the cost of money invested in in-process inventory must be added to the cost of providing the buffer. We see, then, that this problem fits the typical framework of an inventory problem. We have two opposing costs, and we want to minimize the sum of these costs. The model we discuss is based upon the work of Young (1967), Hunt (1956), and Buchan and Koenigsberg (1963).

Suppose we have a two-stage process, as in Fig. 5.21, each stage with an identical Poisson service rate distribution. It can be shown that e, the machine efficiency (the ratio of productive machine time to total operating time), is approximately

$$e = \frac{J+1}{J+2} \tag{5.24}$$

Intuitively, the above relationship is true when $J = 0$. In this case, $e = 0.5$, which says that half the time one of the stages will be idle. Either Stage A is forced idle because Stage B cannot accept the output from Stage A, or Stage B is idle because Stage A has not completed processing the next unit. (Note that we are assuming an unlimited supply of units into Stage A.)

Presented below, without proof, is the equation for expected total cost for this model:

$$E(TC) = C_1[e/(1-e)] + C_2(1-e) \tag{5.25}$$

where

$E(TC)$ = expected total cost considering Poisson production rate distributions
C_1 = per unit inventory cost per unit time period
C_2 = idle machine cost per unit time
e = machine efficiency

The first derivative of Eq. 5.25 can be set to zero and solved for e. Doing so, we obtain

$$e = 1 - \sqrt{C_1/C_2} \tag{5.26}$$

Equations 5.24 and 5.26 may be equated and reduced to obtain the following expression for J_o, the optimum buffer size:

$$J_o = \sqrt{C_2/C_1} - 2 \tag{5.27}$$

For particular values of C_1 and C_2, the optimal buffer size is determined easily by Eq. 5.27. This value can then be used in Eq. 5.24 to obtain e, which in turn can be used in Eq. 5.25 to determine the lowest expected total cost using a buffer size of J_o.

Suppose, for example, that in a particular two-stage process, $C_1 = \$0.10/\text{hr}/$ unit and $C_2 = \$10.00/\text{hr}$. Equation 5.27 gives $J_o = 8$ units. Equation 5.24 gives $e = 0.90$. Table 5.3 shows the $E(TC)$ calculations for several values of J. The results are plotted in Fig. 5.22. The opposing costs and their relationships are clearly visible. We should remember that since C_1 and C_2 are expressed on an

Table 5.3 Calculations for $E(TC)$, Eq. 5.25

J	$e = \frac{J+1}{J+2}$	$1 - e$	$\frac{e}{1-e}$	$C_1 \frac{e}{1-e}$	$C_2(1-e)$	TC
0	0.5	0.5	1.0	0.10	5.0	5.10
1	0.667	0.333	2.0	0.20	3.33	3.53
2	0.75	0.25	3.0	0.30	2.50	2.80
3	0.80	0.20	4.0	0.40	2.00	2.40
4	0.833	0.167	5.0	0.50	1.67	2.17
5	0.857	0.143	6.0	0.60	1.43	2.03
6	0.875	0.125	7.0	0.70	1.25	1.95
7	0.889	0.111	8.0	0.80	1.11	1.91
8	0.900	0.100	9.0	0.90	1.00	1.90
9	0.910	0.091	10.0	1.00	0.91	1.91
10	0.917	0.083	11.0	1.10	0.83	1.93
11	0.922	0.078	12.0	1.20	0.78	1.98

hourly basis, the expected total cost curve in Fig. 5.22 is also on an hourly basis.

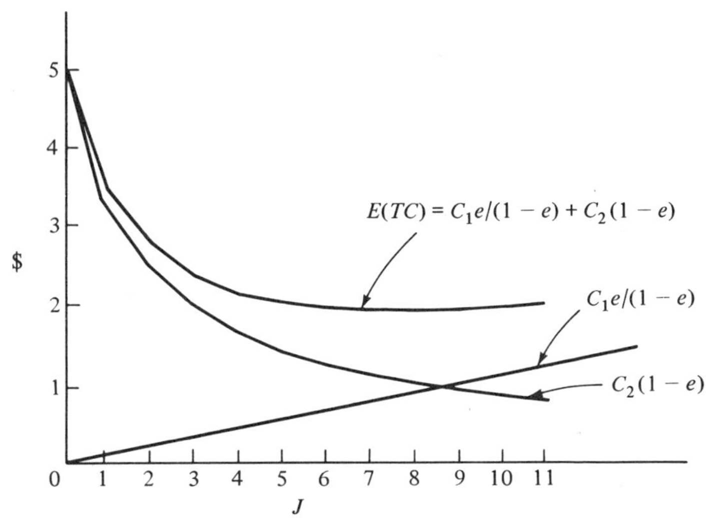

Figure 5.22 $E(TC)$ Curve for $C_1 = \$.10$ and $C_2 = \$10.00$

It is shown in Young (1967) that the above results can be extended to any number (e.g., M) of stages in a production process as long as the service rates of all stages are identical Poisson random variables. For M stages, however, Eq. 5.27 is changed to:

$$J_0 = M(\sqrt{C_2/C_1} - 1) - \sqrt{C_2/C_1} \tag{5.28}$$

If either Eq. 5.26 or 5.27 yields a negative value, J_0 is set equal to zero. Young (1967) also gives a model for the case in which all service times are normally distributed.

Processing Rates Adjustable and Unequal. We now consider the case in which the processing rates of two successive production stages are unequal and at least one is adjustable. We assume again that the production rates at the two stages are Poisson distributed random variables, X_A and X_B, with means μ_A and μ_B for Stages A and B, respectively. Define N as the number of units in the buffer plus the number of units being worked on at Stage B.* Note that when $N = 0$, Stage B is idle, and when $N > 0$, Stage B is working. For the case in which $\mu_A < \mu_B$, it can be shown by elementary queueing theory that the probability function of N is:

$$P_N = (1 - \mu_A/\mu_B)(\mu_A/\mu_B)^N \qquad \mu_A < \mu_B \tag{5.29}$$

Theoretically there is no upper limit on the size of N. In practice, we must restrict the number of units to that which we can accommodate in a buffer at one time. Let us denote the buffer size as J. Clearly, when $N = J + 1$ (i.e., when the buffer is full and 1 unit is in service at Stage B), Stage A must go idle, since there is no place to put its output.

Our approach for this case is to specify acceptable proportions of the total operation time that we are willing to allow Stages A and B to be idle. We define the following notation:

α_A = acceptable proportion of (idle time)/(total operation time) for Stage A.

α_B = acceptable proportion of (idle time)/(total operation time) for Stage B.

The values for α_A and α_B, which must be specified for a given situation, are arrived at by considering the relative seriousness of the two types of idle time conditions.

To begin our analysis, we observe the following relationship:

$$\alpha_B = P(N = 0) = P_0 = (1 - \mu_A/\mu_B)$$

from which

$$\mu_A/\mu_B = 1 - \alpha_B \tag{5.30}$$

The relationship in Eq. 5.30 can now be used to adjust either μ_A or μ_B or both for the desired α_B value. Once the proper values of μ_A and μ_B are determined, the complete probability function of N is determined from Eq. 5.29. It

*Either 0 or 1 unit is being worked on at Stage B at any particular time.

is now a simple matter to determine the buffer size J that satisfies the desired value for α_A. We simply find the largest integer value of N such that the sum of the probabilities of N values greater than J is not greater than α_A. Symbolically,

$$J = \max\,[N - 1], \text{ such that } \sum_{i=N+1}^{\infty} P(N_i) \leq \alpha_A$$

$$= \max\,[N - 1], \text{ such that } \sum_{i=0}^{N} P(N_i) > 1 - \alpha_A \qquad (5.31)$$

Let us consider an example. The production rates of Stages A and B are independent Poisson random variables. Suppose μ_A is fixed at 10 units per hour, but μ_B is adjustable. An analysis of pertinent costs leads us to specify $\alpha_A = 0.10$ and $\alpha_B = 0.10$. Our problem is to determine μ_B and J such that the other conditions are satisfied.

We can determine μ_B directly from Eq. 5.30.

$$\mu_B = \mu_A/(1 - \alpha_B) = 10/(0.9) = 11.1 \text{ units/hour}$$

We now can determine the complete distribution of N from Eq. 5.29:

N	0	1	2	. . .	18	19	20	etc.
P_N	0.1000	0.0900	0.0810	. . .	0.0150	0.0135	0.0121	etc.
$\sum_{i=0}^{N} (P_N)_i$	0.1000	0.1900	0.2710	. . .	0.8773	0.8908	0.9029	etc.

The probability that $N \leq 20 = 0.9029$, which satisfies the desired α_A value of 0.10. However, since N includes 1 unit in service at Stage B, the actual buffer size J is 19 units.

The approaches to the in-process inventory problem that we have discussed are but two of several such models available. Neither of the two should be regarded as appropriate for all situations. More research is needed on the in-process inventory problem so that useful transfer functions may be developed.

GRAPHICAL AND TABULAR TRANSFER FUNCTIONS

Most manufacturing firms must maintain inventories of thousands of individual items. It is very time consuming to apply economic lot size formulas, such as Eqs. 5.4 and 5.15, to each individual item.

It is quite easy to develop ordering tables for various combinations of procurement costs and holding costs. It is then simply a matter of looking up the correct order quantity for a given usage or demand value.

Consider, for example, Eq. 5.4, the basic lot size ordering formula. This formula can be written as:

$$Q_0 = \sqrt{\frac{2(PC)}{CC}}\sqrt{D}$$

It is often the case that the same values of PC and CC are applicable to a large number of items. Thus, we can form several common PC/CC ratios, leaving only D in the expression. An ordering table can then be constructed for each PC/CC ratio.

Suppose that the values PC = \$5.00 per order and CC = \$0.60 per unit per year are applicable to a large number of items. The expression for order size becomes

$$Q_0 = \sqrt{\frac{2(5.00)}{0.60}}\sqrt{D} = 4.09\sqrt{D}$$

We now can construct an ordering table, such as Table 5.4, for various values of D. A similar table would be constructed for all common PC/CC ratios.

Table 5.4 Ordering Table for PC = \$5.00 and CC = \$0.60, or Any Other Values of PC and CC Whose Ratio is 5.0/0.6 = 8.33

D	Q_0	D	Q_0
100	41	10,000	409
500	91	20,000	579
1,000	129	50,000	910
2,000	183	100,000	1293
5,000	289	200,000	1830

Note that the function $Q_0 = 4.09\sqrt{D}$ can be plotted, as in Fig. 5.23. Whereas an ordering table such as Table 5.4 can include only a limited number

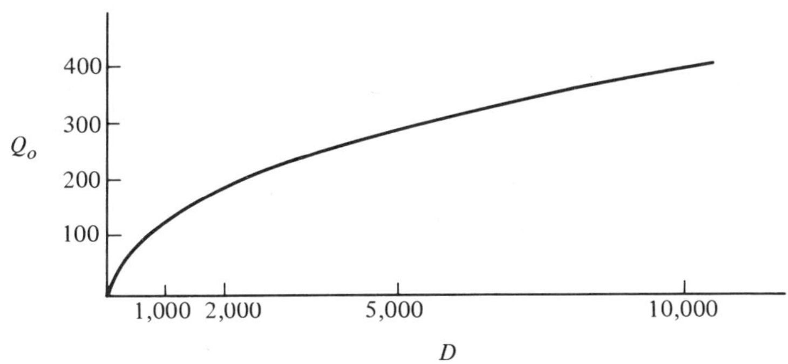

Figure 5.23 Graph of Q_0 as a function of D when PC/CC = 8.33

of values for D, the graph can be used for any value of D. A similar graph would be constructed for all common PC/CC ratios. A better graph can be obtained for large values of D by using log paper.

There are many other graphical and tabular aids that can be used as transfer functions for making inventory operating decisions. Magee (1967, pp 63, 64), for example, illustrates an "ordering chart" and a nomogram, both of which would be helpful, especially in a manual inventory system.

FURTHER CONSIDERATIONS

The several inventory transfer functions developed in the previous sections involve certain implicit assumptions which are sometimes unrealistic in practice. In most of our models, for example, D was used to express the expected annual demand for or usage of an item. We assumed that the usage rate would be relatively uniform throughout the year with only random fluctuations. This assumption may be valid for many situations, but by no means would it be valid for many other situations.

Another assumption implied in our models is that we have no upper limitation on storage space for items in inventory. This is usually not the case, and it is quite common for an organization to be forced to order smaller lot sizes than those determined from the Q_0 and MQ_0 formulas.

A similar assumption pertains to an upper limit on inventory investment. Again, we may be forced to order smaller lot sizes in order to stay within the maximum permissible inventory investment.

Other considerations also may not be accounted for in the Q_0 and MQ_0 calculations. These three, however, illustrate the adjustments that sometimes are necessary in our inventory system. Each of the three assumptions is treated in the following sections.

Time Phasing from Operations Planning. An operations plan, as developed by the methods discussed in Chapter 4, commonly calls for uneven production of certain items over a planning period, say a year. The requirements for all purchased parts and manufactured parts that go into these items are also uneven over the planning period. Our inventory procedures must be compatible with the time-phased requirements of the operations plan.

Consider Production Plan 2 shown in Fig. 4.11. This plan calls for a production rate of 120 units per month during the first 6 months of the year and 60 units per month during the last 6 months. Now, suppose that 5 units of a certain raw material used in the production of this item are needed for each unit of the finished product. This results in a requirement of 600 units per month during the first 6 months and 300 units per month during the last 6 months. If there are no other requirements for this particular raw material, we can determine one Q_0 value from Eq. 5.4 for the first 6 months, and another Q_0 value for the last 6 months.

Suppose $PC = \$10.00$/order, and $CC = \$0.20$/unit/year for this item. For the first 6 months, $D' = 3600$ units; for the second 6 months, $D'' = 1800$ units. The carrying cost and the demand must be for the same time span. Therefore, we must change CC to $0.10/unit/half year to be compatible with the two 6-month planning periods. Equation 5.4 gives us

$$Q_0' = \sqrt{\frac{2(10.00)(3600)}{0.10}}$$

$$= 850 \text{ units for the first 6 months}$$

$$Q_0'' = \sqrt{\frac{2(10.00)(1800)}{0.10}}$$

$$= 600 \text{ units for the second 6 months}$$

This same approach can be used for a shorter time period, provided the carrying cost and demand are for the same time period. Unequal time periods can also be used. For example, the production plan might call for a certain production rate for the first 3 months, another rate for the next 6 months, and still another rate for the last 3 months of the year.

The above approach is only an approximation and will not, in general, yield a least cost inventory policy for a year. A more precise formulation is explained in Wagner and Whitin (1958, pp. 89–96).

Physical Space Limitations. Let us now consider the problem of a limited amount of storage space for our inventories. This condition will often force us to deviate from the calculated Q_0 and MQ_0 values for individual items. However, we would still like to minimize total cost, subject to the space limitation.

If we have n items, our total cost equation is:

$$TC = \sum_{i=1}^{n} [PC_i(D_i/Q_i) + CC_i(Q_i/2)] \tag{5.32}$$

We define K as the total cubic feet of our storage facilities and k_i as the cubic foot requirement for storing 1 unit of item i. If there is a possibility that orders for all items will arrive at one time, then our physical space restriction can be expressed in terms of the order sizes Q_i as follows:

$$\sum_{i=1}^{n} k_i Q_i \leq K \tag{5.33}$$

If we can assume that the orders arrive in a prescribed fashion such that the total amount in storage at one time remains approximately the same, then the space restriction takes on the following form:

$$\sum_{i=1}^{n} (k_i Q_i)/2 \leq K \tag{5.34}$$

We consider here only the case expressed by Eq. 5.33 and assume that we want our warehouse completely full upon the receipt of our orders. This assumption changes Eq. 5.33 to an equality. Upon rearrangement, we have

$$\sum_{i=1}^{n} k_i Q_i - K = 0 \tag{5.35}$$

We now employ the mathematical technique of Lagrangian multipliers to optimize our total cost function within the limitations of the space restriction. An operator λ is multiplied by Eq. 5.35 and added to the total cost function, Eq. 5.32. Note that this addition is permissible since we are simply adding a zero quantity to the right-hand side. The result is

$$TC = \sum_{i=1}^{n} PC_i(D_i/Q_i) + \sum_{i=1}^{n} CC_i(Q_i/2) + \lambda(\sum_{i=1}^{n} k_i Q_i - K) \tag{5.36}$$

To minimize the original total cost expression subject to the space restriction, we must minimize TC over both Q_i and λ. We accomplish this by taking the partial derivative of TC in Eq. 5.36 with respect to Q_i.

$$\frac{\partial TC}{\partial Q_i} = -\frac{(PC_i)(D_i)}{Q_i^2} + \frac{CC_i}{2} + \lambda k_i \tag{5.37}$$

We now set this partial to zero and solve for Q_i,

$$Q_i = \sqrt{\frac{2PC_i D_i}{CC_i + 2\lambda k_i}} \tag{5.38}$$

We now must find those values of Q_i from Eq. 5.38 which simultaneously satisfy the space restriction of Eq. 5.35 by "trying" various values of λ in Eq. 5.38 until a satisfactory solution is found.

Let us illustrate the method by considering two products with the following characteristics:

	D	PC	CC	k
Product 1	5000	25	5	1 ft³
Product 2	1000	25	10	2 ft³

We assume that total storage space is limited to 300 cubic feet. Our usual lot size formula, disregarding the restriction, gives the following values:

$$Q_1 = \sqrt{\frac{2(5000)(25)}{5}} = 224 \text{ units}$$

$$Q_2 = \sqrt{\frac{2(1000)(25)}{10}} = 71 \text{ units}$$

Space required for Product 1 = 224 cubic feet.

Space required for Product 2 = 142 cubic feet.

Now, since total space is limited to 300 cubic feet, Q_1 and Q_2 must be reduced. We want to reduce Q_1 and Q_2 in such a way that total cost, expressed by Eq. 5.32, is minimized. This is accomplished in an iterative solution procedure, as follows:

1. Specify a value for λ.
2. Use this value in Eq. (5.38) to obtain the Q_i values.
3. Determine if the relation in Eq. 5.35 is satisfied.
   a. If yes, the solution is complete.
   b. If no, increase or decrease λ as required, and return to Step 2.

This iterative procedure is continued until Eq. 5.38 yields Q_i values which also satisfy the relation expressed in Eq. 5.35.

For our example, let us start with $\lambda = 1$. Equation 5.38 gives $Q_1 = 189$ and $Q_2 = 60$. These quantities would require $189 + (2)(60) = 309$ cubic feet of storage. This amount exceeds our K value of 300 cubic feet. Both order quantities need to be reduced further. Close examination of Eq. 5.38 reveals that to decrease Q_i we must increase λ, and vice versa.

Let us now try $\lambda = 1.5$. Equation 5.38 now gives $Q_1 = 177$ and $Q_2 = 56$. These quantities would require $177 + (2)(56) = 289$ cubic feet of storage. We have reduced Q_1 and Q_2 too much, thus indicating that some value of λ between 1 and 1.5 would be appropriate.

Let us now try $\lambda = 1.25$. We now get $Q_1 = 182$ and $Q_2 = 58$. These quantities would require $182 + 2(58) = 298$ cubic feet, which is very close to the total capacity of 300 cubic feet.

The above procedure would be clumsy if performed manually for a large number of products. It would be quite easy, however, to perform the necessary calculations on a digital computer.

Budgetary Constraints. Just as there is often a limitation on storage space, quite frequently there is also a limitation on the amount of money that can be tied up in inventory at one time. We define C as the total number of dollars that can be invested in inventory at one time and c_i as the number of dollars per unit that item i costs. We express the budgetary constraint as:

$$\sum_{i=1}^{n} c_i Q_i/2 - C = 0 \tag{5.39}$$

This zero-valued term is then multiplied by the Lagrangian λ and added to the expression for total cost, Eq. 5.32. The new total cost equation is then partially differentiated with respect to Q_i, set to zero, and solved in terms of Q_i. The result is:

$$Q_i = \sqrt{\frac{2PC_iD_i}{CC_i + \lambda c_i}} \tag{5.40}$$

The same iterative solution procedure described in the previous section is also applicable to this case.

Feedback and Corrective Action

In the previous section we discussed several inventory transfer functions available for use in an operations planning and control system. Many other inventory models can be found in inventory texts and professional journals. A particular control system may require several different inventory transfer functions. The choice of a set of inventory transfer functions depends on the type of operation (make-to-order or make-to-stock), the availability of materials and suppliers, the seriousness of stockouts, the demand pattern, the accuracy of the forecasting transfer function, and many other factors.

Regardless of the transfer functions selected, they cannot be expected to operate flawlessly with no monitoring of their performance. It is imperative, therefore, that we continuously evaluate the performance of the complete set of inventory transfer functions. In this section we discuss the concept of feedback and corrective action for the inventory function.

SOURCES OF FEEDBACK*

The successful operation of the inventory planning and control function is greatly dependent upon accurate and timely feedback. Several sources of feedback are discussed briefly in the following paragraphs. This feedback is used to update the two information files for the inventory function, P.III and VS.III.

Vendors. Vendors provide feedback concerning the availability, or non-availability, of materials. A vendor might indicate that a shipment will be delayed, or that a partial shipment will be sent now with the balance of the order to follow shortly.

Sales. Sales provides feedback on actual demand, the expected effects of

*It would be helpful to refer frequently to Fig. II.1 while reading this section.

sales compaigns, excessive inventory buildups at warehouses and retailers, and other intangible information related to the market.

Operations. From Operations we receive feedback on actual production, actual manufacturing lead times, and expected completion dates of orders.

Engineering. The inventory function receives from Engineering feedback concerning new materials and parts, instructions to discontinue old materials and parts, and actual percent defective of both purchased and manufactured items.

Purchasing. The Purchasing function provides feedback concerning new suppliers, price changes, quantity discounts, unusual market conditions, and actual lead times.

Financial Control. From Financial Control and accounting, we receive feedback concerning modifications to inventory investment policy and current costs (such as holding cost and ordering costs) used in our inventory transfer functions.

Physical Count. Periodically, a physical count of all inventory items must be taken to account for errors in recording inventory transactions, pilferage, breakage, spoilage, etc.

SHORT-TERM CORRECTIVE ACTION

Corrective action must be taken whenever actual occurrences deviate significantly from those that we had planned. When this happens with a manufactured item, we can often expedite the production order by requesting a higher priority. For purchased items we can put a tracer (follow-up letter or telephone call) on an item whose inventory level has fallen dangerously low. In fact, we might even go to a different supplier.

Another short-term corrective action sometimes appropriate is to substitute another material or part for one whose stock level has been depleted. We can sometimes purchase a part we normally manufacture or subcontract part of our work.

In the case of stockouts of finished goods, we can sometimes convince the customer to wait for the product. There will be times, however, when despite all we can do, we will run out of raw materials, in-process inventory, and finished goods. In such cases the appropriate course of action usually must be made using human judgment as the transfer function.

LONG-TERM CORRECTIVE ACTION

We must continuously monitor the effectiveness of our inventory transfer functions and adjust the parameters whenever system performance is significantly different from performance criteria. For example, we must keep accurate data on lead times from vendors, so that any change can be reflected in a new

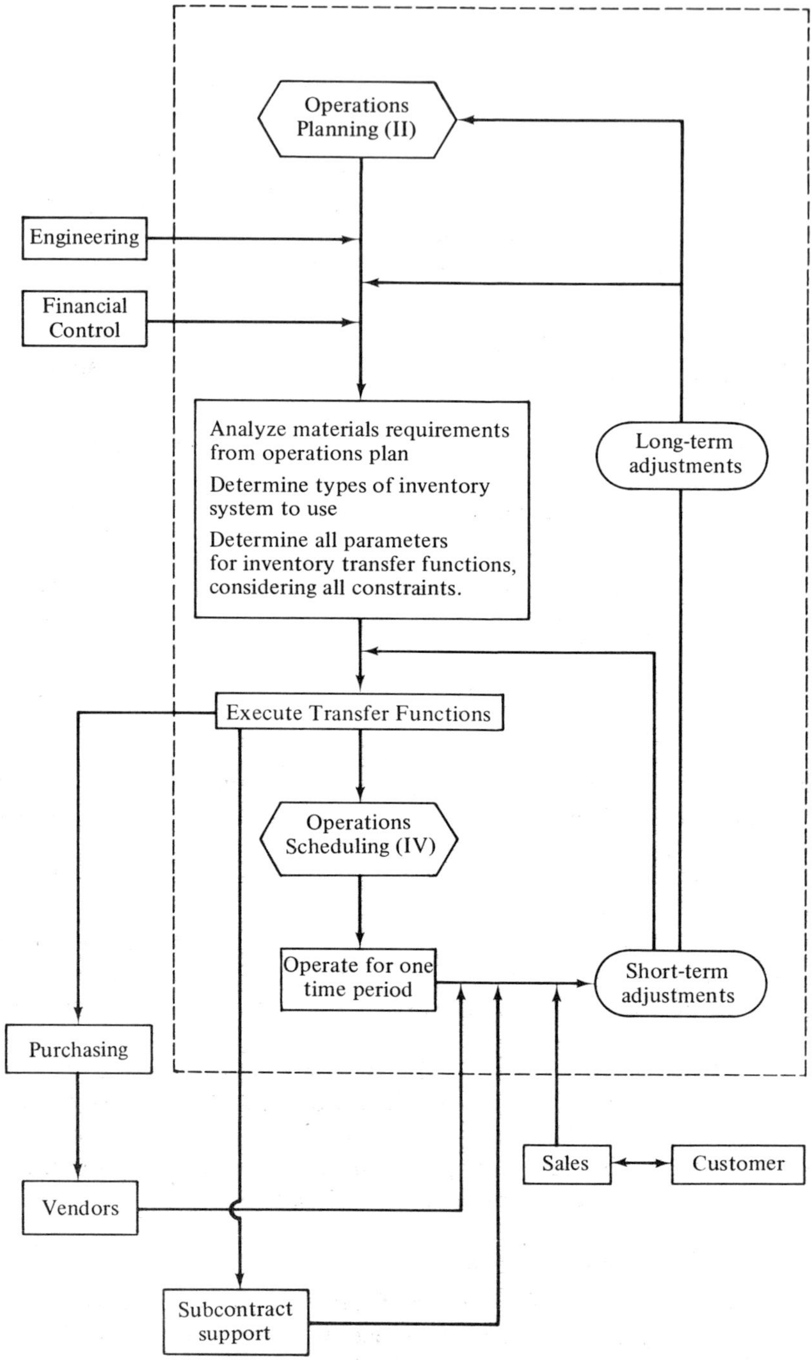

Figure 5.24 Interfaces for Inventory Function

safety stock level. If too many stockouts have occurred, safety stock should probably be increased or time between orders should be decreased. Conversely, if stockouts never occur, then safety stock levels can be reduced somewhat.

Changes in demand levels large enough to necessitate a change in the operations plan also necessitate a change in the parameters for the transfer functions. Remember, all inventory transfer functions depend greatly on an accurate estimate of demand.

Finally, it might become advantageous to change the type of inventory system, fixed order size or fixed order interval, being used. We recall that it is entirely feasible to use elements of both systems simultaneously to control inventories for all our items.

Interfaces

Inventory planning and control interacts with many other functions of the enterprise. These interactions were discussed in detail in previous sections of this chapter. Figure 5.24 is included to help us consolidate the previous discussions of the interfaces with the inventory function. The dashed line boundary separates elements of the operations planning and control system from those outside the system.

For the most part, Fig. 5.24 needs no further discussion or elaboration. We should explain, however, that the block entitled "Operate for one time period" does not necessarily mean a day or a week. The inventory function usually operates on an essentially continuous basis. "One time period," therefore, may be a very short period.

REFERENCES

Brown, R. G. 1967. *Decision Rules for Inventory Management.* New York: Holt, Rinehart and Winston.

Buchen, J.; and Koenigsberg, E. 1963. *Scientific Inventory Management.* Englewood Cliffs, N. J.: Prentice-Hall, Inc.

Buffa, E. S. 1968. *Production-Inventory Systems Planning and Control.* Homewood, Ill.: Richard D. Irwin, Inc.

Eilon, S. 1962. *Elements of Production Planning and Control.* New York: MacMillan Company.

Hunt, G. C. 1956. Sequential Arrays of Waiting Lines. *Operations Research* 4.

Magee, J. F.; and Boodman, D. M. 1967. *Production Planning and Inventory Control*, Second Ed. New York: McGraw-Hill.

Starr, M. K.; and Miller, D. W. 1962. *Inventory Control: Theory and Practice.* Englewood Cliffs, N. J.: Prentice-Hall, Inc.

Wagner, H. M.; and Whitin, T. M. 1958. Dynamic Version of the Economic Lot Size Model. *Management Science* 5: 89-96.

Young, H. H. 1967. Optimization Models for Production Lines. *Journal of Industrial Engineering*, 18: 70-78.

SOURCES OF ADDITIONAL TRANSFER FUNCTIONS

Listed below are several additional sources of inventory transfer functions. The list is intended, not as an exhaustive bibliography on the inventory function, but as a representative sample of the inventory techniques available.

DeMatteis, J. J. 1968. An Economic Lot-Sizing Technique, I: The Part-Period Algorithm. *IBM Systems Journal* 7: 30.

Falkner, C. H. 1969. Optimal Ordering Policies for a Continuous Time, Deterministic Inventory Model. *Management Science*, 15: 672.

Hausman, W. H. 1969. Minimizing Customer Line Items Backordered in Inventory Control. *Management Science* 15: B-628.

Herron, D. P. 1967. Inventory Management for Minimum Cost. *Management Science*, 14: B-219.

Mendoza, A. G. 1968. An Economic Lot-Sizing Technique, II: Mathematical Analysis of the Part-Period Algorithm. *IBM Systems Journal* 7: 39.

Pierce, J. F. 1968. A Multi-Item Economic Lot-Sizing Problem. *IBM Systems Journal*, 7: 47.

Sargent, R. G.; and Bradley, H. E. 1969. A 'Variable S' Inventory Model. *Management Science* 15: 716.

Saundercook, J. F. 1969. ABC Inventory Classification Systems. *Production and Inventory Management*, 10: 23.

Schussel, G. 1968. Job-Shop Lot Release Sizes. *Management Science* 14: B-449.

Zangwill, W. I. 1969. A Backlogging Model and Multi-Echelon Model of a Dynamic Economic Lot Size Production System—A Network Approach. *Management Science* 15: 506.

6 Operations Scheduling

Concepts

Demand Forecasting (Chap. 3) provides an estimate of demand for each product during each operating time period. Operations Planning (Chap. 4) develops a time-phased production plan, in which resources are matched to production requirements and are smoothed over the planning period. Inventory Planning and Control (Chap. 5) regulates the flow of materials, parts, and products into, through, and out of the production system. This regulatory action results in the placement of purchase orders for purchased items and production orders for manufactured items. The operations scheduling function must now fit the production orders into the existing work loads of the various departments or operating facilities.

At this point we should distinguish between the terms *scheduling* and *sequencing*. We use sequencing to mean the determination of the *order* (first, second, etc.) in which tasks waiting at a work center are to be performed. These tasks become scheduled when a clock time has been specified for their beginnings and endings.

Still another term that occasionally causes confusion is *loading*. We use this term to mean the initial assignment of a task to a work center. Loading precedes sequencing which in turn precedes scheduling.

PURPOSE OF OPERATIONS SCHEDULING

The purpose of operations scheduling is to assign specific operations to specific operating facilities with specific start and end times indicated. A production plan spread sheet (see Fig. 4.12) may indicate that a particular facility will operate on two shifts during a certain operating period. The production plan is a gross planning document prepared far in advance. It makes no attempt to specify, for example, that Operation No. 10 on Job No. 296 will be processed on Facility No. 19 beginning at 9:32 A.M. on Tuesday, July 8, and ending at 4:05 P.M. on Thursday, July 10. Detailed activity assignments such as this are made by the operations scheduling component of our control system.

Operations scheduling attempts to assign work to the required facilities in such a way that all the various costs associated with manufacturing are minimized. These costs are caused by factors such as in-process inventory, idle men and equipment (for whatever reasons), overtime, orders completed behind schedule, etc.

To fully appreciate the magnitude and importance of the scheduling function, we must first attempt to visualize the complexity of large manufacturing operations. Orders (regular production orders from Inventory Control and special orders through Sales) are received periodically. Individual jobs are assigned to specific facilities. Certain precedence restrictions must be observed. As jobs progress through the facilities, they often must compete with other jobs for the same scarce resources. This competition results in conflicts on various facilities. Machines break down, some workers fail to appear and others perfrom below or above standard, tools break or wear out, materials are defective, and machines go idle waiting for work from preceding work centers. Orders are cancelled, reduced, or increased. Raw materials fail to arrive when expected. Sales drop suddenly or increase sharply. Marketing begins a crash advertising campaign which causes a sudden increase in demand, after which demand will oscillate sharply until the system is allowed to stabilize again. Engineering introduces product modifications which alter standard processing times, set-up times, operation sequences, operator instructions, etc.

It is very difficult to exercise close control over a dynamic environment such as that described above. Not only are there many factors contributing to the dynamic nature of the system; furthermore, they interact with each other in a mathematically complex manner. In spite of the difficulties, however, the scheduling function must be performed. The objectives of operations scheduling are:

- High percentage of orders completed on time.
- High utilization of facilities and workers.
- Low in-process inventory.
- Low overtime.
- Low stockouts of manufactured items.

These objectives must be accomplished within the over-all framework specified by the operations plan.

The scheduling period, i.e., the period of time over which a schedule is considered fixed (as far as the scheduling group is concerned), should be as short as possible. The range may vary from a few hours to a week, rarely longer. During this time there can be no equipment, personnel, or shift changes. Only emergency production orders are accepted.

An ideal scheduling period would be zero; i.e., orders would be received and assigned continuously, work in process would be reviewed continuously, and the schedule would be revised continuously in response to continuously changing conditions within the plant. This situation is not possible, however, due to time delays resulting from recording, transmitting, processing, and interpreting production data on the shop floor. At best, we can have only a good idea of the current shop status on a short-term basis. We must make our scheduling decisions on this basis.

TYPES OF SCHEDULING PROBLEMS

The nature and complexity of the scheduling function vary greatly among different manufacturing systems. Several techniques have been developed for making decisions. Some are very sophisticated mathematical procedures, while others are purely common sense approaches. Often several techniques must be used in combination to perform the scheduling function for a particular plant.

In the continuous manufacture of standardized products, most of the scheduling problems are resolved when the production line is designed. The sequence of manufacturing operations is fixed, the line is balanced, and the basic production rate is determined. During a scheduling period such as 1 week, the only scheduling decisions that can be made are whether to operate the process at all and, if so, for how many hours, whether to use overtime, whether to increase or decrease the number of shifts, whether to subcontract part of the work, and whether to change the basic production rate by reducing or increasing the labor force and then rebalancing the production line accordingly.

At the other extreme of scheduling complexity is the pure job shop. A very large variety of job-types are received at unpredictable times. The job shop must maintain many different kinds of raw materials since it does not know what it will be asked to manufacture next. It must also maintain general purpose machines and employ skilled operators capable of performing many diverse jobs. The scheduling function must allocate the shop's available resources to the various jobs coming in.

Even though the five objectives of scheduling discussed previously also apply to job shops, the results are usually the opposite. It is common for orders to be completed late. It is almost impossible to obtain high utilization of facilities and workers. High in-process inventories are common. Overtime work is often necessary. Finally, stockouts of raw materials are common.

In the job-shop system none of the scheduling decisions is designed into the manufacturing process because a unique manufacturing process must be designed for each job coming into the shop. This design consists of the determination of the sequence of general purpose machines necessary to manufacture the product.

Between the extremes of continuous manufacturing and the pure job shop is another type of manufacturing, called intermittent production. In this type of manufacturing, only a known set of products, perhaps with modifications, are relevant. Often many of the products are similar in nature and are processed on the same facilities. The sequence of manufacturing operations for each product is usually fixed, except for special orders involving custom modifications. The demand for each product is usually known fairly well.

Although it is possible for an intermittent production system to produce only to customer order (such a system would have many of the characteristics of the job-shop system), the more usual case is to fill customer orders from inventory. Finished goods inventories are replenished periodically from batch runs. Economic batch sizes were discussed in Chap. 5. Production orders received from Inventory Control usually call for a quantity equal to the economic batch size. These are the orders which must now be scheduled into the existing shop load.

The scheduling transfer functions developed in this chapter are applicable to both intermittent and job-shop production situations. We continue to emphasize intermittent production, however, as we have done throughout the book.

BASIC PROBLEM IN OPERATIONS SCHEDULING

Several difficulties are encountered when we attempt to develop a detailed schedule in which specific tasks are assigned to specific facilities on a detailed time basis. These difficulties can be illustrated by considering an example.

Suppose we receive three production orders that are to be scheduled onto three of our facilities. Each order consists of several operations, or tasks. The operation sequence (the order in which operations must be performed on the several facilities) for each production order has been determined and is considered fixed. The three production orders are shown in Table 6.1. The operations are listed in the required sequence for each order. For each operation, the process time in hours and the facility number to which the operation must be assigned are shown.

We have 13 operations to be scheduled to the three facilities. Our objective is to schedule the operations such that the production orders are completed as soon as possible. We must observe the operation sequences shown in Table 6.1 for each order.

Most scheduling periods begin with previously placed production orders already scheduled onto certain facilities. Suppose, for example, that Facilities 1,

Table 6.1 Three Production Orders to be scheduled onto Facilities 1, 7, and 12

Production Order Number	Operation Sequence	Process Time	Facility Number
96	*C*	10	12
	D	20	1
	A	10	7
	B	20	12
	E	10	1
122	*A*	10	7
	B	20	12
	D	10	1
	C	10	7
201	*B*	20	1
	C	10	7
	A	10	12
	D	20	1

7, and 12 already have scheduled the operations shown in Fig. 6.1. This figure is simply a bar chart on which the horizontal scale represents time and the vertical scale has no meaningful unit of measurement. Time zero on the horizontal scale represents the beginning of the scheduling period.

We see that Facility 1 has two previously assigned operations, each requir-

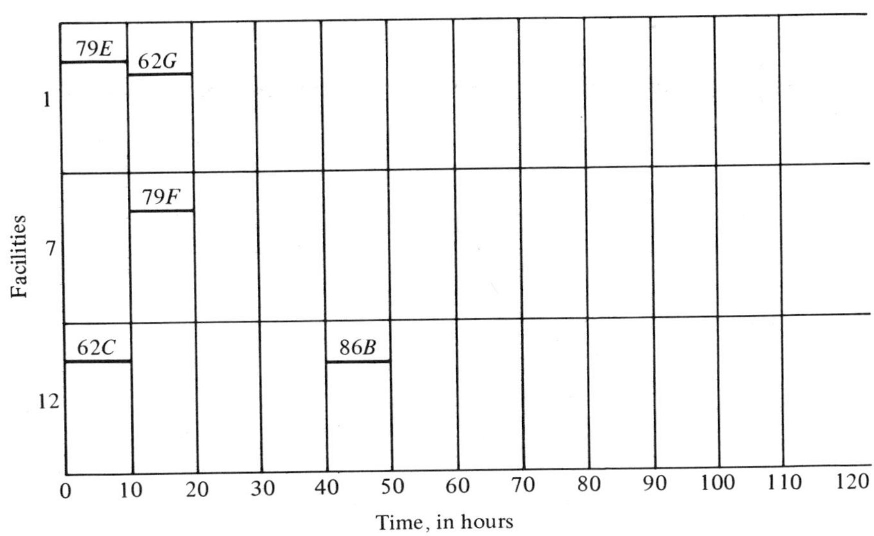

Figure 6.1 Schedule Bar Chart showing existing Work Load prior to new Operations

ing 10 hours of process time. The notation 79*E* means "operation *E* on production order 79." We usually try to schedule new operations onto the facilities such that the previously scheduled operations remain unchanged. This is not necessary, however, and we may very well wish to give higher priority to certain new operations than to certain existing ones.

Let us now place our new operations onto our schedule bar chart. Initially we ignore existing operations and simply show each new operation in the desired position. The results, called the initial load, are shown in Fig. 6.2. Note that the required operation sequences shown in Table 6.1 have been observed.

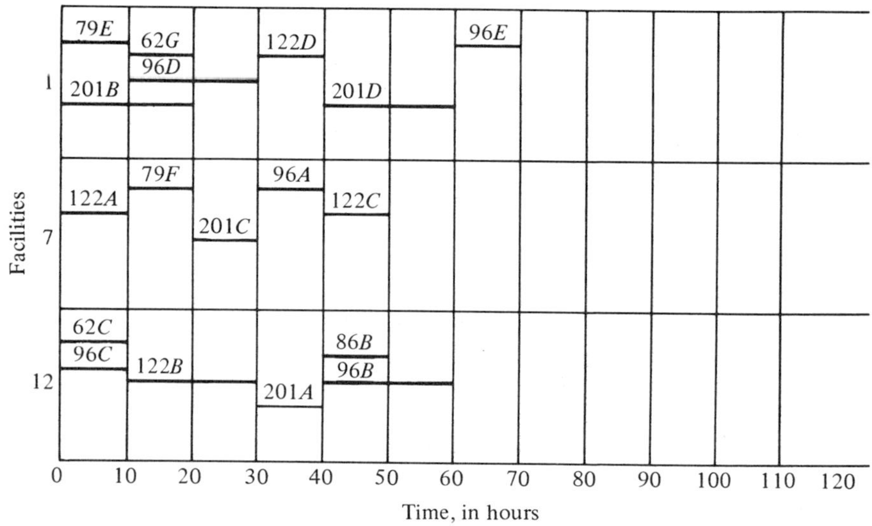

Figure 6.2 Schedule Bar Chart showing "Initial Load," consisting of previously scheduled Operations and new Operations

Let us define the relationship between operations and facilities to be such that only one operation at a time may be performed on a facility. We see in Fig. 6.2 that more than one operation is assigned to the same time interval. This condition is known as a schedule conflict. A decision must be made as to which operation to perform first, which second, etc. This process is called resolving the conflicts.

When all conflicts have been resolved (when we have eliminated all operation overlaps on all facilities) within the constraints of the original operations sequence for each production order, we have generated a feasible schedule. A feasible schedule is one that will work. It may not be the best schedule, or even a good one, but it can be executed.

The process of resolving schedule conflicts is more complex than it first appears to be, primarily because of the interdependencies between facilities caused by the required operations sequences. We see in Fig. 6.2 that no conflicts exist on Facility 7. It is tempting to conclude, therefore, that the five operations can be performed during the time intervals to which they are now assigned. We quickly see, however, that schedule resolutions in Facility 1 will force movement of some of the operations in Facility 7 due to the required operations sequence.

Regardless of how we choose to resolve a particular conflict, one of the overlapping operations must be displaced to the right, thus causing a time delay. When this operation is delayed, the starting time of the following operation on the same production order (but at a different facility) is delayed by a corresponding amount. For example, when we resolve the first schedule conflict at Facility 1 (the four operations assigned to times 0 to 30), either 201*B* or 96*D*, or perhaps both, are delayed. Suppose we arrange the four operations in the following order, with time shown in parentheses: 79*E* (0—10), 62*G* (10—20), 201*B* (20—40), 96*D* (40—60). Operation 201*B*, originally assigned to begin at time 0, now is scheduled to begin at time 20 and end at time 40. From Table 6.1 we see that the next operation on production order 201 is 201*C*. It cannot begin at Facility 7 until 201*B* is completed at Facility 1, which is time 40 rather than time 20 as originally assigned. Similar adjustments are required in all remaining operations on production order 201. Operation 96*D* is delayed even more. All operations following *D* on production order 96 must be displaced 30 time units.

A question that naturally arises at this point is how should the conflicts be resolved? Even with a schedule as small as that shown in Fig. 6.2, the complexity of the scheduling problem is almost overwhelming. Intuitively, we might be tempted to try all possible conflict resolutions and select the best one. It can be shown that $(7!)(5!)(6!) = 435{,}456{,}000$ different resolutions are theoretically possible for this small problem. Complete enumeration of all possible solutions is clearly not a workable procedure.

Another basic question that needs to be asked at this point is what is a good schedule? How would we know one if we saw one? By what criterion do we measure the "goodness" of a schedule? We might be seeking the schedule which results in the most operations being completed during the scheduling period. We might be trying to minimize idle time on our facilities. We could state several equally plausible criteria by which to measure our schedule. It is important to recognize that a schedule that satisfies one criterion does not necessarily satisfy other criteria. For example, our schedule could be loaded so heavily that very little idle time results. However, this schedule may be contrary to the other criterion of maximizing the number of operations completed. The criterion used should relate directly to the objective we are trying to accomplish.

We defer to the Transfer Functions section further discussion of scheduling procedures. The primary purpose of this section has been accomplished by bringing out the complex nature of the scheduling problem. And yet, the total problem is still more complex. Also to be considered are the possibility of variable process times (rather than known amounts, as we assumed above), nonavailability of workers or materials, equipment breakdowns, emergency orders, and other factors.

SCHEDULING DECISIONS

From the preceding discussion we now can state explicitly the decisions that the scheduling function must render (see Fig. 6.3).

Load Facilities. Production orders are extended into specific tasks. Manpower, machine and material requirements are determined. Tasks are assigned to facilities. Where alternate assignments are permitted, consideration is given to existing work loads on alternative facilities. All production orders, planned and unplanned, must go through this step.

Evaluate Work Load. Aggregate requirements for manpower, machine time, and materials are determined for each facility and then compared to the corresponding capacities. Depending upon the result, decisions are made regarding change of work force size, change of production rate, amount of overtime, change in number of shifts, and whether to use subcontract support. All these decisions must conform to the over-all operations plan, or approval must be obtained for significant deviations.

Raw material and subassembly requirements are checked. If sufficient quantities to produce the requested order size are not in inventory, a decision must be made whether to produce a partial order. In all cases, materials and subassemblies allocated to approved production orders are no longer available for other uses.

Sequence Tasks at Each Facility. All tasks now assigned to each facility must be arranged in the order in which they are to be accomplished. Precedence relations among the tasks from the same production order must be observed. Any priorities in effect must also be observed. Only those tasks for which sufficient raw materials exist are sequenced.

Develop a Detailed Schedule. All of the above decisions now are put on a time scale. The schedule must be manipulated (by one of the transfer functions discussed later) until no time conflicts exist. When the schedule is free of all conflicts it is ready for release to Dispatching.

Figure 6.3 represents the total process of arriving at the scheduling decisions discussed in this section. This procedure must be executed prior to each scheduling period. Note how very important it is to have accurate and timely feedback of status information from the shop floor.

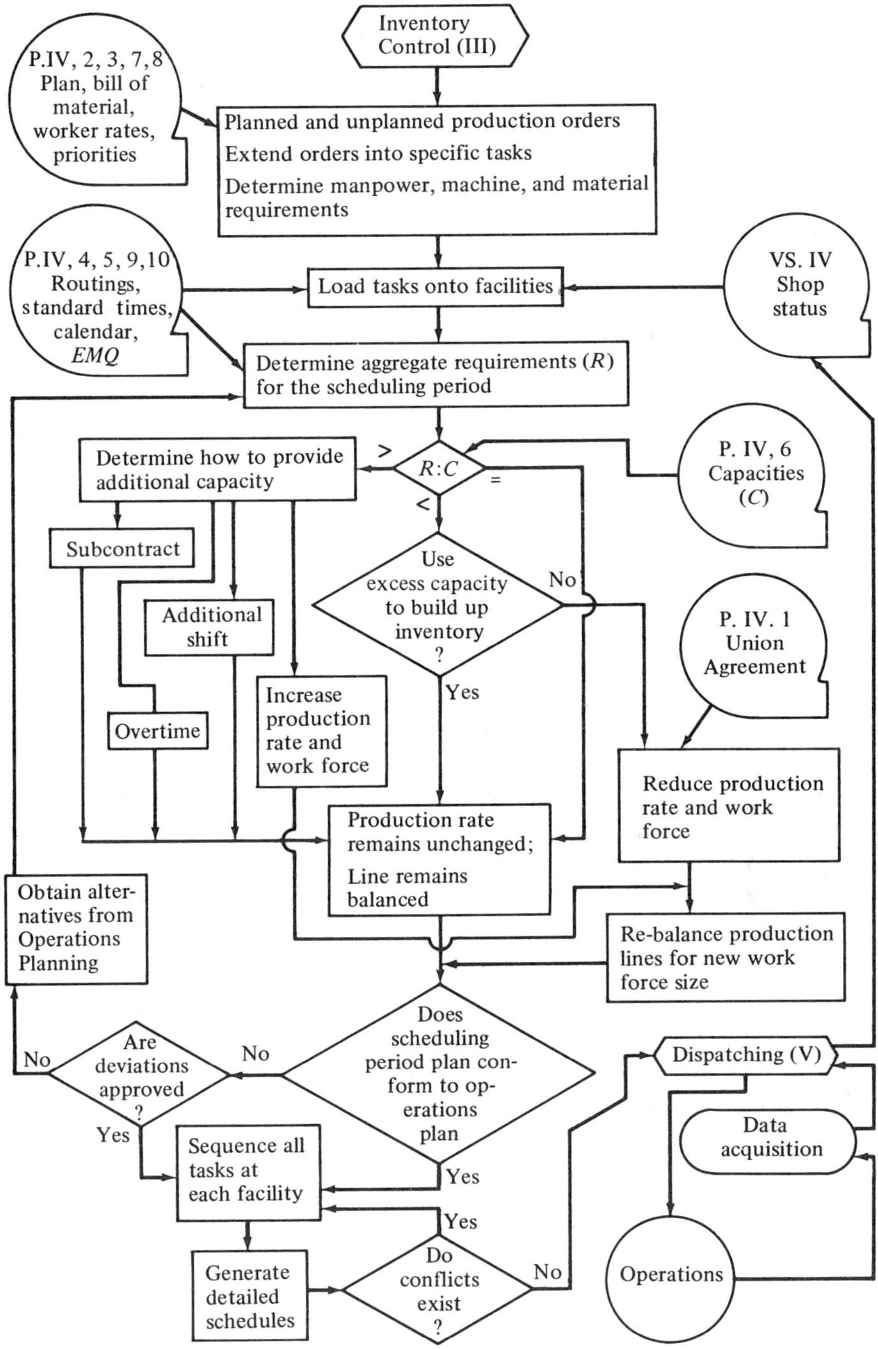

Figure 6.3 Overview of the Operations Scheduling Function

Inputs

The operations scheduling function requires information from several sources. In our usual manner, we present the required information according to the two basic types, parameters and variables. Much of the information required for the short-term scheduling period is similar to (but more detailed than) the information required for operations planning.

PARAMETER FILE FOR SCHEDULING (P.IV)

P.IV.1 Union-management contract on overtime, layoff policies, holidays, employee allowances, etc. Source: management.

P.IV.2 Time-phased operations plan. Source: Operations Planning.

P.IV.3 Bill of materials for each order. Source: Engineering.

P.IV.4 Standard and alternative routings of orders through the manufacturing facilities. Includes current line-balances with associated cycle times and production rates. Source: Engineering and Operations Planning.

P.IV.5 Process times and set-up times for all components to be manufactured and for all alternative routings. Source: Engineering.

P.IV.6 Capabilities and capacities of all facilities. Source: Engineering.

P.IV.7 Skills and production rates of all employees. Source: Engineering.

P.IV.8 All priority rules and priority assignments. Source: management.

P.IV.9 Detailed company calendar for the scheduling period, including holidays or any special situation. Source: management.

P.IV.10 Economic manufacturing quantities (*EMQ*). Source: Inventory Planning.

VARIABLE STATUS FILE FOR SCHEDULING (VS.IV)

VS.IV.1 Progress of production orders in process: percent complete, number of units completed, quality records, actual completion times, material shortages, due dates, remaining process time.

VS.IV.2 Status of production facilities: queue sizes, occurrences of tool breakages or equipment breakdowns, idle man and machine time, scrap and rework records, number of set-ups.

VS.IV.3 Operator performance: absenteeism, number of workers on vacation, availability of particular skills, productivity records.

UPDATING THE FILES

The scheduling function simply cannot be performed properly without current and accurate information. Perhaps no other function in our control system is as sensitive to the accuracy and timeliness of updated information files.

The parameter file is kept current in much the same manner for scheduling as it is for other functions. Particular emphasis should be placed upon such things as changes in materials or methods, new or modified equipment, and new skills acquired by individual workers.

The values in the variable status file are kept current through a well-designed data acquisitions and progress control system. Traditionally, this has been the weakest component in the control system. This problem is discussed further in Chap. 7.

Transfer Functions

APPROACHES TO SCHEDULING

A great number of techniques and procedures have been developed to perform the scheduling function. These approaches vary greatly in sophistication and generality. At one extreme are the straight-forward, common sense approaches, applicable only to a small number of situations, perhaps just one company. At the other extreme are the highly sophisticated approaches intended to be very general and thus applicable to a wide set of scheduling situations. In fact, these approaches are so limited by restrictive assumptions that they can be applied directly in very few real world situations.

Between the two extremes are only a few basic scheduling techniques both useful and generally applicable. Most scheduling methods actually in use are still very much "cut and try" procedures. Much development remains to be done in this area.

The several approaches to scheduling may be categorized into four general groups: analytical, iterative, heuristic, and charting.

Analytical Approach. Several attempts have been made to structure the scheduling problem in a formal mathematical model. This aim has been accomplished only for trivial problems involving at most three machines. A recent book by Conway, Maxwell, and Miller (1967) contains much material that should contribute to a general scheduling theory.

Iterative Approach. In this approach, all possible combinations of task sequences at each facility are tried and the best combination is chosen. Conceptually, this approach is good, but it is not practical because of the tremendous amount of time and computational effort required. In general there are

$$\prod_{i=1}^{m} [k!]_i$$

theoretically possible combinations of task sequences, where k is the number of tasks assigned to the ith facility and m is the number of facilities.

Heuristic Approach. In this approach, logical decision rules or computer simulation models are used to arrive at a schedule. Whereas the optimum is usually not obtained with this approach, consistently good schedules are obtained with reasonable amounts of computational effort and time.

Charting Approach. This approach has been used in all sizes and types of plants and usually is in some form of bar chart called the Gantt Chart.

We limit our treatment of scheduling transfer functions primarily to the heuristic and charting approaches, which have worked in practice because of their simplicity and flexibility. The analytical and iterative approaches have not been used extensively in industry, although they have considerable merit from an academic and theoretical point of view.

It appears that the best practical method of performing the scheduling function is still the human scheduler aided by charts, decision rules, and perhaps a computer simulation model.

Scheduling Assumptions. There are several assumptions usually made in most scheduling techniques. The degree to which these assumptions restrict the usefulness of the generated schedule is different for different manufacturing systems. In most cases, a useful schedule results despite the assumptions. In any case, it is important that we recognize what the assumptions are. The more common ones are:

- Processing times for operations are deterministic and known.
- Set-up times are deterministic and known.
- Facilities and operations are defined such that no overlap of operations on facilities is allowed.
- Processing times are independent of the order in which they are performed.
- Each operation, once started, must be completed before another operation can be started on that facility.
- Facilities never break down.
- Due dates of orders are known and fixed.
- Operation sequences are fixed.
- Sufficient space is available for storing orders waiting to be processed on the several facilities.

TABULAR AND GRAPHICAL TRANSFER FUNCTIONS

Tabular and graphical methods have been used for many years in the loading of work orders onto operating facilities and in the detailed scheduling of work operations. These methods are still useful transfer functions, particularly for relatively small manufacturing operations. These methods also provide a useful framework within which the basic concepts of loading and scheduling may be discussed.

Loading Charts. When production orders are issued by Inventory Control, they require the expenditure of our available resources: machine time, manpower, and materials. Our resources are limited for a given time period; this limitation is called *capacity*. As new orders are received a certain amount of the available capacity must be reserved for the completion of the orders. It is important that we maintain accurate records of capacity reserved for existing production orders at each facility. The difference between total capacity and reserved capacity at a facility is the amount of *remaining capacity* for the time period concerned.

Perhaps the simplest method of maintaining current facility load records is to enter each new production order into a table such as that shown in Table 6.2 (Scheele, Westerman, and Wimmert; 1960). The table must be revised just prior to a scheduling period. As work is completed on each operation, the remaining work, in hours, is entered into the table. In this way, the total load on each facility is determined on a periodic basis.

Table 6.2 Loading Chart: shows Total Hours of assigned but uncompleted Work at each Facility

Production Order Number	Facility 1	Facility 2	Facility 3
78		10	7
79	20		
84	16		5
88		12	
89	24	8	
104			6
etc.			
Total Hours Remaining	60	30	18

The loading chart in Table 6.2 is useful for reflecting the total amount of each facility; a major weakness is that it does not show *when* the various operations are to be performed. For example, we do not know what proportion, if any, of the 30 hours of work assigned to Facility 2 must be performed during the next scheduling period. This weakness can be overcome by maintaining a capacity-load chart, such as the one shown in Fig. 6.4. Such a chart would be required for each facility. The information going into the chart is derived from the detailed schedule, the generation of which is discussed in the next section. We see, then, that loading and scheduling constitute a circular decision process.

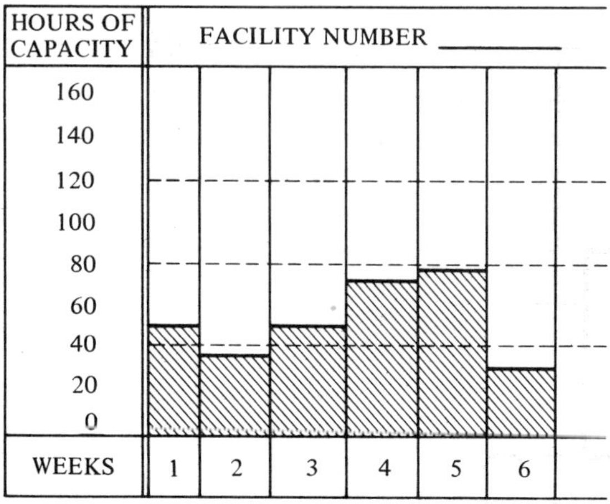

Figure 6.4 Capacity-Load Chart, showing actual hours of work assigned to this Facility for several future time periods

There are many possible methods of portraying facility loads. In general, the most useful ones are those that show capacity and load on the same form.

Note that in the preceding discussion of loading, it was assumed that the assignment of operations to facilities was predetermined. Whenever operations can be assigned to more than one facility, some means must be provided to make the assignments in the proper manner. In general, we want to assign operations to facilities that can perform the operations in the least amount of time. However, we must also consider the relative loads of two or more alternative facilities. A procedure for making such assignments is discussed in the section on decision rules and heuristic procedures.

Gantt Charts. One of the oldest techniques available for sequencing and scheduling operations to facilities is the Gantt Chart. The Gantt Chart is quite useful in showing planned work activities versus actual accomplishments on the same time scale. The schedule bar chart in Fig. 6.2 can be considered an elementary form of a Gantt Chart. There are numerous possible variations of the Gantt Chart. Many are color coded to indicate certain conditions such as material shortages, machine breakdowns, etc.

The basic Gantt scheduling chart is shown in Fig. 6.5. Operations assigned to each facility are sequenced into a feasible schedule. Actual progress is indicated on the same chart to show any deviations from planned operation process times. The status of the facilities can be read from the chart. Operations 44*A* and 65*D* have been completed. Operation 65*E* is exactly on schedule as of now. Operation 44*C* is a half day behind schedule. Operation 86*B* is 2 days ahead of schedule. Operations 86*D* and 103*C* are scheduled but have not yet begun. We should recognize that the time scale may be in any appropriate time units—hours, days, weeks, or months.

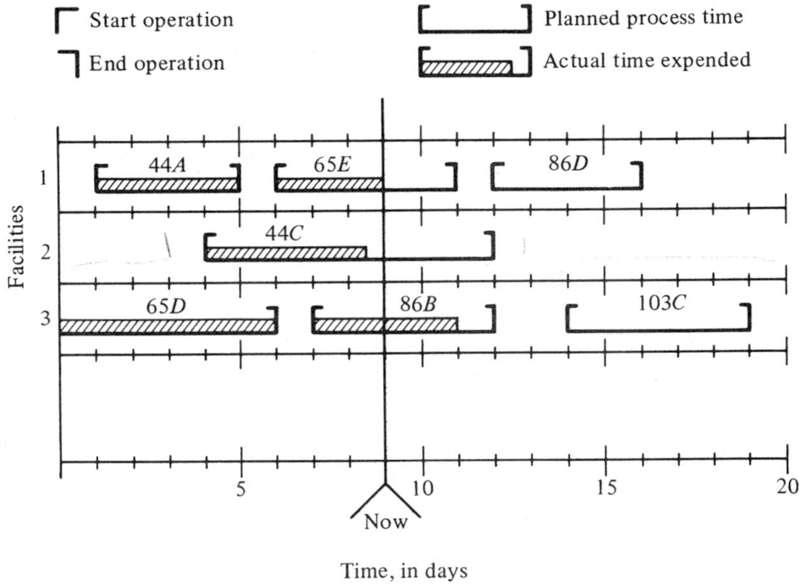

Figure 6.5 Gantt Scheduling Chart

When the Gantt Chart is used to sequence several operations resulting from new production orders, a good rule to follow is to sequence the most heavily loaded facility first, the second most heavily loaded facility second, etc. Although a fairly good schedule can be generated in this way, it will not be optimal under any criterion except by accident.

The Gantt scheduling chart must be updated very frequently in order to reflect the changing status of the shop. Certain jobs will take longer than expected, while others will be completed ahead of schedule. Machines will break down, workers will be absent, and materials will run out. All these factors make it difficult to keep the Gantt Chart updated on a timely basis. Nevertheless, the Gantt Chart should not be overlooked as an aid for loading, sequencing, and scheduling.

Production Flow Chart. Another graphical transfer function often useful in scheduling is the production flow chart. These charts are discussed in Chap. 4, and examples are shown in Figs. 4.9 and 4.10.

Cathode Ray Tube Displays. Although the authors are unaware of any such application, it seems reasonable that a cathode ray tube connected remotely to a digital computer could be made to project a Gantt Chart-type schedule showing current shop status, existing load, etc. New operations could be loaded onto a particular facility with a light-pen input device, and the resulting effects on the remainder of the facilities could be noted immediately. The ability to interact with the computer in this way would be of great assistance in developing a good schedule. It might very well be that such an approach could overcome much of the awkwardness of the Gantt Chart.

DECISION RULES AND HEURISTIC PROCEDURES

The charting techniques described in the previous section become very clumsy in large manufacturing operations. Let us recall the example used in the first part of this chapter in which 18 operations were assigned to three facilities. The initial load is shown in Fig. 6.2. Even for such a small load, we found the task of scheduling the operations quite formidable. If we were faced with the problem of scheduling, say, 10,000 operations to 100 facilities, we would quickly become buried in the details of the problem.

In an effort to obtain practical scheduling procedures, much time has been directed toward the development of logical decision rules. Most of these rules are based upon an underlying heuristic rationale; most of them are designed for computer application. We discuss some of these rules and procedures in the following paragraphs.

Index Method for Loading. Sometimes, several different facilities can perform similar operations. If this is the case, we must decide which operations should be assigned to which facilities. This situation occurs almost exclusively in job-shop (make-to-order) systems.

Suppose, for example, that we have five operations to be performed and that three facilities are available for performing these operations. A particular facility can perform a particular operation faster than can other facilities, whereas it may perform other operations slower than can the other facilities. The processing times for the several possible combinations are shown in Table 6.3. Where no value is entered, that facility cannot perform the particular operation. The hours available for new assignments are shown on the bottom row for each facility.

Table 6.3 Processing Times in hours for five Operations on three Facilities

Operation Number	Facility 1	Facility 2	Facility 3
22*C*	100	150	125
27*A*	200	100	220
44*G*	25	50	20
32*B*	40	30	—
51*E*	60	50	70
Hours Available	160	110	150

Ideally, we would like to assign each operation to the particular facility which can perform the operation in the shortest period of time. In our example, we would like operations 27*A*, 32*B*, and 51*E* to be performed on Facility 2. This is not possible, however, since the total processing time for the three operations is 180 hours and only 110 hours are available for the scheduling period. When an operation cannot be performed on the fastest facility, we would like to assign it to the facility with the next lowest processing time. The Index Method of loading provides a systematic procedure for assigning the operations to the facilities in accordance with the above reasoning. This method will not generally give an optimum solution in terms of minimum total processing times. It will, however, give reasonably good loads. For large problems, it will usually give better results than can be obtained by intuition or visual inspection.

A step-by-step procedure for using the Index Method follows:

Step 1. Given a set of operations to be performed within a scheduling period, and the hours available on each facility during that period, construct a table such as Table 6.3.

Step 2. Determine the efficiency of each facility for each operation. The facility with the lowest processing time for a given operation is assigned an index of 1.00. The facility with the next lowest processing time for that operation is assigned an index value equal to the ratio of the processing time on this facility to the processing time on the best facility. Continue until all facilities have been assigned an index value for each operation. Enter these index values into a table such as Table 6.4.

Step 3. Operations are now assigned to facilities on the basis of lowest index, if sufficient time is available. If sufficient time is not available,

Table 6.4 Index Method for Loading

Operation Number	Facility 1		Facility 2		Facility 3	
	Hours	Index	Hours	Index	Hours	Index
22*C*	(100)	1.00	150	1.50	125	1.25
27*A*	200	2.00	(100)	1.00	220	2.20
44*G*	25	1.25	50	2.50	(20)	1.0
32*B*	(40)	1.33	30	1.00	—	—
51*E*	60	1.20	50	1.00	(70)	1.40
Hours Available	160		110		150	
Hours Assigned	100 40 140		100 100		20 70 90	

assign operations to the facility having the next lowest index, again if sufficient time is available. Continue until all operations have been assigned to facilities. When an operation is assigned, parentheses are placed around the number of hours required on that particular facility and time is entered in the bottom row.

Step 4. (Optional) It may be possible to improve the load by visual inspection and manual manipulation. It may be desirable, for example, to spread the work out over the several facilities, particularly if the Index Method loaded some facilities much more heavily than others.

The last step is included to indicate the necessity for exercising human judgment in conjunction with heuristic procedures. Heuristic procedures are not optimizing procedures; they are aids to human judgment and are not to be followed blindly.

Note that the Index Method could be used with costs rather than with processing times in Tables 6.3 and 6.4. All pertinent costs, such as set-up time, processing time, labor, scrap, etc., would be included. The same procedure would apply.

Decision Rules for Scheduling. Considerable attention has been given in recent years to the development of logical decision rules for scheduling operations to facilities. Examples of research studies directed toward these developments can be found in Conway (1964); Conway, Maxwell, and Miller (1967); and Muth and Thompson (1963).

We are almost forced to use some kind of decision rule for scheduling. We have already discussed the complexity of the scheduling problem and the fact that little progress has been made toward analytical solutions. Decision rules provide a practical way of arriving at good schedules, but they do not, in general, generate optimal schedules.

Before it is meaningful to discuss specific decision rules, it is first necessary to discuss the objectives of the scheduling function. In the discussion associated with Fig. 6.2, we asked, "What is a good schedule?" Perhaps we can approach this question by speculating on what the ideal schedule would be.

Clearly, an ideal schedule would be one which resulted in: (1) all production orders being completed by their due dates, (2) zero idle time on each facility, and (3) no work-in-process inventory building up between facilities. Each of these factors has an associated cost that is unwanted but usually unavoidable.

If we cannot generate an ideal schedule, then we want to generate one as close to ideal as possible. Our objective, therefore, is to generate schedules that minimize the sum of the costs caused by lateness, idle time, and work-in-process inventory.

Many decision rules that attempt to accomplish the objective just stated have been developed. One study (Conway 1964) investigated 92 different rules, many of which were combinations of others. Presented in the following paragraphs are some typical decision rules. They were chosen to demonstrate the types of rules that have been considered. Most of them are designed to try to minimize only one of the three cost factors mentioned earlier.

First Come First Served. Operations are placed in line as they arrive at a facility and are processed in sequential order. This rule is probably the most commonly used in practice. It is by far the easiest to implement and performs quite satisfactorily for many situations.

Last Come First Served. This rule is very similar to the first come first served rule. It is useful in situations in which operations are stacked while waiting for processing. The top operation is done next because it is the most convenient. There is a danger, however, that operations near the bottom of the stack will wait for an excessive amount of time before getting processed.

Shortest Operation First. From the operations waiting to be processed at a facility, this rule selects the operations having the shortest processing time to be performed next. This rule tends to minimize idle time on facilities, particularly when work must flow through several sequential facilities. An obvious disadvantage is that long operations are penalized and can get hung up in the system for a long time. This problem can be overcome by imposing a limit on the amount of time an operation is allowed to wait at one facility.

Longest Operation First. From the operations waiting to be processed at a facility, this rule selects the operation having the longest processing time to be performed next. This rule is useful when the plant is operating at or over capacity. It may also be possible to use this rule as a guide in subcontracting or refusing smaller or less profitable jobs.

Least Time to Due Date. The operation having the earliest planned due date is processed next. This rule tends to minimize over-all lateness of orders. (Note that orders from all customers are given equal priority, which may be undesirable.)

Minimum Slack Time Per Operation. For each operation waiting to be processed at a facility, the remaining processing time (at other facilities) of this order is subtracted from the total time remaining before the planned due date. This sum is then divided by the number of operations following this operation on this total order. The operation having the smallest such value is assigned next. This rule was found to work very well in a large job shop (Le Grande 1963).

Random Selection. The next operation to be processed at a facility is chosen randomly from those waiting. This rule does not perform well in most situations.

Choosing and Maintaining a Decision Rule. Most of the rules discussed above work better for one kind of production environment than for another. The choice of a decision rule for a particular situation depends upon the criterion being used, the size of the shop, the nature of order arrivals, the size of the orders, the similarity of the products ordered, and other like factors.

As conditions change, the rule may need to be changed or at least modified. The performance of the rule should be monitored carefully by maintaining periodic records on such factors as percent of the orders completed by their due dates, idle time on facilities, and accumulation of work-in-process inventory.

Resolving Schedule Conflicts with Decision Rules. Let us now consider how schedule conflicts, such as those shown in Fig. 6.2, are resolved by a scheduling decision rule. We illustrate the method by resolving the conflicts shown in Fig. 6.2 using the first come first served rule.

A convenient procedure for resolving schedule conflicts is to establish a time index T and move it from left to right across the schedule bar chart: all conflicts on all facilities must be resolved before T is incremented. Conflicts are resolved according to the particular decision rule being used.

Referring to Fig. 6.2, we see that a conflict exists in Facility 1 at $T = 10$. Operations 79*E* and 201*B* are both occupying the space 0–10, whereas Facility 1 can work on only one operation at a time. Since 79*E* was a carry-over from a previous schedule assignment, it is selected to be performed first by the first come first served rule. Operation 201*B* is moved 10 time units to the right so as to clear the space 0–10 of conflicts on Facility 1. This movement is shown in Fig. 6.6. The dashed line to the left of 201*B* indicates how far it was moved.

We must immediately investigate what effect this movement has had upon the other operations of production order number 201. From Table 6.1 we see that three other operations must follow 201*B*. These must also be moved to the right 10 units. These movements are also shown on Fig. 6.6.

We now determine whether any conflicts exist at other facilities at $T = 10$. No conflict exists at Facility 7, but there is a conflict between 62*C* and 96*C* at Facility 12. The resolution of this conflict and subsequent movements of other operations following 96*C* are shown in Fig. 6.7. We now have a conflict-free schedule through $T = 10$.

T is now increased to the time value at which the next conflict exists in any facility. In this case, conflicts exist in Facility 1 and Facility 12 at $T = 20$. The resolutions are shown in Fig. 6.8.

This process is continued until all conflicts have been resolved. Figures 6.9 through 6.11 show the remaining iterations.

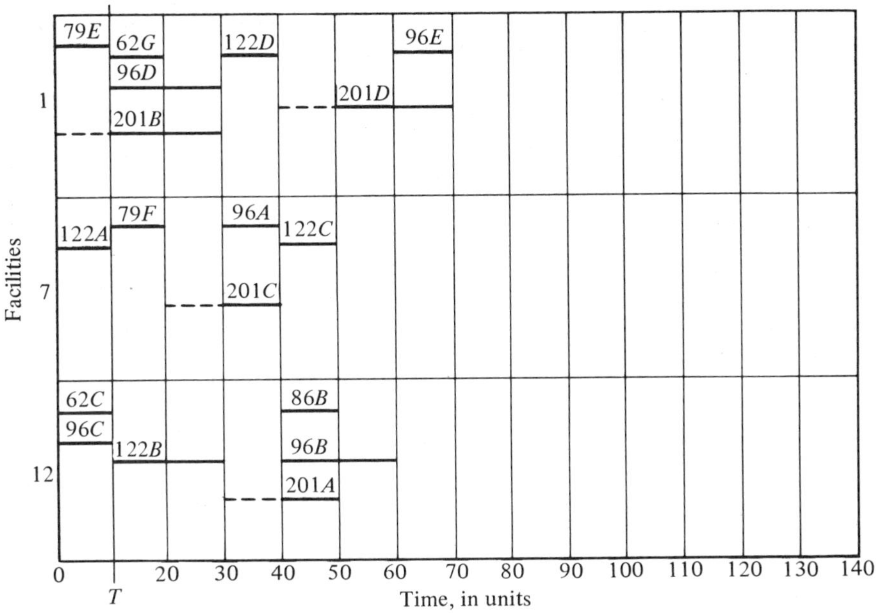

Figure 6.6 First Iteration in resolving Schedule Conflicts from Initial Load in Figure 6.2. Only the Conflict in F1 is resolved

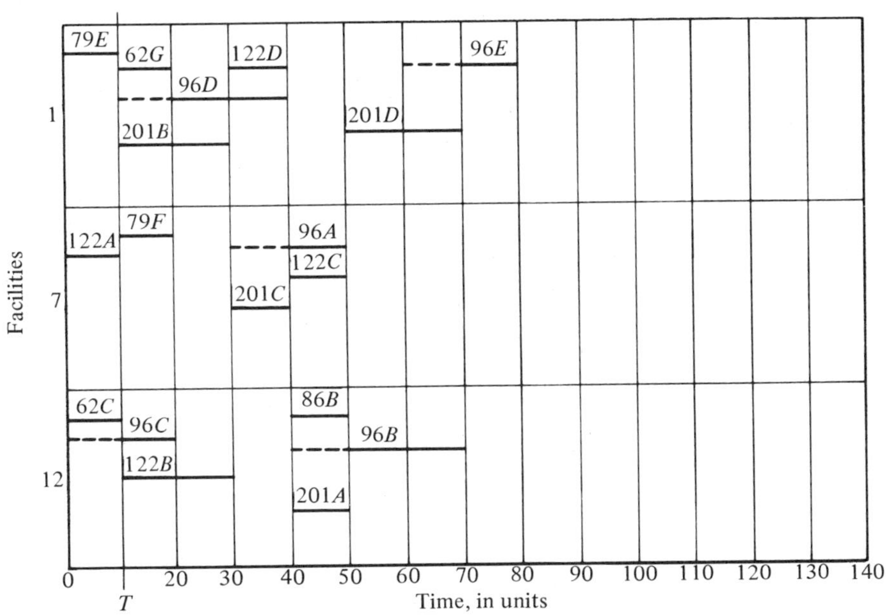

Figure 6.7 Second Iteration. All Schedule Conflicts at $T = 10$ are resolved

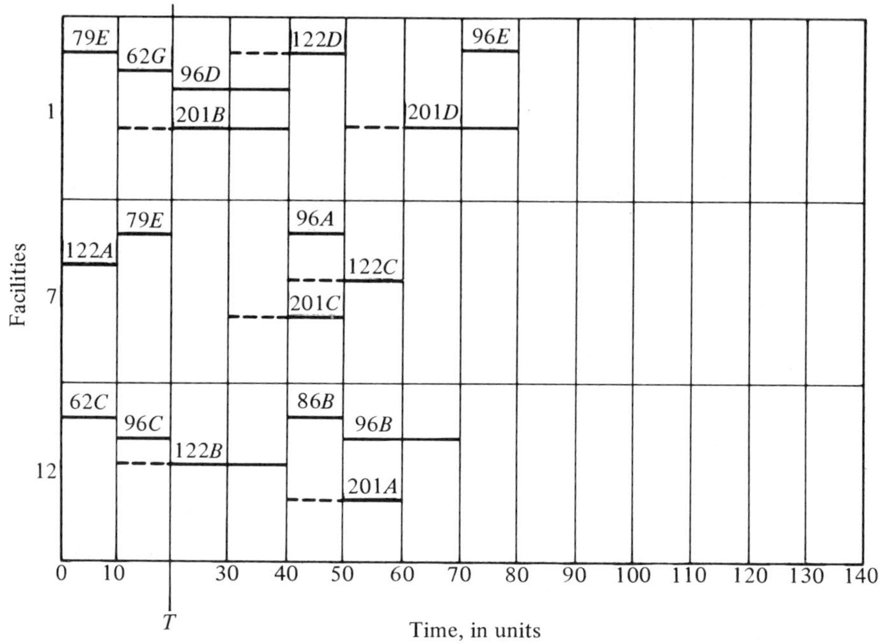

Figure 6.8 Third Iteration. All Schedule Conflicts at $T = 20$ are resolved

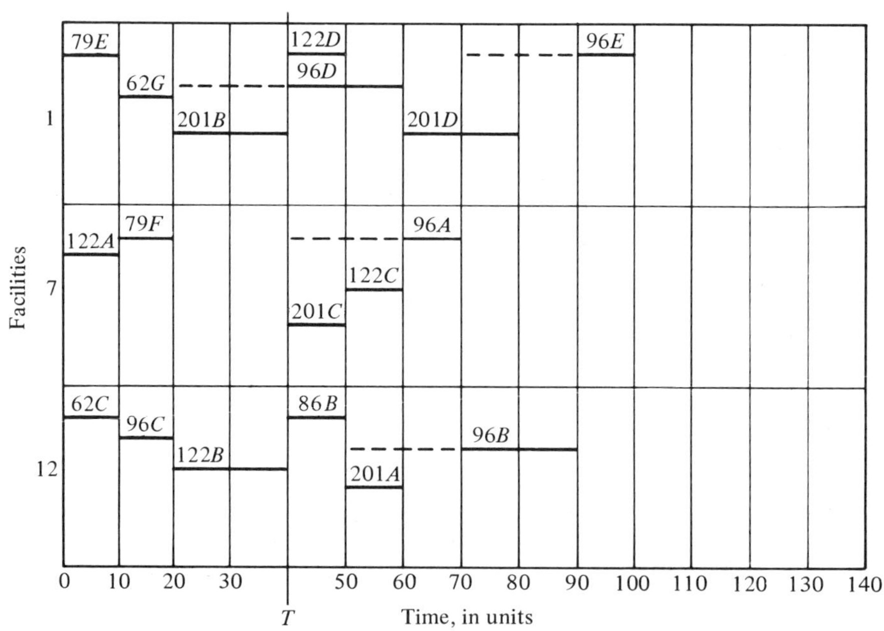

Figure 6.9 Fourth Iteration. All Conflicts at $T = 40$ are resolved

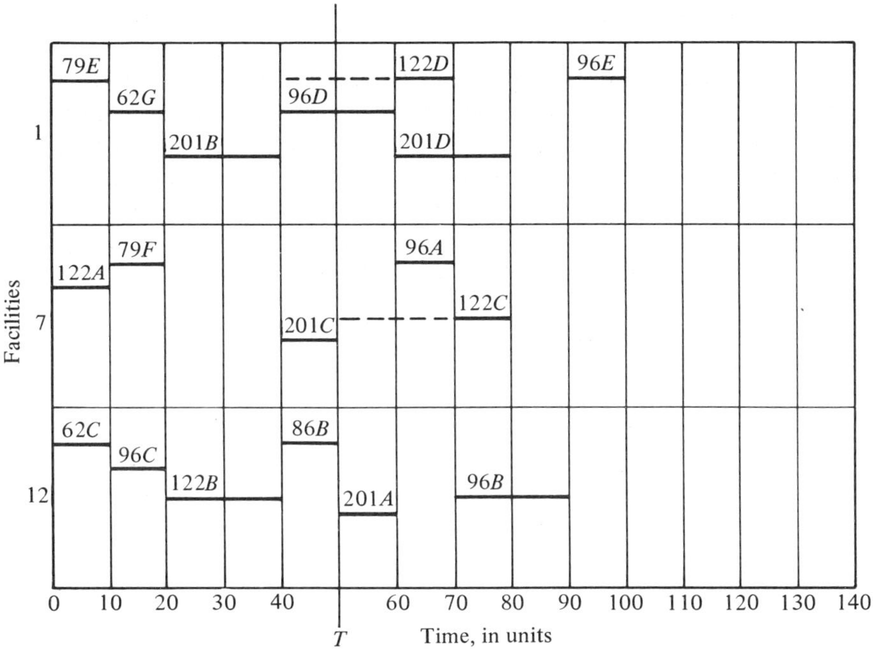

Figure 6.10 Fifth Iteration. All Conflicts at $T = 50$ are resolved

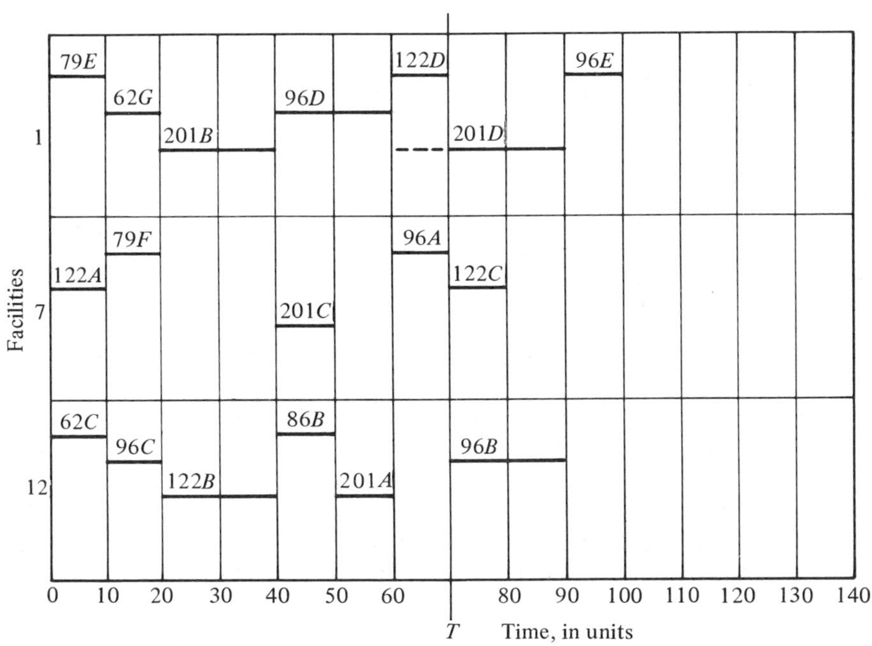

Figure 6.11 Final Iteration. All Schedule Conflicts are resolved

The final schedule in Fig. 6.11 represents a feasible schedule, which means that no schedule conflicts exist on any of the facilities and that the technological sequencing for each production order, shown in Table 6.1, has been observed. The feasible schedule is not necessarily optimal, and we have no way of determining how nearly optimal it is. The only measures of performance we have are the amount of schedule delay of the production orders and the amount of idle time on the facilities. Schedule delay for each production order is defined as the difference between the ideal ending time (from Fig. 6.2) and the actual ending time (from Fig. 6.11) for the last operation in each order. These differences are summed to give us the total schedule delay. In our example, the total schedule delay for the three production orders is $30 + 30 + 30 = 90$ time units. Total idle time on the three facilities is $0 + 30 + 10 = 40$ time units. Only idle time between scheduled operations is counted.

The procedure presented above for resolving schedule conflicts can be applied using any decision rule desired. The reader is invited to use the shortest operation first rule with the procedure and compare the results with those obtained with the first come first served rule.

SCHEDULING BY DISPATCHER

The methods of scheduling discussed above are examples of centralized scheduling systems. In such systems, all start and stop times of individual operations at specific facilities are specified by a central scheduling group. A decentralized scheduling system would be one in which operations are loaded onto facilities by a central group and sent via a dispatcher to the foremen in charge of the several facilities. The dispatcher reviews the current workload with the foremen and a decision is made on the spot as to the best sequence for processing the several operations.

This method of scheduling may be fairly effective in manufacturing systems in which the facilities are relatively independent. (This implies that most production orders consist of a very small number of operations.) One advantage of this method is that the foreman can consider similar set-ups and similar raw materials when deciding on the best sequence of operations.

A major disadvantage of the decentralized scheduling method is the lack of conscious effort toward total shop optimization. It can be shown easily that optimum sequences at individual facilities are usually not optimum sequences for the plant as a whole.

Another serious disadvantage of decentralized scheduling is that it usually degenerates into an expediting system. In such systems, expeditors are assigned to push through specific orders. The result is often a popularity contest. Foremen tend to give priority to operations being expedited by personal friends or by the more forceful expeditors. Such a system usually winds up with the several system components working against each other. This method of scheduling is recommended only for very special situations, and even then it must be used with extreme care.

ANALYTICAL METHODS

None of the scheduling transfer functions presented thus far will generate optimal schedules each time they are applied. It is possible to obtain an optimal schedule with any of the methods presented, but this would result more as a matter of chance than by design. What is worse, we have no way of knowing when we do obtain an optimal schedule.

It is sometimes possible to obtain optimal solutions for certain decisions required of the scheduling function. Most situations of this type involve the assignment of work to facilities. We now discuss two methods that yield optimal solutions.

Linear Programming for Loading. Eilon (1960, p. 345) discusses the use of linear programming in loading an array of production orders onto the available facilities for a scheduling period. The objective is to assign the orders to the facilities in such a manner as to minimize the cost of processing time during the period. This is essentially the same problem as was treated by the Index Method, discussed earlier; however, here we allow order splitting.

Let there be n products to be made during the next period. The number of required units of each product are $U_1, U_2, \ldots, U_n$, respectively. There are m facilities on which these products can be manufactured. It is not necessary that all facilities be capable of producing all products. One unit of a product can be completely manufactured on just one facility. The quantity required for a product U_i can be split and produced on several facilities.

If U_1 (the number of required units of product 1) is distributed among the facilities such that U_{11} is the quantity produced on Facility 1, U_{12} is the quantity produced on Facility 2, and so on, then the total requirement for Product 1 can be expressed as

$$U_1 = U_{11} + U_{12} + \cdots + U_{1j} + \cdots + U_{1m} = \sum_{j=1}^{m} U_{1j} \tag{6.1}$$

In a similar manner, the requirements for the other products are distributed among the m facilities. If a certain facility j cannot manufacture a certain product i, then $U_{ij} = 0$.

Let p_{ij} be the production rate at which Facility j can produce one unit of Product i. Obviously, U_{ij}/p_{ij} is the amount of time Facility j spends on Product i. The total production time assigned to Facility j for all products is

$$\frac{U_{1j}}{p_{1j}} + \frac{U_{2j}}{p_{2j}} + \cdots + \frac{U_{ij}}{p_{ij}} + \cdots + \frac{U_{nj}}{p_{nj}} = \sum_{i-1}^{n} \frac{U_{ij}}{p_{ij}} \tag{6.2}$$

The available capacity at each facility is restricted, perhaps from last period's carryover. The total production time assigned to a facility must not exceed its maximum capacity L_j. That is,

$$\sum_{i=1}^{n} \frac{U_{ij}}{p_{ij}} \leq L_j \qquad \text{for all } j \tag{6.3}$$

We must add the usual nonnegativity constraints of linear programming, requiring that all quantities produced be nonnegative.

$$U_{ij} \geq 0 \tag{6.4}$$

Let the cost of producing one unit of Product i on Facility j be c_{ij}. Since we want to minimize the total cost of production, C, our objective function is

$$\text{minimize } C = \sum_{i=1}^{m} \sum_{j=1}^{n} c_{ij} U_{ij} \tag{6.5}$$

subject to the constraints expressed in Eqs. 6.1, 6.3, and 6.4. The problem is now in proper form for solution by the simplex method of linear programming.

Eilon (1962, p. 348) extends the above procedure to consider the use of overtime and subcontract support when sufficient capacity is not available on regular time. Since the use of overtime and subcontract support usually involves a premium cost, this factor must be added to the problem. For example, consider Facility 2. We denote its maximum capacity on regular time by L_2^1; its maximum capacity on overtime is denoted by L_2^2; and the maximum amount of contract support obtainable is L_2^3. The costs of producing one unit by regular time, overtime, and subcontracting are, respectively, c_{ij}^1, c_{ij}^2, and c_{ij}^3. The number of units produced by the three methods are, respectively, U_{ij}^1, U_{ij}^2, and U_{ij}^3.

The original problem is now expanded to include the two new means of obtaining production. We see then that we are simply adding more rows to the original problem. Essentially, we may regard overtime and subcontracting as additional "facilities," each with its own cost coefficients and capacity limitations. Note that we could expand the problem in this same manner for all other facilities in the shop.

Johnson's Sequencing Model. In 1954, Johnson (p. 61) presented a sequencing algorithm that generates optimal sequences for the special case of n operations processed on two facilities, Facility 1 and Facility 2, in the same order. The generated sequences are optimal in the sense that total elapsed time for processing all n jobs is minimized.

Suppose that we have five operations, all of which must be processed first on

Table 6.5 Processing Times for five Operations to be processed through two Facilities

Operation Number	Processing Time	
	Facility 1	Facility 2
1	5	2
2	1	6
3	9	7
4	8	8
5	10	4

Facility 1 and second on Facility 2. No job splitting is allowed. The operations and their processing times on the two facilities are shown in Table 6.5.

Johnson's algorithm consists of the following steps:

Step 1. Locate the smallest remaining processing time in table (Table 6.5).

Step 2. If the minimum time is on Facility 1, assign that operation to be performed next. If the minimum time is on Facility 2, assign that operation to be performed just before the last operation whose minimum time was on Facility 2.

Step 3. Remove the operation from the list of unassigned operations.

Step 4. Repeat Steps 1 through 3 until all operations have been sequenced.

Step 2 might be slightly confusing until we have performed this algorithm a few times. Essentially, in Step 2 we are working from both ends toward the middle. In the case of a tie in Step 2, it really does not matter how we make the decision as long as we are consistent. In such a situation, we arbitrarily sequence the operation as though its minimum time occurred on Facility 1.

The minimum processing time in Table 6.5 is 1 for Operation 2 on Facility 1. According to the algorithm, we specify that Operation 2 be performed first and remove it from the list of unassigned operations. The next lowest time is 2 for Operation 1 on Facility 2. Operation 2 is to be performed last, that is, fifth in the sequence. Continuing in this manner, we obtain the optimal sequence 2, 4, 3, 5, 1.

It is instructive to illustrate this sequence on a schedule bar chart, as in Fig. 6.12(a). We see from Fig. 6.12(a) that total elapsed time is 30 time units. Figure 6.12(b) shows the sequence resulting from the first come first served decision rule. The total elapsed time resulting from this rule is 34.

There can be found two-stage processes in industry to which Johnson's method may be applied directly. It can also be used in shops containing several facilities, among which two facilities constitute a bottleneck in the flow of product through the shop (Elmaghraby 1966, p. 258).

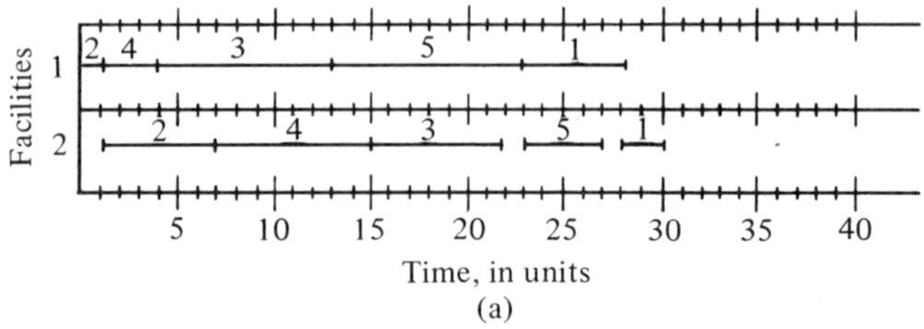

Figure 6.12a Optimal Sequence using Johnson's Algorithm. Total Time = 30

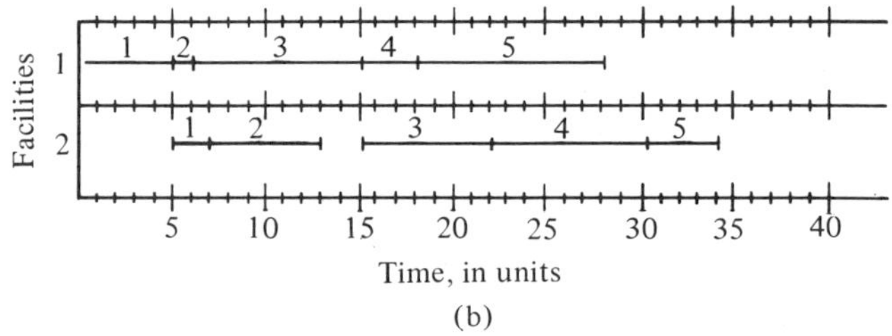

Figure 6.12b Sequence resulting from First Come, First Served Decision Rule. Total Time = 34

There have been several other algorithms developed which yield optimal sequences for restricted cases. No practical solution procedure has yet been presented which generates optimal sequences for the *m* machines, *n* jobs case.

SCHEDULING PROJECT ACTIVITY

There are many activities in industry which are performed on an irregular basis, perhaps only once. Such activities as constructing an addition to a plant, building a plant, building a bridge, rebuilding a large piece of equipment, developing a particular missile, etc., are performed only once. Such activities are called Project activities.

Project activities are not new. The building of the pyramids would certainly fall into this category, as would the development of the V2 rocket in Germany before and during World War II. The distinguishing characteristic of most project activities is the very large number of individual operations that must be performed to complete the project. There are usually many complex relationships between the individual operations, the scheduling of which becomes an extremely difficult task.

Until the late 1950's, most projects were scheduled by a complicated maze of Gantt Charts. This method proved to be more and more inadequate as the size and cost of projects increased. In the late 1950's, two groups working independently on different projects developed strikingly similar network methods that greatly facilitated the task of planning, scheduling, and controlling large projects.

A joint effort by the du Pont Company and the Univac Division of Remington Rand Corporation led to the widely used Critical Path Method (CPM). At about the same time, the Navy was searching for a better way of planning and controlling the Polaris program. The Navy Special Projects Office, working with the management consulting firm of Booz, Allen and Hamilton, developed the Project Evaluation and Review Technique (PERT), a network technique quite similar to CPM. In general, PERT has been used more widely in defense and space projects, whereas CPM has been used more widely in the construction industry and for other large industrial projects. A scheduling algorithm for the PERT method is given in Appendix B.

Feedback and Corrective Action

We discussed in the previous section several scheduling transfer functions available for use in an operations planning and control system. Many other scheduling methods can be found in texts such as Conway, Maxwell, and Miller (1967) and Muth and Thompson (1963) and in professional journals. A particular control system may require the use of several different scheduling transfer functions. The scheduling transfer functions chosen must be compatible with transfer functions chosen in the areas of Forecasting, Operations Planning, and Inventory Planning and Control.

We must provide in the design of our control system the means of monitoring the performance of the scheduling transfer functions selected. In this section we discuss the concept of feedback and corrective action for the scheduling function.

SOURCES OF FEEDBACK

Accurate and timely feedback is very important for all components of our control system and especially so for the scheduling function because random deviations occur much more frequently at this level of the system and the degree of control over detailed components of the system is much greater.

Information requirements for the scheduling function can be compared to those of a pilot landing a jetliner at night at a busy airport. The pilot receives information from the tower, from a large array of instruments, and perhaps from other members of the crew. The requirements for timely and accurate information are obvious. The need for timely and accurate information for the scheduling function is just as urgent. Specific sources of feedback are discussed in the following paragraphs. This feedback is used to update the two information files for the scheduling function, P.IV and VS.IV.

Operations. Operations provides the biggest and most important set of feedback for scheduling. The exact current status of production orders, facilities, and workers is required. Specific information is presented as VS.IV.1, VS.IV.2, and VS.IV.3 on page 214.

Engineering. The scheduling function receives from Engineering feedback concerning new bills of materials, new manufacturing methods and routings, new process and set-up times resulting from methods improvement or refined historical estimates, revised production rates of all employees, updated scrap rates on the several facilities, etc.

Operations Planning. The Operations Planning function provides feedback regarding changes in the time-phased operations plan and in the line balances.

Management. Scheduling receives from management any change in the union-management contract that affects the scheduling of work. Management also provides its desired priority assignments for certain customers.

Personnel. Personnel must indicate the availability of workers to accommodate changes in work force size.

Maintenance. An important feedback to scheduling comes from Maintenance. Just as actual processing times can vary from those planned, so can the time for performing preventive maintenance. In some shops, maintenance is scheduled right along with production operations. Unplanned maintenance is also an important feedback. Accurate information is needed on the frequency and cause of breakdowns.

SHORT-TERM CORRECTIVE ACTION

Adjustments in the schedule must be made quite frequently in response to the dynamic events occurring on the shop floor. This component is perhaps the most critical in our entire control system. Corrective action is necessary because processing times are random variables, worker productivity varies, machines break down, materials are depleted, product and material quality varies, and so on. On the spot decisions are frequently required to maintain plant activity. Such decisions are usually made with human judgment acting as the transfer function. More will be said about this in Chap. 7.

LONG-TERM CORRECTIVE ACTION

We must continuously monitor the effectiveness of our scheduling transfer functions to determine whether our scheduling objectives of high facility utilization, low in-process inventories, and few late order deliveries are being accomplished. We can keep records on manufacturing lead times, how long various size orders stay in the system, in-process inventories, idle facility time, time between machine breakdowns, repair times, etc. Based upon our analysis of such factors, we may decide to seek improvements in our scheduling system.

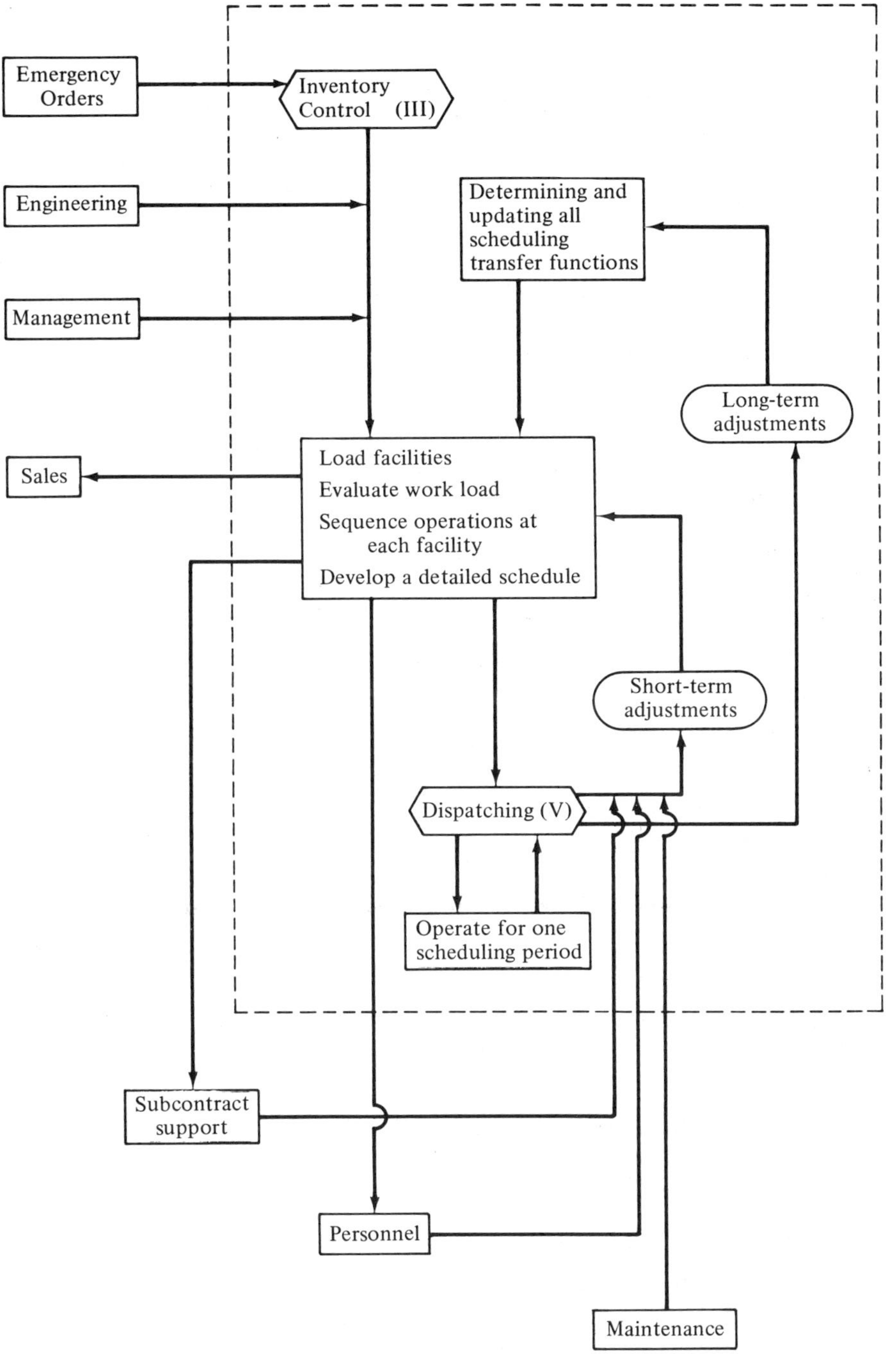

Figure 6.13 Interfaces for Scheduling Function

Interfaces

The scheduling function interacts with very few functions outside the boundaries of the operations planning and control system. The interfaces with management, maintenance, and engineering are discussed in the previous section. In addition, we should mention that scheduling provides Sales with estimates of delivery dates of orders and with manufacturing lead times.

Figure 6.13 is included to help us consolidate the previous discussions of the interfaces with the scheduling function. Note that while the scheduling period is often 1 week, corrective action is taken on a continuous basis, usually under the coordination of the Dispatching function.

REFERENCES

Conway, R. W. 1964. An Experimental Investigation of Priority Assignment in a Job Shop. RAND Report RM-3789-PR, RAND Corporation (Santa Monica, California).

Conway, R. W.; Maxwell, W. L.; and Miller, L. W. 1967. *Theory of Scheduling*. Reading, Mass.: Addison-Wesley Publishing Company.

Eilon, S. 1962. *Elements of Production Planning and Control*. New York: Macmillan Company.

Elmaghraby, S. E. 1966. *The Design of Production Systems*. New York: Reinhold Pub. Corp.

Johnson, S. M. 1954. Optimal Two- and Three-Stage Production Schedules with Set-Up Times Included. *Naval Research Logistics Quarterly* 1: 61-68.

LeGrande, E. 1963. The Development of a Factory Simulation System Using Actual Operating Data, *Management Technology* 3.

Muth, J. F.; and Thompson, G. L. (Eds.) 1963. *Industrial Scheduling*. Englewood Cliffs, N.J.: Prentice-Hall, Inc.

Scheele, E. D.; Westerman, W. L.; and Wimmert, R. J. 1960. *Principles and Design of Production Control Systems*. Englewood Cliffs, N.J.; Prentice-Hall, Inc.

SOURCES OF ADDITIONAL TRANSFER FUNCTIONS

Listed below are several additional sources of scheduling transfer functions. The list, not intended as an exhaustive bibliography on the scheduling function, is a representative sample of the scheduling techniques available.

Bakshi, M. S.; and Arora, S. R. 1969. The Sequencing Problem. *Management Science* 4: B-247.

Eilon, S. 1969. Multi-Product Scheduling in a Chemical Plant. *Management Science*, 15: B-267.

Fabrycky, W. J.; and Shamlin, J. E. 1966. A Probability Based Sequencing Algorithm. *Journal of Industrial Engineering* 17: 308.

Fox, P. D.; and Kriebel, C. H. 1967. An Empirical Study of Scheduling Decision Behavior. *Journal of Industrial Engineering* 18: 354.

Glassey, C. R. 1967. An Algorithm for a Machine Loading Problem. *Journal of Industrial Engineering* 18: 585.

Marchbanks, J. L. 1966. Daily Automatic Rescheduling Technique. *Journal of Industrial Engineering* 17: 119.

Muth, J. F.; and Thompson; G. L. (Eds.) 1963. *Industrial Scheduling*. Englewood Cliffs, N.J.: Prentice-Hall, Inc.

O'Brien, J. J. (Ed.) 1969. *Scheduling Handbook*. New York: McGraw-Hill Book Company.

Spradlin, B. C.; and Pierce, D. A. 1967. Production Scheduling Under a Learning Effect by Dynamic Programming. *Journal of Industrial Engineering*, 18: 219.

Vergin, R. C. 1966. Production Scheduling Under Seasonal Demand. *Journal of Industrial Engineering*, 17: 260.

7 Dispatching and Progress Control

Concepts

PURPOSE

In the preceding chapters we have concentrated primarily on the way decisions are made in an Operations Planning and Control System. For example, we have talked about inventory decisions, forecasting decisions, scheduling decisions, and the like. Now we must consider how all these decisions are implemented; that is, how they are transferred into action. Somehow, the decisions must be relayed from the production control office to the foremen and workers on the shop floor; material has to be issued; instructions have to be gotten to workers; parts must move from one operation to the next. As work is accomplished, we must have a means of recording progress, comparing it to planned progress, and then taking whatever corrective action that seems appropriate.

We could say that the primary purpose of dispatching and progress control is to close the loop in our Operations Planning and Control System. To accomplish this, we must provide ways to detect when a deviation has occurred, analyze what has happened, decide what corrective action would be appropriate, implement the corrective action by incorporating it into subsequent scheduling decisions, and finally, feed back progress information to the planning section to improve future planning.

BASIC PROBLEM IN DISPATCHING AND PROGRESS CONTROL

It is meaningful to divide the activities performed by this function into five categories:

1. Production initialization: triggering production and keeping it moving.
2. Data acquisition: recording, collecting, and transmitting production information (actual events that occur on the shop floor).
3. Performance evaluation: comparing results to expected results and performance to standard performance.
4. Short-term corrective action: decisions necessary for solving immediate problems; keeping the wheels of production in motion.
5. MOE evaluation and long-term corrective action: maintaining cumulative performance records on important measures of effectiveness (MOE) within the system, adjusting pertinent transfer function parameters when appropriate, and identifying cause-effect relationships with regard to significant deviations from system control limits.

The basic problem is to design a control system that will provide input to the operating facilities, monitor events as they occur, compare actual outcomes to the range of outcomes that reasonably could be expected, and then take appropriate action on both a short-term and a long-term basis. The short-term corrective action affects the next set of inputs to the operating facilities, and the long-term corrective action affects the manner in which production activities are planned over a long period, e.g., 1 year.

We want our control system to react responsively to significant deviations from standard system performance and to changes in important noncontrollable variables, such as demand trends, lead times, etc. At the same time, however, we do not want the control system to overreact to chance occurrences that have no long-term significance to the production system. Achieving this delicate balance of responsiveness is a difficult task, but it is a factor that greatly influences the stability and effectiveness of the control system.

SYSTEM DESIGN CONSIDERATIONS

There are perhaps fewer methods available for helping us design this part of our control system than for any of the other functions. This lack of research may explain in part why so little attention has been given to this important function in journal articles and texts. The following factors should be considered when designing a control system:

1. What data should be gathered?
2. By what specific means should data be gathered, transmitted, processed, and stored? (The degree of sophistication of the system must be justified economically.)
3. To what degree of centralization will this part of the system be designed?

4. By what methods will planned performance, allowable variations, and actual occurrences be portrayed (Gantt charts, CRT displays, etc.)?
5. It is helpful to establish decision categories for short-term corrective action, with specific assignment of organizational responsibility for each category. In establishing these categories, the "management by exception" principle should be employed so that decisions are made at the lowest possible organizational level.
6. What type of priority system, if any, should be used for resolving schedule conflicts in the shop?
7. What specific measures of effectiveness should be monitored? (Which are responsive to changes in basic conditions?)
8. For which specific random variables should we maintain current probability distributions?
9. How frequently should we update transfer function parameters?

Note that the above factors are stated in terms of questions. The answers to these questions are decisions. In Chap. 2, such decisions were called structural decisions, because they pertain to the specific nature of the control system itself. When we make decisions such as those above and when we select a particular transfer function to include in our control system, we are literally deciding how to decide.

Structural decisions can also be cast into our transfer function concept. Doing so gains us very little, however, since transfer functions for structural decisions consist almost entirely of human judgment at the present time. Computer simulation is another transfer function that shows much promise for helping arrive at structural decisions. We pursue this possibility further in Chap. 8.

DISPATCHING AND PROGRESS CONTROL DECISIONS

We now turn our attention to the specific decisions made in the dispatching and progress control function. As is true for the other functions of the total system, the decisions we discuss in this section must be rendered, whether formally or informally, logically or emotionally, and regardless of whether a particular company recognizes that the decisions are being made at all. Again, we discuss the decisions according to the five activity categories presented earlier.

Production Initialization. The following steps are accomplished for each Production Order that appears on the official shop schedule:

1. Translate schedule into sets of Production Order papers (one set for each production order on the schedule):

- Blueprints and drawings
- Operations instructions
- Schedule and route sheet
- Route cards (these identify and stay with the order as it proceeds through the various processing operations). The route cards can be devised so that they become feedback on production progress.

2. Issue orders to release materials, tools, jigs, and fixtures. Raw materials and subassemblies are reserved for each production order on the schedule.
3. Arrange for change-over from old production items to new ones on the schedule; include maintenance services.
4. Arrange for materials handling, and coordinate the movement of orders from one operation to the next.

Data Acquisition. We have considered the flow of information from the several operations planning and control functions to the operating facilities. We now must consider information flow in the other direction; i.e., feedback.

It is necessary to acquire feedback because there will almost always be deviations from our plan. Some of the major causes of deviations are variation in demand, variable operation times, variable worker performance, equipment failures, variable quality of purchased and manufactured items, variable lead times from suppliers, rush orders, cancelled orders, labor problems, and unforeseen schedule conflicts resulting from the complex interrelationships between operations being processed through the several facilities.

Specific decisions for the data acquisition activity fall into two categories:

1. Reporting planned events: estimating percent completion on started but unfinished operations; whether a planned event has occurred (when *is* an operation considered complete?); whether to report required variations in the procedure.
2. Reporting unplanned events: whether, when, and how to report particular unplanned events (minor breakdowns, short work stoppages, unusual set-up time, etc.); are their impacts on over-all operations significant enough to report them; if so, should they be reported immediately, once a day, once a week, etc.; in what form should they be reported?

Performance Evaluation. Regardless of how sophisticated or unsophisticated our data acquisition system is, we must make proper use of data relayed from the shop floor. In general, we must compare actual performance to planned performance, detect significant deviations from the production plan, and determine appropriate corrective action.

Specific decisions required of the performance evaluation activity are:

1. Whether an actual occurrence outside the control limits (outside the range of expected variation) is an isolated random event or an event that signifies a significant change that may be repeated.
2. Whether several consecutive occurrences on one side of nominal performance but still within the control limits constitute a general shift in the total cause system for that variable.
3. What caused the occurrence in (1) above to fall outside the control limits, and what caused the run mentioned in (2) above?
4. Whether conditions have changed enough to warrant a change in the nominal performance values and a recalculation of the control limits.

Short-Term Corrective Action. Once it has been determined that there is a significant difference between planned and actual events, it is necessary to implement the appropriate corrective action. At this point in the system we are more concerned with "what happened and what do we do about it" than with determining why the difference occurred and what should be done to prevent future differences. The latter decision is discussed in the next section.

Many problems that occur in Operations can be solved quickly and informally by on-the-spot decisions. Such information never enters the formal channels of the control system. Other problems can be solved simply by bringing them to the attention of the maintenance group. Still others require formal attention by the decision-making mechanism of the control system. Specific decisions that fall into this latter category are:

- What course of action to follow when production falls behind schedule—use overtime, operate an additional shift, use subcontract support, buy components that we normally fabricate, delay certain jobs, refuse further orders, or a combination of two or more of the above alternatives.
- Whether particular orders should be given the attention of an expeditor to push them through production.
- Given the current status of the production system, how should the schedule be modified to best handle the new situation? What priorities should be assigned to operations in conflict and upon what basis are the priority assignments made?
- What course of action to follow when product quality drops to an unacceptable level. This decision is basic to the quality control system and thus is not pursued further here.

MOE Evaluation and Long-Term Corrective Action. We have discussed means of effecting corrective action on a short-term basis. The types of decisions we discussed had to be made quickly in order to alleviate current production problems. In day-to-day operations, there is not enough time to perform an in-depth analysis of the causes of the deviations from the production plan. Unless we make such an analysis, however, we throw away much valuable information that could possibly lead to solutions that would prevent the recurrence of many problems and that could increase greatly the effectiveness of our over-all planning and scheduling procedures.

Specific decisions that must be rendered by this activity are:

- Whether the numerical value of a parameter has changed (e.g., average process times, average lead time, average time between breakdowns, average idle time, average repair time, percent defective of incoming materials, percent defective of manufactured items, average investment in inventory, etc.).
- Whether other parameters affected by the above changes should be changed accordingly.
- Determining cause-effect relationships.
- Classification of deviations according to their causes.
- Whether the dispatching and progress control function is performing satisfactorily. (We are suggesting that this function monitor itself.)

Figure 7.1 represents the entire process of arriving at the decisions discussed in this section. The short-term corrective action loop is executed essentially continuously, while the long-term correction loop is executed less frequently but is related to all functions in the control system. Note that, in the long-term correction loop there is included a monitoring loop for the dispatching and progress control function itself. The fact that this function monitors its own performance indicates that our over-all control system is an adaptive control system. This term applies to systems that monitor their own performances, learn from their own mistakes, and adjust their own parameters accordingly. This term generally is used to describe completely automatic systems not subject to human intervention. In our case, humans are extremely important components of our system, particularly for monitoring activities.

Inputs

The dispatching and progress control function requires much more detailed and specific information than the other functions in our control system. Note that the Variable Status File is relatively more important in this function than in the others.

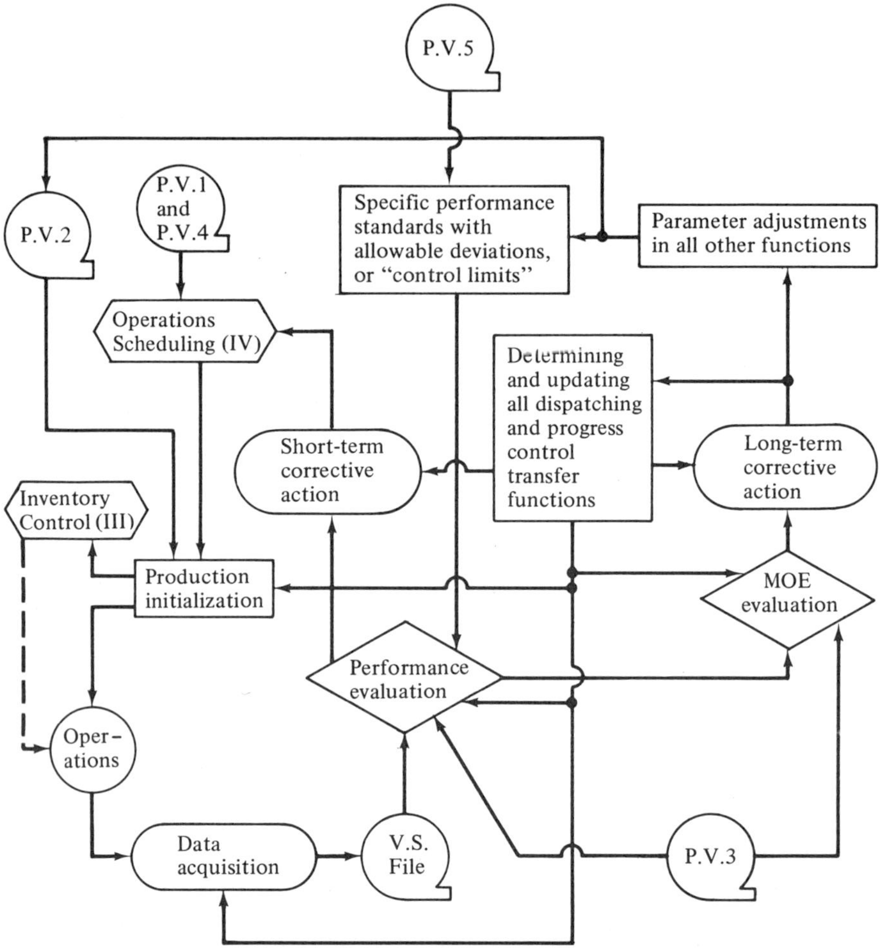

Figure 7.1 Overview of the Dispatching and Progress Control Function

PARAMETER FILE FOR DISPATCHING AND PROGRESS CONTROL (P. V)

P.V.1 Operations schedule. Source: Operations Scheduling.

P.V.2 Blueprints, drawings, operation instructions, route sheets, route cards, and other engineering data. Source: Engineering.

P.V.3 Performance standards and allowable deviations, both short- and long-term basis. Source: Forecasting, Operations Planning, Inventory Planning, Operations Scheduling, and Engineering.

P.V.4 All priorities and decision rules. Source: management and industrial engineering.

P.V.5 Parameters of probability distributions being maintained for significant variables in the system. Source: continuous recalculations.

VARIABLE STATUS FILE FOR DISPATCHING AND PROGRESS CONTROL (VS. V)

The variable status file for this function for short-term corrective action is identical to the variable status file for Operations Scheduling, VS. IV (see page 216). In addition, this file must acquire actual values of all variables whose performances are being monitored. A large number of such variables are indicated in Table 7.1.

UPDATING THE FILES

One of the major activities of dispatching and progress control is data acquisition. This activity, by its very nature, results in updating the Variable Status File. The Parameter File is more difficult to keep updated, since a judgment has to be made as to whether each parameter has changed. The accuracy with which the parameters are kept updated is fundamental to the effectiveness of the entire control system.

At this point we find ourselves face-to-face with the most serious weakness in many operations planning and control systems. No system can be controlled effectively until specific performance standards and allowable deviations have been established; and once established, they must be monitored and revised when necessary. Effective monitoring and revision of standards and allowable deviations cannot be accomplished without a well-designed data acquisition system.

Transfer Functions

In general, transfer functions and formal decision processes are not as well developed for the dispatching and progress control function as they are for other functions. This is particularly true for the design of integrated information systems. While some progress has been made toward formalizing the study of information systems,* most procedures are highly pragmatic approaches tailored for specific systems. Nevertheless, adequately designed information systems can be obtained using the existing procedures provided the designer first uncovers the decision structure necessary to accomplish the objectives of the system. The information system should be designed to satisfy the decision requirements, not vice versa.

We discuss transfer functions according to the five categories previously specified: production initialization, data acquisition, performance evaluation, short-term corrective action, and MOE evaluation and long-term corrective action.

*In particular, see Brooks (1964); Homer (1962); and Lieberman (1956).

PRODUCTION INITIALIZATION

The process of implementing production orders is shown in Fig. 7.2. Some of the transfer functions available for the required production initialization decisions are discussed in the following sections.

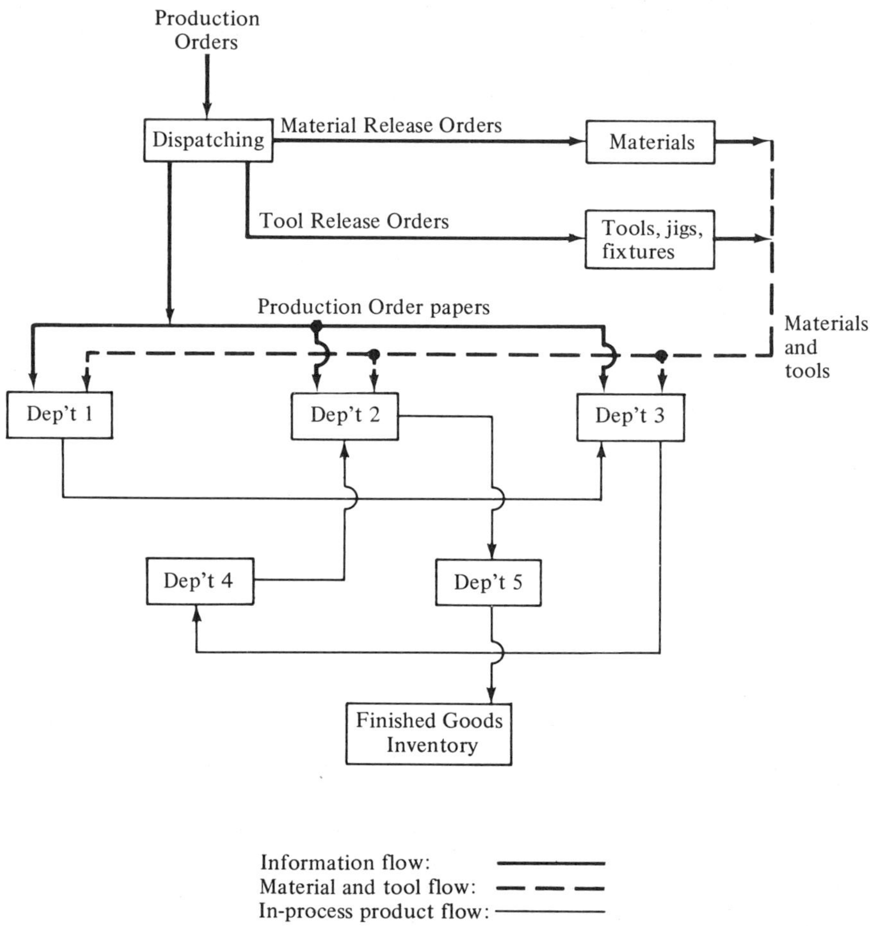

Figure 7.2 The implementation of Production Orders

Computer Programs. It is often very beneficial to develop a computer program capable of extending activities on the official schedule into sets of production order papers, especially when the generation of production order papers follows the same general steps for all or large groups of the manufactured items. The general steps of such a procedure would be as follows:

Step 1. The schedule is the input to the program. Each operation on the schedule has a unique identification.

Step 2. The program then identifies all the items required for the production order papers for each operation, blueprints and drawings, operation instructions, route sheets, etc. These are listed so that the dispatcher can gather them for distribution to the shop floor.

Step 3. The program punches a deck of route cards for each production order. These cards, which stay with the order as it progresses through the various facilities, are used as feedback in reporting production progress.

Step 4. The program can print out any special instructions regarding set-ups, materials handling, etc.

Step 5. The computer can automatically issue orders to release materials, tools, jigs, and fixtures. It can reserve raw materials and subassemblies needed for each production order on the schedule. Here, we need a strong interface between the functions of dispatching and Inventory Control, particularly if the entire control system is highly automated and computerized. Again, the advantages of centralized data files are obvious.

Note that the suggested procedure requires a large amount of current, accurate information. A major problem in such a system is to verify the accuracy of the information and to keep it updated.

Gozinto. In Chap. 4, we discussed Gozinto, a matrix method for determining material requirements and many other types of information. This same procedure would be quite useful in performing the dispatching function. The development of specific procedures for using the Gozinto concept at the operating level is left as an exercise for the reader. The procedure outlined in Chap. 4 for the planning level can be used as a guideline.

Graphical and Mechanical Aids. In a large plant consisting of many facilities and in which hundreds of production orders are in process at all times, coordination of production activities is a formidable task. One of the early approaches to this problem was the use of a "dispatch board," an example of which is shown in Fig. 7.3. There are many possible variations of the basic system, and several related mechanical devices are available commercially. It would even be possible to display the pertinent information on a cathode ray tube (CRT) console connected remotely to a computer. Certain concepts of an advanced computer system are discussed in the next section.

Transfer Functions for Decentralized Control Systems. Some production control systems employ a decentralized dispatching function to avoid the

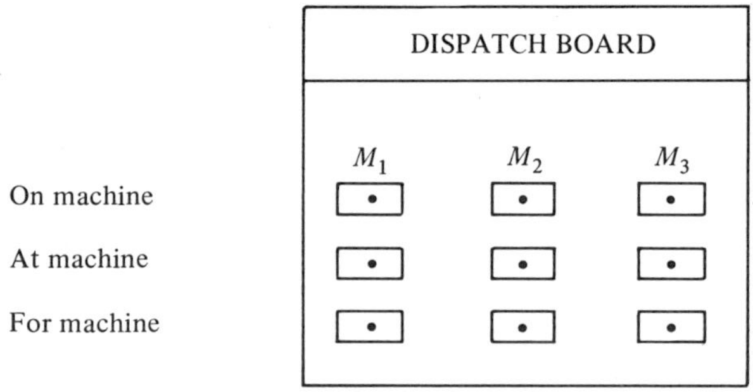

For machine: all the operation tickets for work to be performed on this machine

At machine: when work has been moved to the machine, the ticket is moved up to the "at machine" position

On machine: when the job is actually placed on the machine, the ticket is moved up to the "on machine" position.

Figure 7.3 Dispatch Board for coordinating Work Orders

necessity of a highly complex scheduling system. Under such a system, many of the detailed decisions normally made in the scheduling function (Chap. 6) are deferred to the dispatching function. The transfer functions presented in Chap. 6 then become applicable to the dispatching function. Note that the basic decisions that must be made have not changed; they have only been moved to another component of the control system.

Decentralized dispatching has the advantage of moving decision making to the lowest possible level. It allows the dispatcher and the individual foremen to consider short-term events and to react accordingly. Decentralized dispatching has the serious disadvantage of not considering the total plant when detailed decisions are made. It is difficult to communicate changing conditions between individual facilities.

DATA ACQUISITION

Data acquisition consists of measuring and recording production progress. To do this, we must acquire production data from many points on the shop floors. The specific production data needed are:

- *Progress of work in process:* percent complete, number units completed, quality records, completion times, material shortage, due dates, remaining processing time.
- *Status of production facilities:* queue sizes, occurrences of tool breakages and equipment breakdowns, idle man and machine time, scrap and rework records, number of set-ups.

- *Operator records:* absenteeism, availability of particular skills, productivity records.

The specific method for acquiring and recording this data depends on the type of production, the size of company, and the sophistication of the control system. Traditional data acquisition systems record production progress through such means as having the operator fill out an operation completion report or having the operator place a prepunched card in a collection box when the operation is completed. These reports or cards are collected manually, usually once or twice each day. They are then brought into a central dispatching office, where visual control charts are updated from the collected records. The visual control charts are simply mechanical variations of the familiar Gantt chart. Many versions of these mechanical Gantt charts are commercially available.

In the central dispatching office, the control charts are analyzed and an attempt is made to identify problem areas. When problems arise that threaten the schedule, expeditors are assigned the task of solving the problem, which may involve sending a rush order for raw materials, assigning priorities to work in progress, etc.

In more sophisticated data acquisition systems, production progress is recorded by electromechanical means. Data is entered at input stations dispersed throughout the plant. An input station of this type is shown in Fig. 7.4. All "constant" information is prepunched into a regular 80-column card, perhaps as part of the computerized dispatching system suggested in the previous section. Whenever some event has occurred (such as an operation completion or

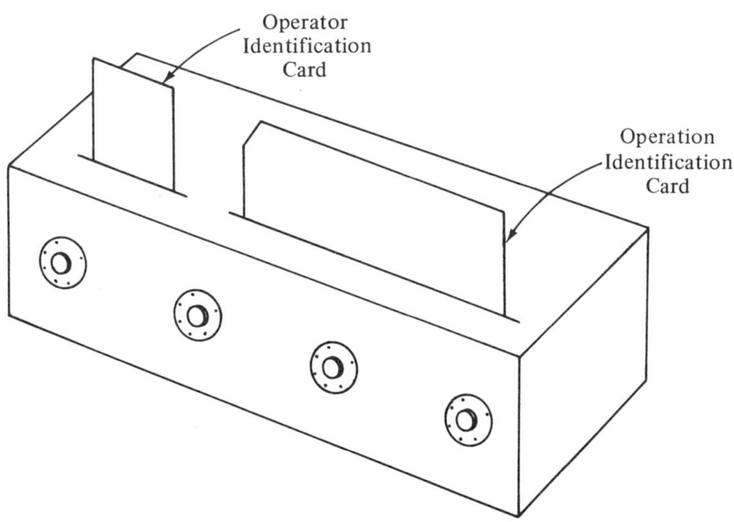

Figure 7.4 Input Station for reporting Production progress

a specified percentage completion), the operator inserts the 80-column card into a slot and his own identification card (often similar to a credit card) into another slot. Variable information, such as processing time, scrap factor, etc., is entered either by turning knobs or setting levers. Some commercial input devices have a third slot that receives information about the machine that performed the activity. This slot is especially useful when one input station is used in conjunction with several machines.

The information is transmitted to a central data receiving unit where it is processed and combined with information from other input stations. The ultimate purpose of this data is to compare actual results to planned results and to provide a basis for subsequent corrective action. These activities are discussed in the following two sections.

The manner in which a sophisticated on-line operations control system functions is illustrated in Fig. 7.5. The system shown collects production data, processes it, sends certain information to other functions, sends summarized data to MOE for long-term corrective action, and returns data to dispatching so that short-term corrective action may be initiated. In addition, the system can reply to inquiries from the shop floor. The remote stations must be capable of receiving the replies and therefore must be equipped with either a typewriter device or a CRT display unit.

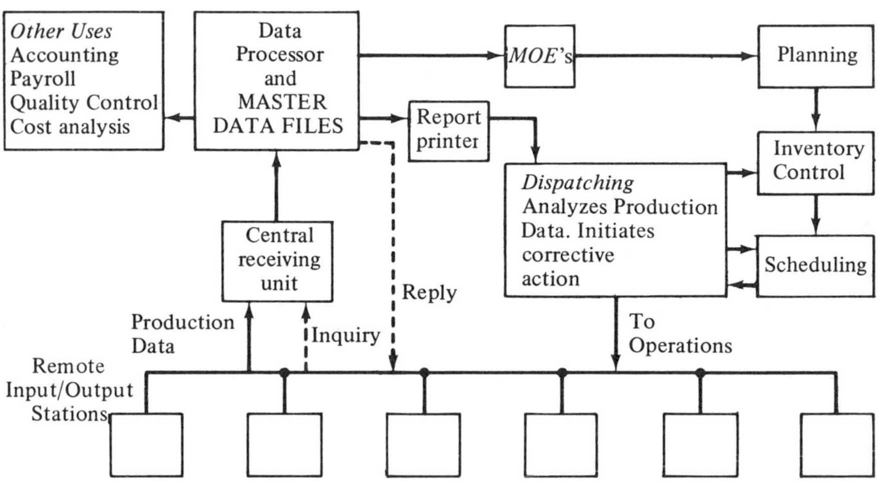

Figure 7.5 Representation of an On-Line Operations Control System with Remote Inquiry capability

Such a system is usually very difficult and expensive to develop and implement. Whether it can be justified depends on whether extremely tight control of operations is necessary. In those cases where such a system can be justified, the following favorable results can be expected:

- Lower in-process inventories
- Fewer stockouts and schedule delays
- Less man and machine idle time
- Lower per unit cost of production
- Fewer unforeseen events, thus reducing reliance on expeditors, and
- Better basis for production planning through feedback of various production data.

PERFORMANCE EVALUATION

Regardless of how sophisticated or unsophisticated our data acquisition system is, we must make proper use of the data relayed from the shop floor. In general, we must compare actual production performance to planned performance, detect significant deviations from the production plan, and determine appropriate corrective action. We concern ourselves in this section only with short-term performance evaluation. In the last section, we look at the performance of the system over a longer period of time and emphasize the determination of cause-effect relationships.

Transfer functions for short-term performance evaluation are often graphical. A very common transfer function is the familiar Gantt scheduling chart shown in Fig. 6.5. A review of Gantt charting would be helpful at this point.

Numerous other graphical aids can be used for short-term performance evaluation. For example, one gross measure of plant performance is the total lateness of all unfinished production orders (see Fig. 7.6). The "goal" for total lateness is a value arrived at by observing the system over many operating periods and noting the amount of total lateness that appears to be "healthy." Note that the goal for total lateness is not zero. The proper amount of total lateness is that which strikes a balance between the cost of late orders and the cost of excessive idle time which occurs when total lateness is very small.

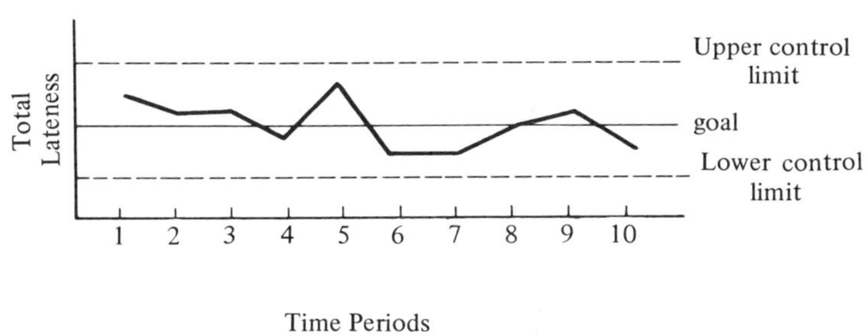

Figure 7.6 Total Lateness Chart for entire Plant or for individual Departments

A total lateness chart for each department could also be kept. Control limits can be determined for this and other measures of performance by computing the standard deviation of past observations and using this figure in somewhat the same manner as is done in constructing quality control charts. Control limits are useful for recognizing when a certain factor has exceeded its normal range of fluctuation. As in quality control procedures, an observation outside the control limits may be simply an isolated random occurrence with no long-lasting effect on the system or it may signify a significant problem area.

Some other data that would help with short-term performance evaluation are:

- Value of in-process inventory.
- Value of raw materials inventory.
- Value of finished goods inventory.
- Number of backorders.
- Amount of idle time for total plant and for each facility.
- Production count for total plant and for each facility.
- Amount of scrap and rework for total plant and for each facility.
- Number of machine adjustments necessary on each machine.
- Number of tools broken for total plant and for each machine.
- Amount of overtime required for total plant and for each machine.

For each of these measures, a chart such as that shown in Fig. 7.6 would be maintained.

In addition to checking for points outside the control limits, we should also be alert to runs; i.e., several consecutive points on one side of the goal, or nominal value. When this event occurs, we should consider the possibility that the system has undergone a basic change. The section concerned with long-term corrective action attempts to identify the cause of the change.

SHORT-TERM CORRECTIVE ACTION

Many problems arise on the shop floor every day whose solutions require very little formal decision making. When a tool breaks, replace it. When a machine drifts out of adjustment, reset it. Corrective action is essentially predetermined by the nature of the problem and the narrow range of reasonable alternatives. Formally, we can say that human judgment is the transfer function in such cases, but in reality very little judgment is required.

A large number of problems arise on the shop floor whose solutions are not predetermined and therefore require formal decision processes. These include such problems as large queues building up at machines, schedule delays, bottlenecks, broken-down machines, emergency orders, material shortages, employee absenteeism, and such things.

Human Judgment. Human judgment is by far the most commonly used transfer function for determining short-term corrective action. In such an approach, decisions are left to the lowest possible level. The machine operator himself may determine which job, of those waiting at his machine, he will perform next. The group foreman may decide to work four hours overtime in order to catch up.

Expeditors. The use of expeditors, while actually a human-judgment type of transfer function, is so commonly used that it is worthy of separate discussion. Late orders are assigned to expeditors, who attempt to push the orders through the facilities.

There are several disadvantages of this method. Different expeditors push different orders and consequently work against each other. The most serious weakness of the expeditor method is that it usually degenerates into a "squeaky wheel"* system. Foremen tend to give priority to operations being expedited by personal friends or by the more forceful expeditors.

Priority Assignment Rules. Managers of medium to large manufacturing plants now recognize that corrective action with regard to rescheduling the shop load usually can be accomplished best by some sort of priority assignment system. Such a system gives consistency to decision making and usually results in a smoother operation over-all.

The particular priority assignment system used in a plant depends upon many factors. Some of these factors are (IBM 1960, p. 31):

- *Type of customer:* large, small, old, new (e.g., high priority for large or old customers).
- *Profitability of the item:* highest priority to most profitable item.
- *Classification of the item:* high priority for a service or spare part and for certain rush orders.
- *Utilization of facilities:* high priority to orders which will more quickly fill following idle work stations.
- *Relative value (investment) of the item:* high priority to high value items.
- *Safety stock of item:* high priority to item with least remaining stock on hand.
- *Due dates:* high priority to item with earliest promised delivery date.

*The wheel that squeaks the loudest gets the grease; the expeditor who pushes hardest gets the attention.

- *Scheduled start date:* high priority to item with earliest scheduled start date.
- *Slack time in remaining operations:* high priority to item with least slack time in remaining operations.
- *Length of operation:* high priority to shortest operation.

Many priority assignment rules have been developed and used in industry based upon one or more of the above factors. We consider two of these, the critical ratio method and the sum of two digits methods.

The critical ratio method (Plossl and Wight 1967, p. 297) is useful for situations in which there is a manufacturing lead time for items produced in lots. (None of the items can be used until the entire lot has been completed.) We need the following information for each item we manufacture:

1. Reorder point quantity.
2. Stock on hand, continuously.
3. Total manufacturing lead time.
4. Remaining manufacturing lead time.

We want to compare the rate at which stock on hand is being used up to the rate at which manufacturing lead time is being used up. We do this by determining two factors:

$$A = \frac{\text{Stock on hand}}{\text{Reorder point quantity}}$$

$$B = \frac{\text{Remaining lead time}}{\text{Total lead time}}$$

$$\text{Critical Ratio} = CR = \frac{A}{B}$$

Let us consider an example. Suppose the reorder point of an item is 600 units and the manufacturing lead time (total time to complete one lot of this item) is 4 weeks. Suppose that 1 week after the order was placed, stock on hand has dropped to 400 units. Then,

$$A = \frac{400}{600} = 0.67$$

$$B = \frac{3}{4} = 0.75$$

$$CR = \frac{A}{B} = \frac{0.67}{0.75} = 0.895$$

We see that stock on hand and lead time are being used up at almost the same rate, so we are not overly concerned at this point. But, suppose for another item, $A = 0.5$, $B = 0.9$; then $CR = 5/9 = 0.555$. Here, stock on hand is being depleted at almost twice the rate that manufacturing lead time is being used up. Therefore, we should be greatly concerned about running completely out of stock before the lot being manufactured is completed.

In general, we can set up guidelines such as the following:

Range	Condition	Action Required
$0.80 \leq CR \leq 1.20$	Acceptable	None at this time
$CR < 0.80$	Too low	Expedite
$CR > 1.20$	Too high	Consider delaying production

Each company would have to set its own ranges based upon its own situation.

A similar approach can be used in determining priority among orders waiting to be processed at a work station. Suppose we have three orders waiting at a work station (W. S.) with the following characteristics:

Order	Due Date	Scheduled Completion On this W. S.
29	84	78
38	90	80
52	85	75

Date: 80

$$\text{Ratio} = \frac{\text{Due Date} - \text{Date}}{\text{Due Date} - \text{W. S. Date}}$$

$$R(29) = \frac{84 - 80}{84 - 78} = \frac{4}{6} = 0.67$$

$$R(38) = \frac{90 - 80}{90 - 80} = \frac{10}{10} = 1.00$$

$$R(52) = \frac{85 - 80}{85 - 75} = \frac{5}{10} = 0.5$$

We see that Order 52 is more critical than Order 29, even though it has a later due date. Order 38 is on schedule, as indicated by the ratio of one. The sequence in which the orders should be processed on this work station is 52, 29, 38.

The sum of digits method (IBM 1960, p. 33) is useful for resolving schedule conflicts at work stations, based upon (1) the type of item and (2) the schedule status. A numerical coding system is established for each of these two factors. High numbers indicate low priorities. A typical coding system for type of item is:

Code	Type of Item	Explanation
9	High volume, rapid production	Large safety stock, little urgency.
7	Batch production	Smaller safety stock, little urgency.
5	One-time production item	Exact quantity is on order, more flexibility in due dates.
4	Service or spare part	Required for immediate customer need.
3		Expedite, authorized by production control with 15 days stock remaining.
2		Expedite, authorized by department manager with 10 days stock remaining.
1	(Highest priority)	Expedite, authorized by plant manager with 5 days stock remaining.

A typical coding system for the schedule status is:

Code	Explanation
9	4 or more days ahead of schedule
8	3 days ahead of schedule
7	2 days ahead of schedule
6	1 day ahead of schedule
5	On schedule
4	1 day behind schedule
3	2 days behind schedule
2	3 days behind schedule
1	4 days behind schedule
0	5 or more days behind schedule

Each order awaiting processing at a work station is assigned one of the codes for each of the two factors. The two codes are added for each order. The order having the lowest sum is most critical and is therefore processed next. The other orders are then sequenced in increasing order of the sums. Note that the sequencing must be performed daily, because not only do new orders join the queue, but the schedule status code (the second factor) changes for each item each day. When two or more orders have the same sum of digits, it is necessary to sequence those orders according to some other decision rule, such as length of operation (top priority to the order having the shortest operation time because it will least delay the other orders in the queue).

There is no reason why the sum of digits method could not use three or four (or even more) factors, rather than the two we discussed. The other factors would have to be meaningful measures of priority, they would have to have a meaningful relationship to the other factors, and their code values would have to be assigned carefully. The thing that prevents us from using many factors is that there is a large cancelling effect when several factors are used simultaneously.

We have discussed but two of many priority assignment methods in use in industry; many others are described in the literature concerning sequencing, scheduling, and dispatching. We should note before leaving the topic that priorities can be applied at two levels:

1. In scheduling; that is, in assigning operations to facilities. This amounts to an a priori resolution of anticipated schedule conflicts as discussed in Chap. 6.
2. In operations; on the shop floor, the queues are sorted according to the established priority rule. This is the level at which the two rules we have just discussed would be applied.

Computer Simulation. It is conceptually feasible to construct a computer simulation model of the manufacturing operations and to experiment with alternative decisions before actually implementing a particular set of corrective actions. The idea here would be to inform the decision maker what would be likely to happen if certain decisions were made. He could then select the best set of decisions from those he considered.

Such a model would require extremely accurate and timely information. It would be very expensive to implement and maintain. While such an on-line crystal ball is not attainable for most companies today, it holds such great promise that it will surely be attempted in the not too distant future. An example of a partial simulation model for very short-term corrective action is described in LeGrande (1963).

MOE EVALUATION AND LONG-TERM CORRECTIVE ACTION

In the previous section we were concerned with effecting corrective action on a short-term basis. The types of decisions we discussed had to be made quickly in order to alleviate current production problems. In day-to-day operations, there is not enough time to evaluate current operations status with respect to our overall operations plan, nor is there time to perform an in-depth analysis of the *causes* of deviations from the plan.

Monitoring the Operations Plan. When an operations plan is originally developed (according to the methods discussed in Chap. 4), we recognize that the plan is only a general framework within which detailed operations are to be performed. We know that day-to-day operations may fluctuate quite a bit and that moderate deviations from the plan are sometimes necessary.

The most common method for monitoring the operations plan is a simple graphical plot that shows both the plan and the actual occurrences. (See Fig. 7.7.) It is often more desirable to plot planned versus actual production as cumulative values. Figure 7.8 is such a graph for the same basic data shown in Fig. 7.7.

Graphs such as Figs. 7.7 and 7.8 also can be kept for other measures of effectiveness, such as the number of hours of overtime planned and actually used,

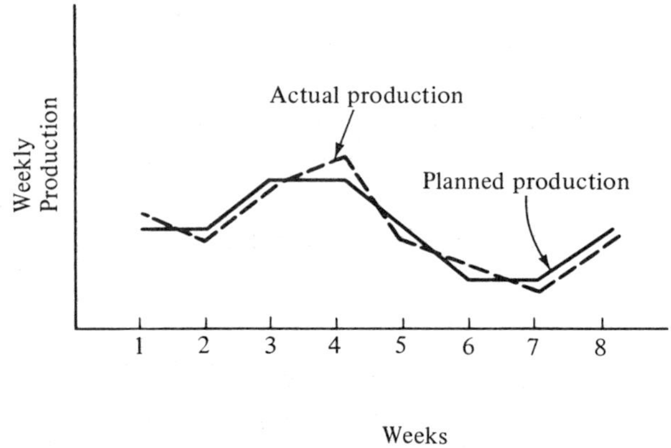

Figure 7.7 Comparison of Planned and Actual Production

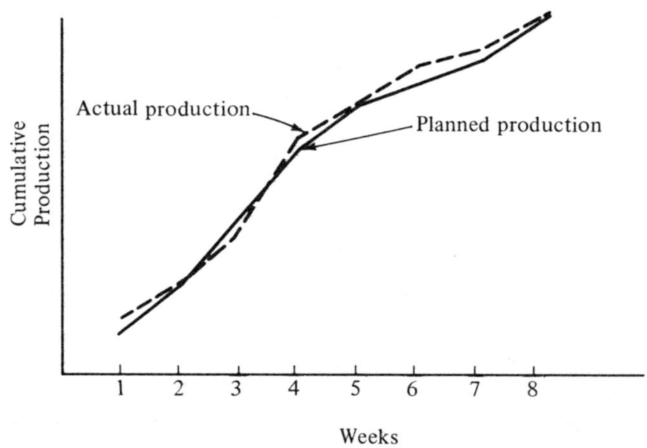

Figure 7.8 Comparison of Cumulative Planned and Actual Production

planned and actual inventory accumulations, etc. We should also recognize that a separate set of graphs can be kept for each major finished good, for each product line, or for total plant production.

For monitoring large order production or project type activity, a percent completion graph, as shown in Fig. 7.9, is very effective. Percent completion is plotted against a reverse time scale. The control limits are intended to portray allowable deviations from planned percent completion. Note that the control limits converge as they approach the delivery date.

Cost control and schedule control for large projects can also be aided by using graphical representations. Figure 7.10 shows two such methods. Again,

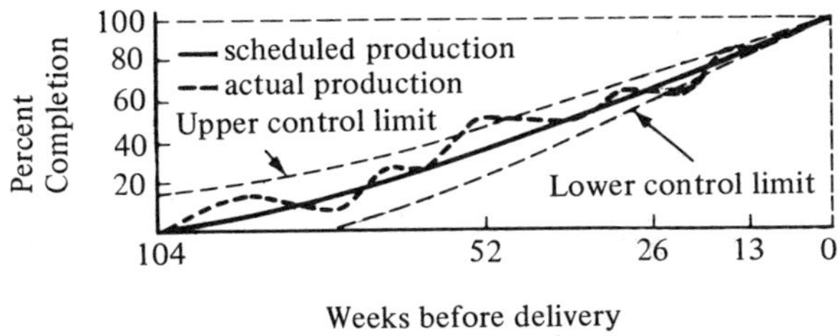

Figure 7.9 Measuring Production in terms of Percent Completion

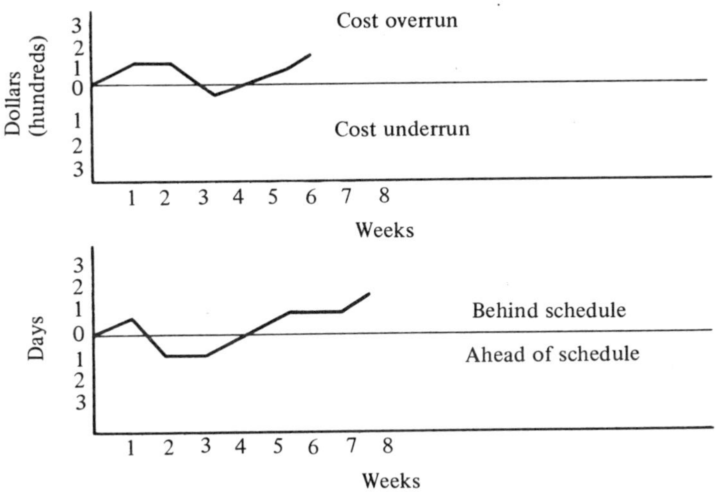

Figure 7.10 Cost and Schedule Progress Charts (adapted from Greene, 1967, p. 97)

we could place control limits on these graphs that would express the permissible fluctuations.

We could devise many other graphs and charts that would help us evaluate performance. For example, we could portray running averages of idle time, work-in-process inventory, raw material inventory, throughput time, per unit production cost, overhead rates, etc. We could then compare last period's actual performance for each of these measures to the running average.

Classifying Deviations. Each significant deviation from our established standards (operations plan, schedule, etc.) should be recorded and its cause analyzed. A few causes usually generate most of the problems. We need to know the relative frequency of causes so that we can direct our attention to the more frequently occurring problems. For example, work stoppages and holdups can be classified according to the following causes (Eilon 1962, p. 411):

- Lack of drawings or instructions.
- Lack of materials.
- Lack or failure of tools.
- Work held up by previous operations.
- Machine breakdown.
- Operator missing or not available.
- Waiting for inspection to approve work or machine setting.

We can classify other deviations according to their causes. The relative frequencies can be cumulated, as described earlier. We must also trace down cause-effect relationships so that we can anticipate and prevent problems. For example, high scrap rate at a work station may be the result of faulty casting (in another department) rather than of faulty operation or tooling at the work station. In general, when a deviation or problem occurs, we should make every effort to associate the cause of the problem with surrounding occurrences.

We can form statistical distributions of many of the variables in the system. For example, two distributions relating to equipment breakdown are shown in Fig. 7.11. We would probably want to generate separate sets of these distributions for the different classifications of equipment, and perhaps for each individual machine.

A few of the other random variables for which we would want to generate distributions are:

- Lead time for each vendor and for each purchased item.
- Quality records for each vendor and for each purchased item.
- Operation times.
- Queue sizes.
- Total throughput time, etc.

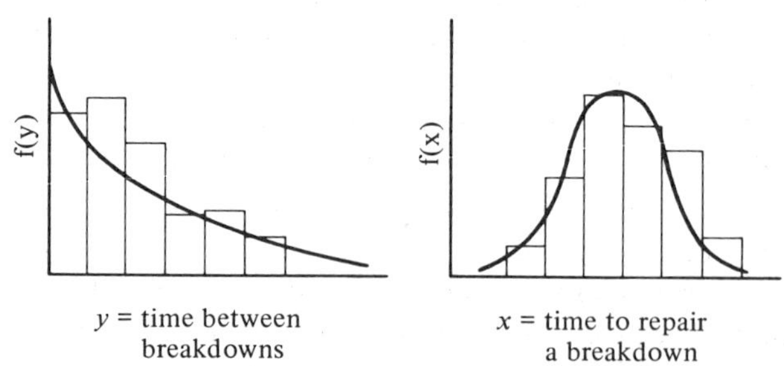

Figure 7.11 Distribution relating to Equipment Breakdown

It is important to emphasize that all of these distributions are generated and kept updated through feedback.

Measures of Effectiveness for the Control System. Suppose that we do all the things discussed in the preceding paragraphs and that we have at our disposal all the timely, accurate information that has been suggested. What is our next step? What do we do with this information? Exactly how do we use it to improve the performance of our control system? Let us engage momentarily in wishful thinking and discuss what we would *like* to be able to do. In other words, let us discuss an idealized control system of which our present topic, MOE evaluation, is the correction loop (we are speaking here of the Planning Control Loop in Fig. 1.3.).

Ideally, we would like to have a mathematical model of our production system that includes all the relationships between the components of the system. Such a model would not only tell us the total system effect of certain corrective actions (such as the effect on Department *B* of adding more capacity in Department *H*); it would determine for us the particular set of corrective actions that would, in some sense, optimize total system performance. Our feedback would determine updated values of parameters in the model. Such a system would operate optimally by continuously correcting itself through feedback.

We can use as an analogy a steel rolling process that automatically adjusts its rollers for the desired thickness. Such a system is illustrated in Fig. 7.12. As the steel sheet is forced through the rollers, the thickness gage measures the thickness of the sheet. This thickness is compared to an established standard. When the measured thickness is outside the acceptable range, an appropriate adjustment is automatically effected.

Returning to our production system, it is natural to ask why we cannot exercise control in much the same precise way that is done in the steel rolling process. There are two reasons why we are not so fortunate. First, in a production system we must deal with dozens of variables, many of which interact in a complex fashion, instead of the one (sheet thickness) controlled in the steel rolling example. Second, in a production system we have more than one criterion to which system performance must be compared. It is sometimes tempting to think that our criterion is always monetary, in which case we would attempt to minimize cost or maximize profit. A little reflection reveals, however, that some variables simply cannot be reduced to monetary terms, while many of those that can be are practically impossible to measure.

It seems certain that the current status of management science is such that we cannot expect to be able to surmount the obstacles described above in the foreseeable future. Faced with this dilemma, we must do the best we can with the tools we have to work with. Specifically, we can associate certain measures of effectiveness (MOE's) with each function of our operations planning and control system. We can also specify gross measures of effectiveness for the entire system.

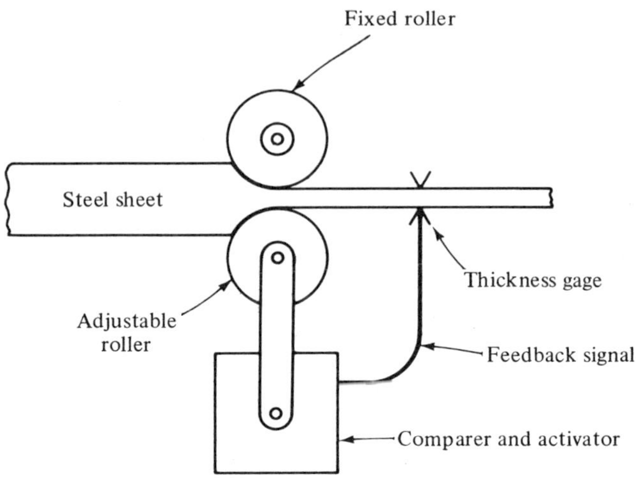

Figure 7.12 Steel Rolling Process that automatically adjusts itself

MOE's for a particular function are arrived at by considering the behavior we know the function should exhibit if it is performing properly. For example, in the inventory function we know that average inventory level must be kept low in order to avoid excessive carrying costs. At the same time, we know that enough inventory must be carried to avoid excessive stockouts and order (or set-up) costs. Three MOE's that we should associate with inventory control are, therefore, average inventory level, number of stockouts occurring, and number of orders placed. In this case, we can also specify MOE's for carrying cost, order cost, and stockout cost. These costs then can be combined into one overall MOE, called total cost. Inventory theory has been developed to the point that for individual items we can determine the optimal reorder points, safety stock, and order quantities. When we attempt to apply this theory to the determination of batch sizes for many items whose production conflicts among work stations, we are unable to obtain generally optimal solutions. The MOE's discussed above are useful for bridging the gap between the functions of inventory and scheduling.

MOE's for other functions in the control system can be established by similar reasoning. For each function we determine the behavior that we would like for it to exhibit with respect to pertinent system variables. Then, as the system operates, we measure the values of the variables each period (week, month, etc.) and call these the computed MOE's. We compare the MOE's to established system performance criteria and then adjust the parameters of the transfer functions accordingly.

An instructive way to represent these concepts is to imagine that the operations control system has a large control room, similar to the control room of a petroleum refinery or an electric power generating plant, which contains many dials. There would be one dial for each MOE in our system. Painted on each dial would be control limits within which the performance of the MOE is considered normal, or acceptable. Below the dial would be a control knob that would permit us to adjust the MOE whenever the needle points to a value outside the control limits.

Shown in Fig. 7.13 are several dials for MOE's that can be associated with the inventory control function. There would be a set of dials for each of the several functions of the control system, as shown in Fig. 7.14.

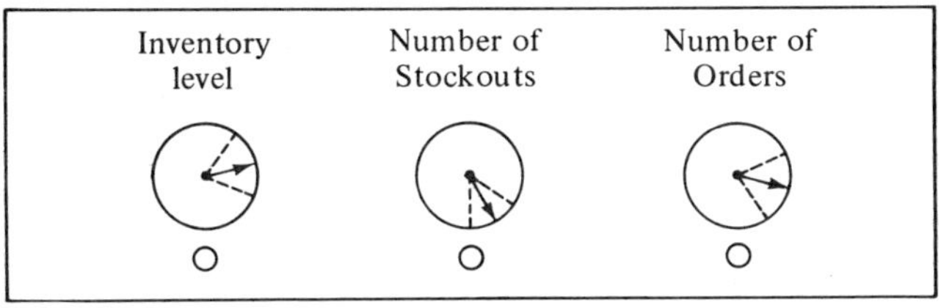

Figure 7.13 MOE's for Inventory Control

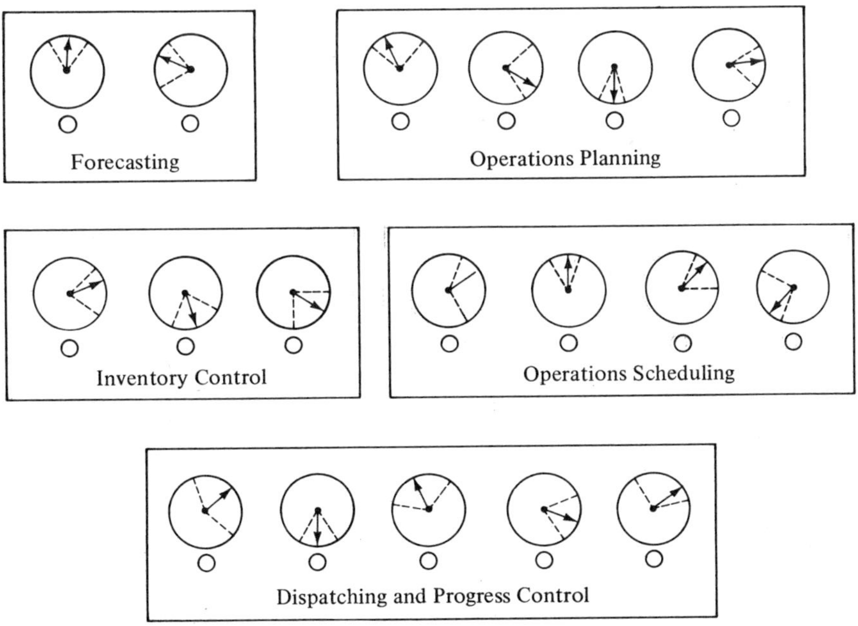

Figure 7.14 MOE's for the entire Control System

Table 7.1 Measures of Effectiveness by Function or Activity

Function or Activity	MOE's	Performance Criterion Compared to
1. Forecasting	1.1 Actual demand for regular market items	1.1 Limit established on forecasted demand
	1.2 Number and size of rush orders	1.2 Corresponding statistical distributions developed from historical data
	1.3 Cancelled orders	1.3 Corresponding statistical distributions developed from historical data
	1.4 Demand for new products	1.4 Market survey; sales analysis
	1.5 Declining demand	1.5 Life cycle projection
2. Operations Planning	2.1 Percent utilization	2.1 Cumulative performance index
	2.2 Backlogged production	2.2 Normal backlog, with upper control limit
	2.3 Operation times	2.3 Standard times
	2.4 Overtime required	2.4 Economic shift change point
	2.5 Subcontract support	2.5 Short-term make-or-buy breakeven point
	2.6 Scrap and rework	2.6 Quality control standards
	2.7 Flow congestion	2.7 Established performance index
	2.8 Inventory levels	2.8 Space allocated for inventory
	2.9 Inventory investment	2.9 Budgetary limitation
	2.10 Skill utilization	2.10 Skill availabilities
	2.11 Capacity utilization trends	2.11 Short-, intermediate-, and long-term capacity maintenance and displacement plans
	2.12 Capacity used by new products	2.12 Engineering standards
	2.13 Amount of delay caused by machine breakdown	2.13 Economic maintenance crew and facilities
	2.14 Process efficiencies	2.14 Standard procedures and work methods; new processes available
	2.15 Production quantities	2.15 Planned production quantities
	2.16 Ending inventories	2.16 Planned ending inventories
	2.17 Overtime required	2.17 Planned overtime
	2.18 Time used in preventative maintenance	2.18 Scheduled preventative maintenance
	2.19 Frequency and duration of equipment failures	2.19 Corresponding statistical distributions
	2.20 Idle time factor	2.20 Average idle time, with upper control limit

	2.21 Manufacturing costs	2.21 Standard costs
	2.22 Overhead burden	2.22 Overhead goal in budget
	2.23 Labor productivity	2.23 Expected productivity
	2.24 Absenteeism	2.24 Amount expected for this period
	2.25 Number of rush orders and other external perturbations	2.25 Anticipated number from historical records
3. Inventory Planning and Control (Some of the MOE's pertain to both individual items and total inventories. Some also pertain to all types of inventory; i.e., purchased parts and material, in-process, and finished goods.)	3.1 Average inventory level and cost	3.1 Standard from *EOQ* analysis
	3.2 Number of stockouts and cost	3.2 Number permitted by management policy
	3.3 Number of orders placed and cost	3.3 Standard from *EOQ* analysis
	3.4 Total cost of 3.1, 3.2, and 3.3	3.4 Expected total cost
	3.5 Inventory value lost in storage	3.5 Cumulative performance index
	3.6 Order and manufacturing lead times	3.6 Corresponding statistical distributions
	3.7 Usage during lead times	3.7 Corresponding statistical distributions
	3.8 Total inventory investment	3.8 Budget constraint
4. Operation Scheduling	4.1 Production Completed	4.1 Scheduled production
	4.2 Completion times	4.2 Scheduled completion times
	4.3 Status of work in process (Percent complete or projected completion date)	4.3 The schedule; due dates
	4.4 Actual operation times	4.4 Standard times
	4.5 Material shortages	4.5 Reorder points
	4.6 Queue sizes	4.6 Economic buffer sizes
	4.7 Manufacturing lead times	4.7 Corresponding satistical distribution
	4.8 Set-up and process changeover times	4.8 Standard times
5. Dispatching and Progress Control	5.1 Number of jobs requiring expediting	5.1 Cumulative performance index
	5.2 Number of holdups due to poor materials handling	5.2 Cumulative performance index
	5.3 Amount of in-process inventory	5.3 Cumulative performance index, budget constraint
	5.4 Measured congestion	5.4 Percent allowance for nonsynchronized facilities
	5.5 Schedule slippage	5.5 The schedule
6. Engineering	6.1 Number of customer complaints	6.1 Product design; material specifications
	6.2 Process effectiveness and cost	6.2 Acceptable performance level
	6.3 Effectiveness of operating procedures	6.3 Acceptable performance level
	6.4 Measured congestion and interference	6.4 Facility design, acceptable performance level

Table 7.1 (Cont'd.)

Function or Activity	MOE's	Performance Criterion Compared to
	6.5 Process time	6.5 Standard times
	6.6 Labor productivity	6.6 Standard worker allowances
	6.7 Quality and scrap records	6.7 Statistical control limits
	6.8 Tool breakage and replacement	6.8 Standard speeds and feeds
7. Purchasing	7.1 Quality records	7.1 Vendor records; acceptable performance level
	7.2 Lead times	7.2 Corresponding statistical distribution for each vendor
	7.3 Material costs	7.3 Vendors' quotations; quantity discounts
8. Financial Control	8.1 Aggregate inventory investment	8.1 Budget allowance
	8.2 Direct labor cost	8.2 Standard for this output level
	8.3 Overhead costs	8.3 Budget allowance for each type
	8.4 Idle man and machine costs	8.4 Upper control limit
	8.5 Scrap and rework costs	8.5 Acceptable standards, with upper control limits
	8.6 Cost of each order	8.6 Estimated cost
	8.7 Shift change costs	8.7 Production plan
	8.8 Marketing cost	8.8 Budget allowance; sales increase
	8.9 Total production costs	8.9 Estimated total costs
9. Total System	9.1 Profit	9.1 Projected profit
	9.2 Return on investment	9.2 Acceptable performance
	9.3 Share of market	9.3 Established goal; competitor's performance
	9.4 Innovation	9.4 Goal; competitor's performance
	9.5 Productivity	9.5 Goal
	9.6 Financial reserves	9.6 Projected reserves
	9.7 Employee turnover	9.7 Goal
	9.8 New customers	9.8 Goal
	9.9 Sales increase	9.9 Goal; market analysis
	9.10 Backlog	9.10 Acceptable level
	9.11 Unsatisfied demand	9.11 Acceptable number
	9.12 Forecasted downward trend	9.12 Life cycle of product line

The primary function of the production control department would be to monitor the several dials and to adjust the MOE setting whenever a needle points to a value outside the control limits. We must recognize, of course, that there are many complex relationships between the MOE's. When we adjust one MOE by turning its control knob, we usually affect the setting of other MOE dials. It is important to realize that this is not an optimizing procedure; rather, it is a balancing or tuning procedure by which we attempt to maintain system equilibrium and effectiveness. The extent to which we accomplish this depends on several things: proper specification of MOE's, correct association of MOE's to system performance criteria, inherent system stability from one period to the next, accuracy and timeliness of feedback data, extent of external perturbations, and the extent to which external perturbations can be smoothed.

A set of MOE's that might be appropriate for an operations planning and control system is outlined in Table 7.1. The MOE's are not mutually exclusive; i.e., a particular MOE may be pertinent to several functions. Some of the MOE's are computed weekly, some monthly, some quarterly, and some annually, depending on the planning horizon for the particular function.

Feedback and Corrective Action

The reader will note that in Chaps. 3 through 6, this section discusses the manner in which transfer functions for Forecasting, Operations Planning, Inventory Planning and Control, and Scheduling, respectively, are monitored and updated. Such a step is not necessary in this chapter since the dispatching and progress control function monitors itself. The discussion in Chap. 8 assists in showing how this is accomplished.

Interfaces

The dispatching and progress control function interacts with all other functions within the boundaries of our operations planning and control system and with many functions outside the boundaries. The reason for so many interfaces is the close association of this function with the total information system of the firm.

In Chaps. 3 through 6, this section included a diagram showing the important interfaces. The corresponding diagram for this chapter is shown in Fig. 7.15. Notice that this is simply Fig. II.1 on page 44.

Note that we have ended Part II with the same diagram we started it with. This is hardly coincidental. We began the development of our control system with an overview, in which all the components were related in a general way to each other and to the total system. We then examined each component in minute detail, ending each chapter with a diagram that shows that com-

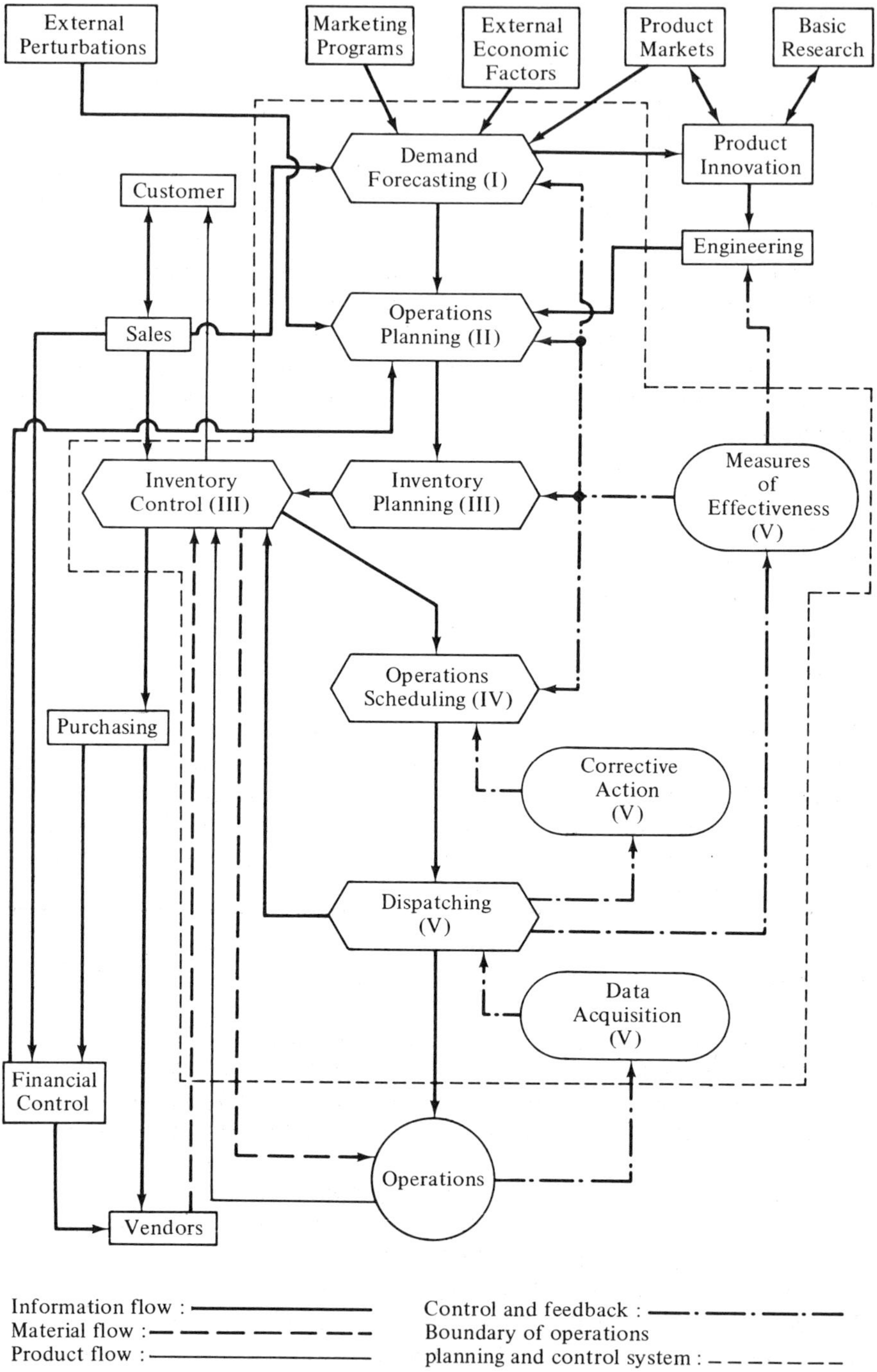

Figure 7.15 Master Flow Diagram: Operations Planning and Control System

ponent's interactions with other elements in the system. Finally, we have put the pieces back together as Fig. 7.15.

Part III is concerned with problems of designing total control systems.

REFERENCES

Brooks, G. H. 1964. *Mathematical Models for Information Processing Systems.* Unpublished Ph. D. Dissertation, Georgia Institute of Technology (Atlanta, Ga.).

Eilon, S. 1962. *Elements of Production Planning and Control.* New York: The Macmillan Company.

Greene, J. H. 1967. *Operations Planning and Control.* Homewood, Ill.: Richard D. Irwin, Inc.

Homer, E. D. 1962. A Generalized Model for Analyzing Management Information Systems. *Management Science* 8, 500–515.

LeGrande, E. 1963. The Development of a Factory Simulation System Using Actual Operating Data. *Management Technology* 3, 1–19.

Lieberman, I. J. 1956. A Mathematical Model for Integrated Business Systems. *Management Science* 2: 327–36.

Plossl, G. W.; and O. W. Wight. 1967. *Production and Inventory Control: Principles and Techniques.* Englewood Cliffs, N. J.: Prentice-Hall, Inc.

IBM Reference Manual E20-8041. 1960. *General Information Manual: Management Operating System for Manufacturing Industries* (White Plains, N. Y.).

SOURCES OF ADDITIONAL TRANSFER FUNCTIONS

Listed below are several additional sources of dispatching transfer functions. The list, not intended as an exhaustive bibliography on the dispatching function, is a representative sample of the dispatching and progress control techniques available.

Beuter, R. J. 1963. A Theory for the Maintenance of Control of the Firm. *Journal of Industrial Engineering* 14: 175.

Conway, R. W. 1965. Priority Dispatching and Work-in-Process Inventory in a Job Shop. *Journal of Industrial Engineering* 16: 123.

Estabrook, L. C. 1966. Product Cost Control Reporting System. *Proceedings of the 1966 AIIE National Conference* (San Francisco).

Joss, E. J.; and Polk, E. J. 1967. Integrated Management Information System. *Journal of Industrial Engineering* 18: 625.

Korn, K. E.; and Lamb, J. H. 1967. Computerized Management of Production Control. *Journal of Industrial Engineering* 18: 677.

IBM Reference Manual E20-0090-0. *Principles of Data Acquisition Systems* (White Plains, N. Y.).

III Control System Design

8 Designing the Control System

In each chapter of Part II, we treated one of the five functions of operations planning and control. Even though we discussed the interfaces of each function, the total control system was discussed only very briefly, at the beginning and at the end of Part II.

There are two chapters in Part III. In this chapter, we discuss certain basic systems concepts, the phases of systems evolution, and several design considerations specific to designing operations planning and control systems.

In Chap. 9, we speculate on the future of operations planning and control systems design.

Systems Concepts

DEFINITION OF TERMS

In recent years, the terms *systems*, *systems engineering*, *systems design*, and other similar words have come into common usage. Unfortunately, these terms tend to mean different things to different people, according largely to the educational and work experience of the person using the term. Therefore, to avoid confusion, we must define these terms.

We can start with the broad dictionary definition of *system*: "A set or arrangement of things so related or connected as to form a unity or organic whole; a set of facts, principles, rules, etc., classified or arranged in a regular orderly form so as to show a logical plan linking the various parts." Since this definition is very broad, it is better for us to adopt a more limited definition: "A system is a set of components related by some form of interaction or interdependence." Then, if we include within "components," the transfer functions, data files, data processing, decisions, and other factors we have discussed, this definition is adequate for our purposes.

Turning to the term *system design*, we must first look at the word *design*. Again, this is a word that is greatly overworked and for which many definitions exist. The most useful understanding of the word for our purposes is: *The bringing together (synthesis) of known concepts and knowledge of the several sciences into a novel, useful and economically feasible configuration in order to satisfy some need.* Now, combining our two definitions, we can say that *systems design* is: *The bringing together of known concepts and knowledge of the several sciences into a novel, useful set of components organized to satisfy a definable user requirement.*

Systems engineering is also a much abused term; many volumes have been, and will be, written about the subject. We avoid this semantic struggle by considering *systems engineering* the entire process, including but not restricted to design, which must take place in order to obtain a functioning system. In later sections of this chapter, we discuss the primary things which must take place to achieve a functioning operations planning and control system.

EVOLUTION OF A SYSTEM

The evolution of a system has four basic phases, with a number of fundamental steps in each phase. The four basic phases are:

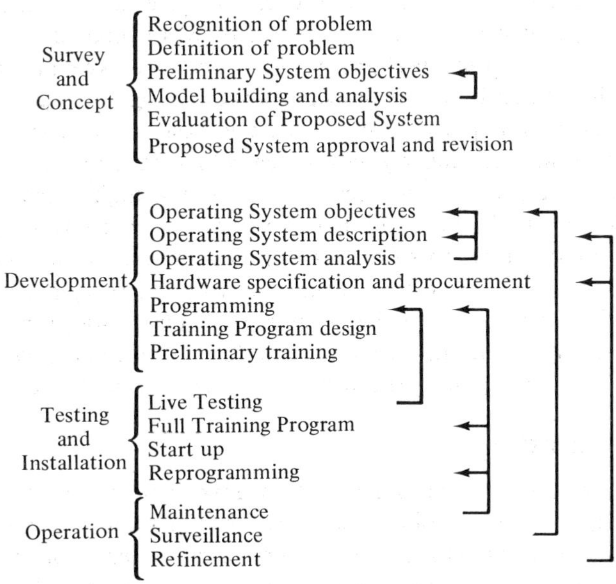

Figure 8.1 Conceptualization of the Design Process

1. Survey and concept.
2. Development.
3. Testing and installation.
4. Operation.

In order to provide the reader a better understanding of this total design process, the four phases and their associated fundamental steps are shown schematically in Fig. 8.1. The four major sections that follow present these four basic phases; subsection headings correspond directly to the fundamental steps of each phase. While studying the next four sections, frequent reference to Fig. 8.1 is suggested.

Survey and Concept

In this phase, we familiarize ourselves with the over-all situation with which we have to deal, establish the conceptual framework for the final system, evaluate our concept in terms of technical and economic feasibility, and establish design objectives for the system. Several steps are essential in this phase.

RECOGNITION OF PROBLEM

This step may appear to be trivial, yet it is nevertheless a very important one. All too frequently, particularly in the area of data processing, some companies have designed and installed very comprehensive systems without any attempt to find out precisely what problem the system was supposed to solve. Such companies were motivated, perhaps, by the notion of "keeping up with the Joneses," by having an impressive computer installation.

It also frequently happens that a system is designed to cure a symptom, rather than to attack the underlying problem which gave rise to the symptom. For example, management may voice concern over a high percentage of late shipments. We would be making a mistake to confine our efforts to the warehouse and shipping areas to solve this symptom when, in fact, the problem might lie in poor communications from the sales districts, poor production planning and control, or even the assignment of unrealistic shipping dates in the face of process limitations. We must, therefore, determine early in the game just what the problem is that we are trying to design a system to handle.

DEFINITION OF PROBLEM

Having learned what our problem is, it is important to state it precisely. In this definition, we should avoid generalities, striving rather to be very specific about what it is that we wish to attack. This careful definition permits us in the next step to specify exactly how we want our system to perform.

PRELIMINARY SYSTEM OBJECTIVES

After our careful, precise statement of the problem, we are now in a position to establish preliminary system objectives. At this point, we view our system rather idealistically, not concerning ourselves too much with either technical or economic feasibility. Rather, we state as precisely as possible what we want the system to accomplish and how we want it to perform.

To be useful, these objectives must be stated in very explicit terms, quantitatively wherever possible. For example, rather than defining an objective as "improving shipment performance," we should specify that, "The system must be capable of making shipments to customers on or before the scheduled shipping date for 95 percent of all orders and 98 percent of all tonnage." Unless we are very explicit, we have no means of measuring the performance of our system after installation, and we will have great difficulty in selecting alternative components in building our system.

MODEL BUILDING AND ANALYSIS

With our problem well defined and the preliminary statement of our objectives prepared, we are now in a position to start the design, or "synthesis," stage. It is important to remember, however, that we are still in the survey and concept stage. We are making only preliminary decisions on a gross basis. In many cases, we include in this model several alternatives for each of the components needed in the final system. For example, if one of the components needed is a forecasting transfer function, we may, at this point, decide only that we will use either the moving average transfer function or the exponentially weighted moving average transfer function.

The general notion involved in this step is that we survey the field for components (transfer functions, data processing techniques and equipment, etc.) that might be useful in our system. We accomplish this survey by referring to our past education and experience, by searching of literature, by examination of similar systems in other enterprises, by examination of the existing system, by consultation with experts, etc. We then fit these components together into a model of our system and analyze this model to see whether these components in this configuration will meet the defined systems objectives. We discard those components which obviously do not meet our needs.

We should also note that in this step, as well as in later developmental steps, we may have to "cycle back" and reappraise our objectives because we may have overshot in defining our objectives and stated one which just cannot be met, considering the existing state of the art and any reasonable short-term development. Note, however, that it is better to overshoot than to underspecify our objectives, as the latter would lead us to less than the best possible system.

We must also be careful in this step to avoid getting too detailed. It is easy to jump ahead to the developmental phase and get involved in the detailed design before we have obtained the over-all concept. It is better in many cases to postulate that we can find a suitable component at a later date than to spend the time now to develop one.

EVALUATION OF PROPOSED SYSTEM

We have now established what might be called the basic framework of our proposed system. The details are far from being worked out, but we have established a skeletal system which we feel to be generally feasible and which will meet our stated systems objectives.

Throughout the prior steps, we have been evaluating the system in terms of meeting objectives, technical feasibility, and to some extent, economic feasibility. We must now do this evaluation formally, with the end in mind of proposing to management that we go ahead with the more detailed design, development, and installation of the system.

We desire as the output of this step the following:

- An understandable (to management) summary of the problem.
- A description of the proposed general system.
- Evidence of technical feasibility.
- An economic evaluation of the system, including:
  a. Estimated time and cost of development and installation.
  b. Estimated operation cost after installation.
  c. Anticipated benefits and savings (or profit improvement) to be derived from the system.
- A plan for the balance of the system development, including design team organization, timing, funding considerations, and a target start-up date. In many cases, it is desirable to present alternative plans, in order to consider time-cost tradeoffs.

It is interesting to note that project planning and control techniques, such as PERT, can and should be used for planning, coordinating, and controlling the design activities. This approach requires that all major activities of the project be separately defined and that their precedence relationships be established. Starr (1964, pp. 117–25) presents an excellent example of the application of the network planning approach to the design of a complete production system, including the control system.

PROPOSED SYSTEM APPROVAL AND REVISION

At this crucial point, we present our proposed system and the evaluation to management for consideration and a go-ahead for development and installation. In practice, we frequently find that we obtain only partial or conditional approval; that is, we are cleared for development subject to some cost criterion or monetary limitation.

We therefore frequently find it necessary at this point to revise our proposed system to conform to imposed constraints. This revision usually takes the form of one or more of the following:

1. Limiting the system to solve only a part of the defined problem.
2. Relaxing the preliminary systems objectives. For example, we may decide to meet only 90 percent of customer orders on time.
3. Settling for a less reliable system.
4. Trading off some installation costs for some higher operating costs.

At any rate, presuming that management does not decide just to live with the existing system, we must at this point firm up exactly what is to be embraced in our system and establish basic plans for its accomplishment.

Development

The initial steps of development tend to repeat, in much greater detail, many of the steps taken during the survey. We are now down to the detailed, painstaking task of designing our operational system. Each of the following steps must be taken, and often repeated over and over, until the results meet our objectives.

OPERATING SYSTEM OBJECTIVES

We now take the preliminary system objectives, developed and perhaps revised during the survey stage, and develop very detailed objectives. Each of the objectives developed earlier is now expanded into perhaps several subobjectives, so that we know in great detail how our system must perform. For example, when we set the objective for meeting scheduled shipping dates during the survey, we were considering all the company's products in a collective sense. During this stage, we set objectives for each product or product line to be realistic with the problems associated with each, but to meet the over-all initial objective.

OPERATING SYSTEM DESCRIPTION

With our detailed objectives specified, we now describe in copious detail all the components to be included in our system. This step involves the selection of specific decisions to be made in the system, the data and data sets to be used, the transfer functions to be used, and the data processing steps, documents, and other procedures to be followed. All these details must be completely specified, described, and related in an orderly manner to make up the system.

This step, the analysis step which follows, and even the previous objectives step, must usually be performed many times. In general, it is a convergent process, in which we eliminate, combine, change sequence, and refine the detailed system components until they meet the system objectives as nearly as possible.

OPERATING SYSTEM ANALYSIS

In this step, we look at the operating system description, together with our detailed objectives, and try to determine whether the system as we are designing it will meet these objectives. At the same time, we concern ourselves with technical feasibility, economic feasibility, and our broader objectives of solving the specific problem involved. As was pointed out in the prior step, this step and the two preceding ones are carried out in a repetitive cycle, and it is frequently difficult to specify exactly which step is currently being performed.

HARDWARE SPECIFICATION AND PROCUREMENT

We finally reach a point through repeated operation of the three prior steps where we know, to a high degree of certainty, exactly what we want concerning decisions, data, data sets, and transfer functions. At this point, we must specify the hardware which will be used to handle these components. At this point (and usually through all of the prior steps), we very much need the services of an expert in this field, either from within the organization or on a consulting basis.

It is beyond the scope of this book to deal with the problem of equipment selection. In general, however, we can say that it is a matter of engineering economic evaluation—we wish to obtain that equipment which will perform the desired operational steps at the lowest possible long-term cost.

PROGRAMMING

In developing operations planning and control systems, we are concerned with programming of two types. When we are dealing with computer-based systems (the majority of operations planning and control systems are of this type), we obviously must program the computer and all associated hardware. This programming is, of course, a highly specialized matter, the details of which must be left to the professional programmers.

As systems designers, however, we must specify in great detail what the programmer must do. This communication is accomplished through the medium of our detailed operating system description and by working closely with and supervising the efforts of the programmer.

We are also concerned with another programming task. For the system to operate effectively, every function of the system must be provided with detailed instructions (and, therefore, a program) to follow in order that the function may be properly performed. We must program the activities of all the organizational areas and the individuals in these areas who are in any way concerned with the system we are designing. This programming takes several forms—

procedures manuals, operating instructions, detailed job design, workplace design and layout, forms design, etc. This part of the programming task must be done as carefully as the equipment programming if the system is to function as designed.

TRAINING PROGRAM DESIGN

Now that we have designed the system and programmed the hardware and other functions, our attention must be turned to training. In general, we are concerned with the training of three groups of people:

1. Operating personnel.
2. Management.
3. Peripheral personnel.

The need for training operating personnel is obvious, in that they are, in fact, components of the system and need to be trained in detail to perform the activities for which we have programmed them. Our new system is usually so different from the old one that they must learn a whole new set of duties. Depending on the organization, the systems designer may get a great deal of assistance in this part of the training program from the training supervisor. However, the technical content of the training must come from the designer. A great deal of assistance may also be had from the foreman or supervisor, but we must bear in mind that he, too, must be trained in matters relating to the new system.

The need for training (or orientation of) management is perhaps less obvious but equally important. It is usually more difficult to accomplish. It is important because, unless the manager knows how the system is supposed to function, he cannot intelligently use it. It is difficult because the demands upon management's time are many and because the ego of management is disinclined to admit to the need of further training.

Training for peripheral personnel might also appropriately be called orientation, since we are only trying to keep them informed. We feel it wise to do this since almost all new developments within an organization may be interpreted as a threat to job security. In most cases, training for these people can be disposed of with an hour or so of discussion by a person from management as to the general objectives of the new system and the way in which it affects (or fails to affect) them.

PRELIMINARY TRAINING

In order both to test our training program and to prepare for the testing of our system, we now undertake preliminary training of key operating personnel who will be active in our system. We train a minimum number of people at this time, but we make every effort to train them as thoroughly as possible. In many cases, the people trained at this time are first-line supervisory personnel who are capable of functioning as operators and who will later assist us in training their personnel.

With our system developed and programmed, and some key personnel trained, we are now ready for the next vital phase.

Testing and Installation

LIVE TESTING

Throughout the developmental stages, we have without a doubt been testing certain aspects of our system as they have been developed. Certainly, the equipment programmers have been making debugging runs, and we have probably tested many other components of our system before deciding to incorporate them.

It is now time, however, to test our entire system, under conditions which approximate as nearly as possible actual operating conditions. We want to debug the whole system, to maximize our probability of a successful start-up.

We do not, at this time, wish to let our system interfere in any sense with the existing, ongoing system. To avoid this possibility and to "shake out" our system, we usually conduct our live test with sets of data that have been concocted to test every feature of our new system. For example, we may concoct a set of 100 sales orders with every conceivable variation and use them repetitively to test the system.

Later on, we may use actual data to test the system, being very careful not to interfere with the ongoing system. Note that we cannot fully test the decision portion of our new system, but only the mechanical portion. If we tried to test the decision portion, we would have to supplant the existing decision system, which we are not yet ready to do.

Based on these live tests, we make modifications to our system, usually in terms of details of programming both equipment and functions. After these modifications are made, we must revise our training program to incorporate the necessary changes.

FULL TRAINING PROGRAM

Our next step is to undertake a full training program for all operating, management, and peripheral personnel, as was discussed in the development phase. We emphasize again that much of the success of the new system will depend upon how well we train the people who will operate and use it.

START-UP

The objective of the start-up operation is, of course, to get our new system on line and to phase out the old one. Just how we accomplish this substitution is not always obvious. We naturally want to get the new system into operation as quickly and as easily as possible. However, despite our careful design and our live testing program, we always have some fear that the new system will not perform adequately at first and that we may have to retreat to the old system for a while so that we can make further modifications. Therefore, we try to begin in a manner that will permit us to revert to the old system temporarily whenever necessary.

There are four basic ways to begin a new system. One, of course, is to merely discontinue the old and start the new at some predesignated time. This method is mandatory for the decision portions of the system, since the new and old systems will differ in the decisions reached, and since only one decision can possibly be implemented. However, this method is to be avoided if at all possible.

Another method is to start the new system, or as much of it as possible, into operation parallel with the existing system, then terminate the existing system when the new system seems capable of continuing successfully. Note again that we cannot use this method for the decision portions.

A third method can be called the step-wise method, wherein some portions of the new system are implemented, then at later dates others are implemented, and so on until the whole system is in operation.

The fourth method embraces the best features of the second and third and uses the first for decision portions only. This method might be termed a step-wise parallel method. In this method, the more mechanical portions (data origination and transmission) are started up, one or two operations at a time, parallel with the existing system. As each new system operation seems to be functioning properly, the counterpart operation of the old system is permitted to lapse. Then, in a few days, another segment of the new system is started, again in parallel, until we progress through the strictly mechanical systems operations to the more intellectual ones. These latter operations, the decisions, must then be started up, but we have much more assurance that they will function properly. Furthermore, by starting up in this manner, we can in general be assured of a fail-safe start-up, in that we can quite easily revert to the old system if necessary to maintain continuity of operation.

REPROGRAMMING

During the start-up operation, we are almost sure to find some weaknesses in our design. As soon as possible after the system is operating relatively smoothly, the necessary changes should be made in both the hardware programming and in the manuals and other instructions which serve to program the other system functions. In some cases, fairly major revisions may be necessary during the early operating period to correct design oversights. Furthermore, during the start-up period, some things may have been changed on a basis of expediency in order to get the system going. We should now reexamine these features with a view to smoothing out rough operation points in our new system.

Operation

From a strict point of view, our work is completed when the system is operating smoothly. Yet, to assure good continuous operation, we need to provide a basis for continuing effective operation of the system. Three steps, or more properly, areas of concern, are important in the ongoing operation. Depending upon the nature of maintenance, surveillance, and refinement, re-cycling may be necessary to one of several steps in the development phase or the testing and installation phase. This possible re-cycling is shown in Fig. 8.1.

MAINTENANCE

In addition to the acknowledged need for maintenance of the equipment and hardware in the system, there is a continuing need for maintenance of the remainder of the system. One obvious point of maintenance, which must be built into an operations planning and control system, is the maintenance of the Parameter File. Less obvious is the need to maintain the programming, both of hardware and functions. The passage of time always brings about minor changes, new personnel, and other factors which tend to degrade the operation of the system unless a mechanism is set up to detect and correct for them. Maintaining a proper level of operator competence is sometimes very difficult, so thought must be given to training of new employees and periodic retraining of old ones in order to maintain system effectiveness.

SURVEILLANCE

Closely associated with maintenance is surveillance. By this, we mean keeping an eye on things, making sure that the system is still doing what it was designed to do and that the function it is performing is a proper one. This area is primarily a management responsibility, but one in which the original designer can be of great assistance. The manager must be provided with some continuing measure of the effectiveness of system performance—some sort of control system on his control system (reference to Fig. 7.1 will reinforce these concepts).

REFINEMENT

The third important area recessary to assure continuing system performance at a high level is that of refinement. The need for refinement may make itself known through either the maintenance or surveillance area. We may find, for example, that a piece of equipment new to the market may be superior to one we are now using and may permit doing easily what we had to do awkwardly in the original design because of equipment limitations. Or, we may read of a new transfer function that we could use to replace one in our original design. In either case, we need to keep aware of opportunities to refine the system and to make these refinements if they are economically attractive.

Design Considerations

Thus far in this chapter, we have discussed certain systems concepts and have explored the basic phases and steps leading from the inception of a system through the design and development activity and finally to the operation of the system. This material has been quite general and would be of assistance to us in designing almost any system. Let us now discuss some considerations pertaining more specifically to operations planning and control system design.

ORGANIZATION FOR SYSTEMS DESIGN AND DEVELOPMENT

As we study the various functions which constitute an operations planning and control system, we are struck by two facts. First, such a system is enormously complex. Second, because the system is so vital to the functioning of the organization, it cuts across many organizational lines. In such a system, properly designed, we cannot say that all of a given function, or set of data, is purely of sales concern or purely of production concern. We note that the transfer functions we select may make a decision which has as great an effect on sales as it does on production. We, therefore, are concerned both with designing a system which will capably serve all these functions and with later convincing many persons in different organizational areas that the system is indeed capable of serving all their unique interests as well as the general interest of the over-all organization.

We can do much to alleviate these concerns if we organize properly the four phases of system evolution. First, we must assure ourselves that we, as systems designers, are in a position to get all the information pertinent to the system to be designed. One effective way to do this is to form, prior to the survey phase, a small guidance or steering committee. Ideally, this committee should consist of three to five persons, including the person who is highest organizationally in each of the several organizational areas which might be embraced by the final system. In a company-wide situation such as we might contemplate here, this might be the director or vice-president for sales, his counterpart in production, and perhaps the control director, especially if data processing activities are under his purview. The chief systems engineer should also be a member of this committee.

The committee need not meet too frequently. For example, it might meet bi-weekly or at the call of one of the members. It should not set up an elaborate reporting structure for the design group, but rather should provide a basis for informal discussion of progress and problems.

There are several benefits which such a committee can bring to a systems development activity.

A basis is provided for obtaining access to information relative to the higher-level decision structure and the data content of this level of activity. These high-level persons can readily provide access to persons at the intermediate level who are knowledgeable about decision structure and data requirements. The committee provides a mechanism for the systems design group to expose the development as it emerges. The committee members, because of their positions on the management hierarchy, can frequently act to solve personnel problems and otherwise assist, particularly as the installation phase is approached. The existence of the committee greatly simplifies the problem of management training. Because these individuals have been a part of the design effort, they will feel a great deal of personal identity with it. They will be personally receptive to the necessary detail relative to the system and will exert their personal influence on intermediate management levels in the direction of being amenable to orientation and training.

We also find that another organizational strategy is useful, both in terms of excellence of system design and in assistance in the implementation step. This strategy is to include, in sizable numbers, persons drawn from the operating groups which will be affected by the system design. Some successful design efforts have used eight to ten such persons to each systems engineer. In choosing such personnel, we should try to obtain the supervisor, assistant supervisor, and a key operating person in each case. In making such a selection, we would look first for a knowledge of the present job, second at some evidence of creativity, and third at educational background.

The roles of personnel such as these, separated temporarily from their regular duties, are to: provide an easily accessible source of fundamental knowledge of the present system, and more importantly, of the detailed problems involved; provide a liaison between the professional systems engineer and the working-level operators; furnish a balance wheel of practical considerations, which when combined with the more theoretical concepts of the systems engineer will give a system more likely to be workable; provide the cadre for preliminary training and perform much of the detailed training of the operating personnel; and assist in overcoming resistance to the new system as it develops and is installed.

Since they have been working closely with the system as it emerges, they will have a very strong feeling of identity with it, which cannot help but assist in getting it across to the organization.

Another important consideration is the use of outside consultants. Many consulting firms have emerged in recent years whose primary business is assisting companies in their efforts to design and implement control systems and management information systems. An advantage of using outside consultants is that often they are highly skilled experts who take a completely objective view of the organization and its requirements. A disadvantage is that, regardless of their experience and training, they cannot know all the details necessary to design a complete control system.

Each company must decide for itself whether to use outside consultants in its system design efforts. A good compromise might be to use both outside consultants and company personnel on the design team.

In summary, we must involve personnel from the functions with which we are dealing if our eventual design is to be a success. We must also include sufficient professional capability, using consultants where such capability is not available within the organization. If we become introspective for a moment, we can see why both are needed—we react best to the new things which we have had a part in formulating and tend to react against new things that are imposed upon us.

OBJECTIVES OF CONTROL SYSTEM DESIGN

We learned in Chap. 4, and emphasized again in Chap. 7, that we must specify explicit objectives for the performance of the operating facilities. We also learned early in this chapter of the necessity for being explicit in stating objectives in designing our control systems. Since an operations planning and control system is a particular type of system, we can state, in a general way, many of the objectives of a given control system. We should recognize that in the design of an actual system, our objectives would have to be stated even more explicitly and related directly to the organization for which the control system is being designed. These general objectives, however, are:

1. *Responsiveness:* We want the control system to respond to significant changes in the total environment but to avoid overreacting to random fluctuations in the normal operating pattern.
2. *Management by exception:* We wish to design a system that generates, accepts, transmits, processes, stores, and distributes information in a timely and accurate manner. The system should automatically consolidate, analyze, and reduce the information as much as possible so that only information actually needed in the transfer functions is presented. To assure this, we, as designers of the control system, should reduce as many decisions as possible to administrative actions. This approach permits management by exception, in which only that information pertaining to exceptional conditions is forwarded to the decision maker.
3. *Optimal performance:* We want to design an integrated operations control system that attempts to optimize the performance of the total system rather than the individual components of the system. The optimization of individual components usually results in suboptimal performance of the total system because of the complex relationships between system components.
4. *Compatibility:* We want our operations control system to be compatible with other systems within the organization. In particular, we wish to minimize duplication and contradiction between major systems. In an ideal total system for the firm, there would be a common data base.
5. *Coordinated framework:* We want to design a smooth-running system in which detailed operations are coordinated within an over-all framework.

We want to avoid squeaky-wheel management, in which production progress is left to the mercy of a frantic expediting system.

6. *Sophistication:* We want to design the control system to the proper degree of sophistication and automation. In general, we should include as much sophistication and automation as can be economically justified. As system designers, we control the degree of system sophistication by our selection of transfer functions for each operations planning and control function. Each transfer function selected should improve total system performance by at least as much as the difference between the cost of this transfer function and the next least costly alternative. The several transfer functions included in a control system will usually be of about the same relative degree of sophistication. For example, there would be no advantage in having a highly sophisticated dispatching and progress control system if the transfer functions for inventory control are relatively gross and inaccurate.
7. *Minimum parameter dependency:* We want our control system to be dependent upon the smallest possible number of parameters, consistent with required system response to pertinent events. In general, the smaller the number of parameters controlled, the fewer complex relationships involved and the fewer adjustments required.
8. *Maintainability:* We want our control system to be able to accept changes (costs, lead times, etc.) in system parameters with a minimum of effort, errors, and confusion.

SELECTION OF TRANSFER FUNCTIONS

In each chapter of Part II, we discussed some of the transfer functions available for rendering the particular decisions pertinent to the operations planning and control function being developed in that chapter. The design of a complete control system for a particular operating environment consists of integrating transfer functions into a working whole.

We may be fortunate enough to find several transfer functions suitable for use in our control system with little or no modification. Then again, we may have to modify certain transfer functions to satisfy the particular requirements of our system or even develop completely new transfer functions.

The process of synthesizing a control system involves a great amount of trial and error, particularly when one first attempts this process. In the past, most designers of operations control systems have relied largely upon intuition, experience, and extensive technical knowledge of the particular operating system. While these characteristics will always be essential for good system designers, it is hoped that the design process will be approached more systematically in future efforts. The following steps are presented as a general procedure for

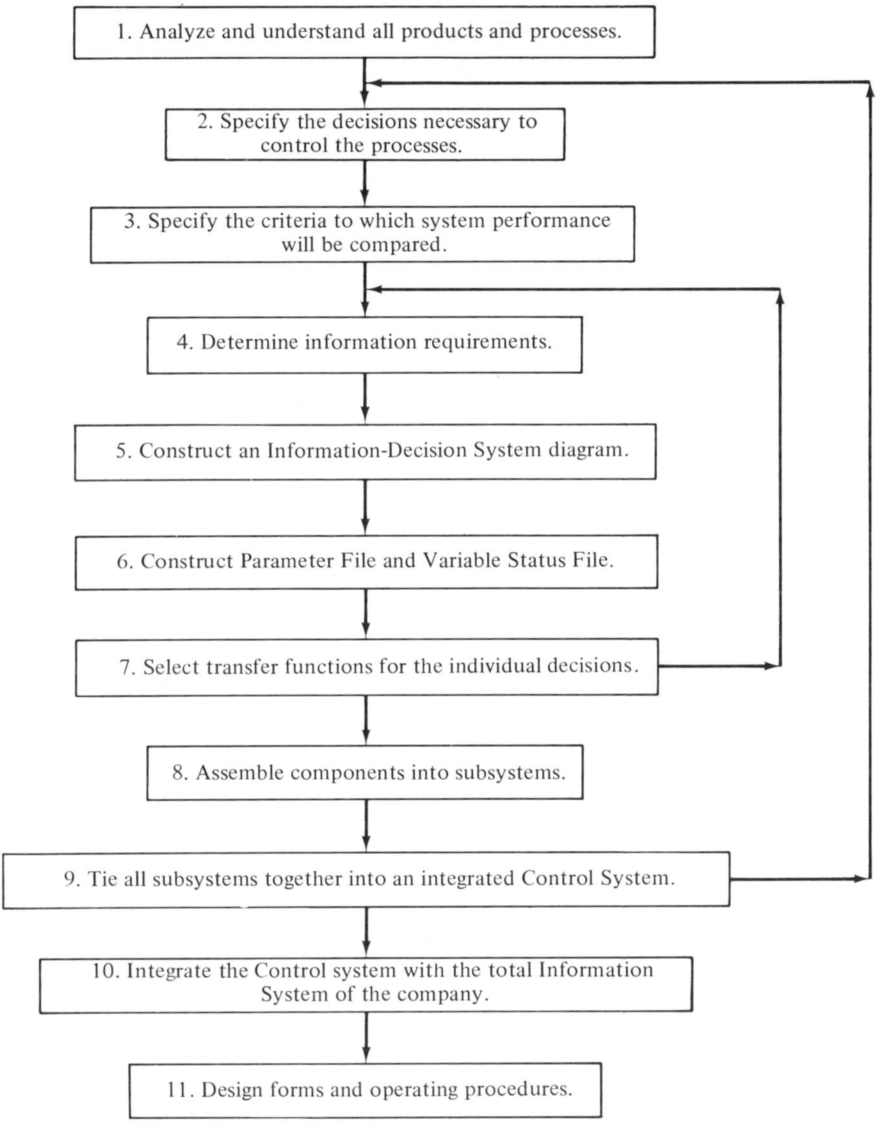

Figure 8.2 The process of selecting Transfer Functions

selecting transfer functions and integrating them into an operations control system. The procedure is shown schematically in Fig. 8.2.

1. Analyze and understand all products, materials, work methods, and all manufacturing and distribution processes.

2. Specify the decisions necessary for controlling the processes and the relationships between decision points.
3. Specify the criteria to which system performance will be compared. (See Table 7.1 for a rather lengthy list of such criteria.) We may not be able to specify numerical values for all criteria at this point, but we must at least state what the criteria are.
4. Determine the information required for making the decisions and for describing system performance so that it may be compared to the criteria specified in Step 3. Designate each information element as a parameter or variable and specify its source.
5. Construct an information-decision system diagram such as that shown in Fig. 2.5. The actual transfer functions (i.e., the decision processes by which the decisions will be rendered) have not been specified at this point.
6. Consolidate all information elements into two common files, the parameter file and the variable status file. It is at this point that we discover the particular information elements required in more than one transfer function.
7. Select transfer functions for the several decisions; keep in mind that the transfer functions selected must be capable of operating together as a total control system. Test each transfer function by the use of simulation or historical data. Repeat Steps 4 through 7 until a suitable transfer function for each individual decision has been selected.
8. Assemble subsystem components of the control system to test the interactions among the several components. This testing is particularly important for the transfer functions. Run these subsystems using simulated or historical data until sets of transfer functions are found which meet the system criteria.
9. Assemble progressively larger subsystems until all the essential components are included in an integrated control system. Continue to test with data, repeating Steps 2 through 9, until all transfer functions are compatible and the total system performs as expected.
10. Integrate the operations planning and control system with the total information system of the company. A common data base for all information requirements is feasible in almost every case and should be strived for.
11. Design forms and operating procedures for all components of the system. Particular emphasis should be placed on the means of acquiring data from various points of origination.

SOME PRACTICAL CONSIDERATIONS

Let us now discuss in greater detail certain practical aspects associated with the process of selecting individual transfer functions and integrating them into a total control system. We follow the same order as in the previous section.

Item 1: It is certainly no easy task to understand "all products, materials, work methods, and all manufacturing and distribution processes." However, we must acquire as complete a picture as we can of the operations system for which we are designing a control system. We must understand our product lines, our markets, the manner of distributing our products, our suppliers, our particular manufacturing methods, etc. In this step an outside consultant is of limited value, whereas the operating supervisors attached to the systems group can be invaluable.

Item 2: Decisions required for controlling processes may be quite difficult to obtain. Decision processes are poorly understood, even by those who are involved in them. This is particularly true of lower-level decisions, i.e., those made on the shop floor. Few such decisions are documented, and few shop foremen or workers can easily describe the decision processes that they go through. Even fewer shop people recognize the dependency relationship between decision points. It is necessary, therefore, for the system design team to dig out the specific decisions and all dependency relationships.

Item 3: Most operating personnel are not accustomed to thinking of shop performance in terms of criteria. They understand, however, such statements as "keep inventory low," "excessive overtime is expensive," "we can't afford many stockouts." It is the design team's job to state criteria explicitly and to eventually specify numerical values for the criteria. It would simplify matters considerably if all criteria could be stated in the same units, such as dollars. If this could be done, a total cost model of the operation could be stated, and the objective of the control system would be to minimize costs over the long run. As brought out in Chap. 7, however, it is usually quite difficult, if not impossible, to express all system variables in monetary terms. Most system variables can be expressed in terms of money, time, or physical measurement (such as the number of stockouts during 1 month). The human component in the control system must then attempt to relate all these "apples, oranges, and pears" and to translate the resulting comparisons into operating decisions.

Another practical consideration with regard to Item 3 is that a logical hierarchy of criteria should be established. That is, a relatively small number of measures of effectiveness should be identified and watched very closely, and only when one of these MOE's goes out of control (see Fig. 7.14) are lower-level MOE's monitored. For example, we could establish control limits on total inventory cost of an item. If this cost exceeds our control limit value, then we look further to identify specifically which factor caused the excessive cost.

Item 4: The difficulty of performing this step is determined largely by how well Item 3 was performed. This step is tedious and time consuming. The important thing is to develop for each decision a complete list of all information needed to make that decision. It is also important to break down the information into its smallest units; i.e., information elements, as described in Chap. 2. Remember that the same basic information element may appear on lists for several decisions (see Figs. 2.1 and 2.2). It is also important to list all information elements needed for each decision, even those that are not available. In this manner, we will be aware of which decisions in our control system must be made with incomplete information.

Item 5: The construction of an information-decision system diagram is rather straightforward, provided that Steps 3 and 4 have been performed completely and accurately. Indeed, the attempt to construct this diagram will help bring out any inconsistencies, inaccuracies, and incompleteness in Steps 3 and 4. In performing this step, it is important not to speculate as to the specific transfer functions needed for making the decisions. At this point, all we want is a general model of the information-decision system for which we are designing a control system.

Item 6: This step is one of the most difficult and time-consuming in the entire design process. It essentially requires the development of a total information system, or what is commonly called an integrated information system. A common data base must be constructed which will serve all information needs of the firm, not just the information needs of the operations planning and control system. Vast volumes of information must be coded and classified according to some over-all scheme. A significant portion of the information required originates in the engineering group. This portion includes information regarding products, processes, materials, quality factors, work methods, time standards, routing sheets, etc. Other important information comes from sales, personnel, accounting, finance, and purchasing. Much of this information is subject to rather frequent change. The information system should be designed such that changes can be made with a minimum of disruption and confusion. This is usually a difficult task to accomplish, but unless information changes can be entered on a regular, timely basis, the control system simply cannot operate as it should. At this point, it is essential to enlist the assistance of one or more specialists in information processing. These specialists may come from the data processing division of the company, from the customer service division of a computer manufacturer, or from a management consulting firm. It is important to recognize that we waited until after the information-decision structure was specified to bring in the information processing specialists. Many firms begin the design of their control systems by calling in these specialists.

Item 7: At this step we must resort to our knowledge of the various transfer functions available. It is very important to essentially forget about the way decisions are made in the old system. We should review the textbooks and professional journals to be sure that all important transfer functions are considered. It would probably be wise at this point to enlist the assistance of an expert in the field, such as a university professor who is carrying on an active research program. Researchers usually keep abreast of the latest developments of transfer functions. It may very well be that no existing transfer function is suitable for particular decisions. In such instances, we must either adapt existing transfer functions or develop new ones to meet our requirements. Let us not forget that many decisions in our system will require the use of human judgment, particularly those short-term decisions made on the shop floor and those regarding whether system parameters should be changed. It is probably safe to say that human judgment is the most commonly used transfer function in decisions involving monitoring activities. Still another practical consideration in this step is that many decision processes discussed in Chaps. 3 through 7 are designed to render several individual decisions. A good example of such a transfer function is the Gozinto procedure, developed in Chap. 4. Conceptually, then, a single decision process might very well perform several of the decision blocks on our information-decision system diagram developed in Step 5. We see again how very important it is that the transfer functions which we select be capable of operating together as a total system. Note that Steps 4 through 7 form a cyclic process in that, after performing these steps, we may have to go back and redefine our information-decision system in terms of the particular transfer functions we select in Step 7. Steps 4 through 7 are repeated until all decision elements in our information-decision system diagram are accounted for. At this point we still do not know whether the system will "work" well enough to satisfy all requirements. That is, we have not actually tested the compatibility of the several transfer functions; we only know that all decisions have been accounted for.

Item 8: At this point we begin testing the compatibility of the transfer functions. One approach is to construct a subsystem for each of the five major functions discussed in Part II—Forecasting, Planning, Inventory, Scheduling, and Dispatching and Progress Control. Each subsystem would contain all the transfer functions needed to arrive at all decisions required of that particular function. The subsystem for Forecasting, for example, might include a transfer function for each of several district warehouses as well as one for the central factory. These transfer functions would have to be mutually compatible. We recall from Part II that many decisions are input to other decisions. The necessity for compatibility in such cases is obvious. At this step it is a good idea to list alternative subsystem configurations for each of the five functions. This leaves us more flexibility in later steps and in the selection of a final system design.

Item 9: At this step we test the compatibility of the subsystems. That is, the set of transfer functions selected to perform the forecasting function must be compatible with the requirements of the planning function. The set of transfer functions selected to perform the Planning function must, in turn, be compatible with the requirements of the Inventory function. In a similar way, all transfer functions must be compatible. In addition, all decision processes within the operations planning and control system must be compatible with the other major systems of the firm. These needs were discussed in the interfaces section of each chapter in Part II. Note at this point that we have acquired a much more thorough understanding of the system and its information-decision requirements than we had at the beginning of the design process. It is appropriate, therefore, to cycle back through Steps 2 through 9 in an attempt to refine our system definition, its decision processes, and its information requirements. Having done this, we are in a position to make our final selection of control system components, especially the transfer functions.

Item 10: No further elaboration on this step is given here. Glans, Grad, Holstein, Meyers and Schmidt (1968) is an excellent text covering the design of total management systems.

Item 11: The text cited above is again suggested for greater detail on this step. Let us say, however, that, in general, the goal should be to originate the data in mechanized or machine-readable form as a by-product of the basic data origination. An example would be in the preparation of the original sales order, where the typing could be done at a remote console connected to a computer, so that the data we wish is automatically recorded, rather than keypunched after the fact. Note that this step is also closely related to the step of equipment selection.

System Documentation

In a previous section, we discussed the programming of equipment and personnel. Because detailed documentation is so important to the success of a new system, we offer in this section specific suggestions for documenting a designed system.

A complete operations planning and control system is very extensive, with numerous components related in complex ways. Usually several people, perhaps several teams of people, are involved in designing the new system. Their efforts cannot possibly be coordinated in any meaningful way without complete and consistent documentation as the design develops. Many of the people who will operate the system will be relatively unskilled and uninformed insofar as technical knowledge of the control system is concerned. Detailed procedures manuals must be prepared for the operating personnel. System documentation is essential in preparing accurate procedures manuals. The control system must be designed to accommodate changes. Again, complete documentation is essential for incorporating changes into the system accurately. The changes can be of two primary types: changes in information contained in the common data files and changes to the control system itself.

The particular type of documentation used varies greatly from one company to another. While it is impossible to specify detailed documentation requirements, we can at least discuss general guidelines that would be helpful in documenting the designed control system.

Systems documentation should reflect the needs of its users. Since the needs of users vary greatly, it follows that different levels, or classes, of documentation would be beneficial. It seems logical to identify three classes of documentation: (1) information system documentation, (2) control system documentation, and (3) a control system procedures manual.

INFORMATION SYSTEM DOCUMENTATION

This class of documentation has to do with the total information system of the company. It is concerned with specific data formats, data file construction, specific data processing procedures, computer program flow charts, and the like. Details concerning the complete management information system are beyond the scope of this book. We should point out, however, that the operations planning and control system is an extremely important component in the total information system. The documentation format should conform to the standard format used for the total information system. Perhaps the most complete treatment to date of the design of total management information systems is presented by Glans *et al.* (1968).

CONTROL SYSTEM DOCUMENTATION

This particular class of documentation pertains to the actual control system itself. It is intended for the use of those persons who maintain the control system. These people would normally be in the production control department, if such a department exists by that name.

A suggested over-all format for this documentation would be the format used for the first seven chapters of this text. If such a format is used, the sections would be:

Section 1: Description of the Production System. This section would describe the company, its products, the production processes, materials flow, the distribution system, and the relationship of the operations planning and control system to sales, purchasing, engineering, marketing, etc. It would be appropriate in this section to state the over-all objectives of this department and to show that the departmental objectives are consistent with over-all company objectives.

Section 2: The Information-Decision Structure. Specify the decision structure of the production system and the information required to support the decision structure. The concepts of Chapter 2 would be followed in writing this section. This section would contain specific information concerning management reports at all levels. Again, the reports should be geared to the decision structure of the firm, not vice versa. This section must tie in very closely with the information system documentation. Detailed information concerning data formats, file structures, etc., would not be repeated here; rather, the appropriate item in the information systems documentation would be cross-referenced as needed. A detailed flow-diagram showing all information flow would be included.

Section 3: Demand Forecasting. This section would present the particular transfer function(s) selected to forecast demand, perhaps with an explanation as to why these transfer functions were selected. Sample calculations of parameters would be included. The method of determining each parameter would be shown. Control limits and other performance criteria would be shown, as well as specific measures of effectiveness. Instructions must be included as to what action to take when the system goes out of control. Methods of long-term corrective action, as well as short-term corrective action, must be specified. Particularly important is the procedure for keeping all parameters updated. Instructions should also be included for changing a transfer function. Since there is a high degree of dependence among transfer functions, we must consider the effect on the rest of the system of changing a transfer function. Finally, this section should discuss interfaces with other functions outside the boundary of the operations planning and control system.

Sections 4 through 7 would treat operations planning, inventory planning and control, operations scheduling, and dispatching and progress control in the same manner as described above for demand forecasting.

The concept of feedback and corrective action should be stressed in Sections 3 through 7. We should show explicitly how feedback from one period's operation is used as input to the decision process for next period.

CONTROL SYSTEM PROCEDURES MANUALS

A set of detailed procedures manuals should be prepared for the technicians and clerks who will be operating the system. An example of such a procedures manual would be the instructions for operating the inventory system. Careful instructions must be distributed regarding the manner in which data is entered into the system. The amount of detail required in the procedures manual depends upon the extent to which system operations have been automated.

An Illustrative Design Example

In Chap. 1, we were introduced to VAST, Inc., a large diversified company, and the plastics fabrication division of it as a means of trying to grasp the magnitude of operations planning and control. Now we wish to continue this same example, to see how some of the techniques and concepts we have learned can be applied in practice. (At this point, it would be helpful to review the narrative material in Chap. 1 relative to VAST, Inc.)

We can now realistically visualize ourselves as systems designers, employed as in-house consultants by VAST's Engineering Department. (We could just as well visualize ourselves as outside consultants, or, in a more limited system, as just working within the Plastic Fabrication Division in a staff systems design capacity). In some manner, perhaps through our own efforts, the management of the Plastics Fabrication Division has become aware of the need for study of the operations planning and control system. We have been summoned to discuss the problem with management.

Having learned by experience and discussion as much about the problem as was previously presented in Chap. 1, we would be ill-advised to give management much advice at this point. Rather, bearing in mind the stages and steps of systems design presented earlier in this chapter, we would probably propose at this time only to survey the situation and would establish with management the general scope of the survey and a tentative date for its completion.

In the case of VAST, Inc., management did indeed seek the assistance of the management information-decision systems group of the Engineering Department. A supervising consultant and two relatively senior engineers then held a series of discussions with management groups, primarily production management at the Toledo level. However, each plant was represented at several of the meetings, and the director of sales and some of his staff were present at the later meetings.

The result of the meetings was authorization for these three systems designers to undertake a survey, in which they would examine all aspects of the eleven sales districts, the Toledo central office, and plant operations at all five plants. They were to take approximately three months in this activity, then report their recommendations to management. They were to have nominal assistance in collecting information, and an assistant to the General Manager was designated as a coordinator between this survey group and the division. Primary emphasis was to be placed on the solvent process product, since the Akron plant seemed to be in the greatest need of immediate assistance. However, sufficient emphasis was to be given the other plants and processes to assure that a system embracing the whole division could be designed, if justified.

Let us now undertake the design of the system. For ease in referring to earlier sections of this chapter, the same outline is followed; frequent reference is made to other sections of this chapter and previous chapters as we need to invoke portions of the concepts and techniques. The various section and subsection headings appear here in italics in the same order as presented earlier.

SURVEY AND CONCEPT—VAST, INC.

Recognition of Problem

During the course of the discussions with management, we noted many references to apparent problem areas. Some of those cited by management were noted in Chap. 1 as lack of control, high and unbalanced inventories, high plant costs, high rates of machine downtime, and poor shipping performance. Our first task, of course, was to uncover the fundamental problems so that we did not waste our time later in solving the wrong problem.

To do this, we needed to visit the principle operating points and to understand all products, materials, work methods, and manufacturing and distribution processes. We also needed to discover the fundamental decision structure and make a first estimate of the data requirements.

In general, it is best to examine a system starting with input and progressing through the system to the output phases. Since it was felt, with justification, that sales provided the basic input, the first visit was to the New York sales district. The three engineers were accompanied on their 3-day visit by the General Manager's assistant and by a supervisor from the Toledo sales order section.

Visit to New York Sales District. The discussions in New York involved every level of worker, but fundamental emphasis was placed on a few critical points. The most important question was, "What decisions need to be made to conduct this phase of the business?" The next important question was, "What information is needed to make these decisions?" And, finally, "Does the present system furnish this information?"

Surprisingly, the single most important decision involved whether a specific sale could be made in response to a customer request. Furthermore, the district had difficulty in making this decision because they had no idea of what was currently in stock at a given plant, what production was scheduled, or what might be on hand but already sold to another customer in another district. For thermoplastic products, each district was given a monthly commitment of each item, but their customer demands often exceeded this amount, and they had no way of knowing whether other districts might be willing to release a portion of their commitment.

As a result, all sales districts maintained almost constant telephone and telegraph contact with the sales order section in Toledo. They literally could not accept an order without clearance from Toledo, although no policy required this. Furthermore, as we shall see, the order section frequently had to contact the production planning groups, and they, in turn, the plant. As a result, the sales district needed from one hour to as long as one week just to tell the customer whether VAST could accept the customer's order.

Another important district decision was that of forecasting sales for ultimate use in operations planning. Two key points were uncovered. First, the districts were forecasting by purely judgmental means, based on the individual salesmen's projections for each product for each customer, with a consolidation and review by the district sales manager. Second, the forecast was not an estimate of demand, but rather an estimate of what the salesmen would like to sell, tempered by a guess as to what would become available through production.

Many of the mechanical details of district operations were also noted. For example, customer inquiries were almost always received by telephone. Then, after obtaining clearance to accept the order, telephone confirmation was made to the customer. This telephone acceptance was followed, with timing dependent on work load, by a typed order confirmation. One copy of this order confirmation was then sent to the producing plant in order to trigger plant shipping action.

We left the district with much information other than this, which we introduce later as it becomes pertinent in the discussion. Our next visit took us to the sales order section in Toledo, the point of reference for all sales districts when clearance to accept an order is needed.

Visit to Toledo Plant: Sales Order Section. Our findings in this section were somewhat dismaying. In general, this group consolidated district forecasts on a monthly basis and kept records relative to inventories and production plans. These latter records were updated monthly in most cases, with some weekly updating. *Their decisions were the same as those of the sales districts:* could a specific sale be made in response to a customer request? In most cases, they had to refer the question to the production planning groups, located one floor below. When they did not refer to production planning they found that the plant frequently did not ship on time; therefore, they referred almost every inquiry unless the plant was clearly overstocked. In essence, the group did only data handling and some little processing, although they felt they made many decisions and proposed getting many elaborate reports from production and the plants. An important point was that this group never handled a sales order as such.

Visit to Toledo Plant: Production Planning Group. The production planning group was our next point of inquiry. Here we did find decisions being made. The primary decisions related to "How much to produce," "When to produce," and "In what quantities to produce." Here, also, an attempt was made to control the level of plant inventories (see Chap. 5).

The principal data inputs to this group were the sales forecasts, inventory on hand at the plant, and data on machine capacities. The group maintained records of products currently in production. Most of the data input was on a weekly or monthly basis. The principal periodic output was a production schedule to the plant on a monthly basis, telling the plant what to produce. This schedule was revised almost daily by teletype or telephone.

On a day-to-day basis, this group handled numerous inquiries from the sales order section. Some of these inquiries could be handled immediately, based on records of inventory and production. A large number could only be answered by calling the plant, since the records were felt to be out-of-date and unreliable in tight situations. Furthermore, this group lacked knowledge of process detail and of day-to-day plant problems, which made it difficult for them to precisely determine production capability. We also noted that this group, like sales order scheduling, never handled a sales order as such.

Visit to Akron Plant. Our survey group next visited the Akron plant for several days. The first few days were spent in becoming familiar with the general process and products and with the general problems of manufacture. Our attention was then devoted to detailed plant procedures in nearly every area.

Because our survey in the sales districts and Toledo had centered on the sales order and the production schedule sent to the plant, we looked first at those groups in the plant which dealt with these documents. We found that two different groups received them and that these two groups were separated both organizationally and physically.

The production order and all other communications from and with the production planning group in Toledo were handled by the plant production control group which was housed in the main plant office and reported to the manufacturing superintendent. Sales orders from the plants were received by the order and invoice group, which was located near the plant warehouse and reported to the accounting superintendent. The shipping group was located in the same office and also reported to accounting.

The production control group had an extremely difficult task. At monthly intervals, they received a production order from the Toledo office, indicating the products to be made. It was not a schedule in any true sense, since it did not indicate any sequence of production or priorities in case of conflict. Furthermore, it was changed, sometimes several times a day, by teletype or telephone as the Toledo group tried to be responsive to day-to-day sales demands. The task of the local group was to somehow establish sequence and priorities in order to schedule the production to machines and to keep the operating efficiency of the plant at a reasonable level. They also had to be concerned with minimizing machine changes, with scheduling machines so that little or no overtime would be incurred by production or maintenance, with assuring that materials would be available (control and procurement of raw materials was an accounting function), and with trying to keep finished product inventory at some reasonable level.

Formal data, both input and output, was sadly lacking, although a great deal of paper work was in evidence. A weekly stock status (inventory) report, of doubtful accuracy, was available. Virtually no communication existed between this group and the order and invoice group.

At the other end of the plant, the order and invoice group received the sales orders from the sales districts. The orders were received by mail, depending on the work loads in the districts, from 2 to 10 days after the sale was made. They also received many teletype orders from the district calling for immediate shipment.

We found that the orders were filed by scheduled ship date. On the day prior to desired shipping, they were removed from the file and given to shipping. Order assemblers then tried to find the proper material, assemble and mark it, and prepare shipping lists. The shipping lists then went back to order and invoicing for the invoicing operation. The Shipping Supervisor then arranged transportation and prepared bills of lading.

We observed that, in a great many cases, the material could not be found. In this case, shipping expeditors started back through the process trying to locate suitable material in process and to urge the appropriate foreman to get it completed. This "pushing" was done with no communication with production control; therefore conflict often arose in the production areas. Only in the case that no such material could be found in process did shipping contact production control, and such contacts were frequently rather heated exchanges.

Contacts with other plant groups disclosed similar problems, all revolving about poor communications, lack of data, and decisions based largely on the personal judgment of the individual. We did find, however, a sincere desire on the part of most of the people to improve the methods of operation.

We also checked at this point some areas of the plant not then directly concerned with the problem, but which would become important in the development of a new system. One of these was plant data processing. This group, part of accounting, had a small computer, used largely for applications such as payroll and labor accounting and for preparation of cost sheets. Almost all their applications were on a weekly or monthly cycle basis, and all their data were acquired from manual records through keypunching.

Recall that we had agreed that the survey would be directed primarily toward the solvent process. We now felt that we had accumulated much of the basic data for our survey, but short visits were made to the sales order scheduling and production control groups of the other product line groups in Toledo and to the other producing plants. In every case, we found a similar situation, differing only in degree.

We now felt that we had recognized the problem and could proceed to the next survey step.

Definition of Problem

As we visited the various locations in the division, talked with the people and observed the operations, the problems began to take shape. It was now necessary to formalize our impressions and to state the problem concisely both for ourselves and for division management. We also note that much of what we observed, such as poor delivery performance, unbalanced inventories, and excessive machine changes, might be considered problems to one small segment of the division, but were really only symptoms to the division as a whole.

We were able to define the problem quite concisely, using the symptoms as evidence of the validity of the problem statement. This definition was:

1. The decision structure of the division, relative to sales, production, and

distribution, was not well defined or coordinated. Responsibility for many decisions was placed with the wrong organizational group.

2. The information system relative to sales, production, and distribution was inadequate to support these activities, in terms of timeliness, accuracy, and sufficiency of data.
3. Methods of making decisions were unreliable and were based largely on the judgment of the individual making the decision.

Preliminary System Objectives

Having defined our problem, we were in a position to state our preliminary system objectives. (The reader should try to relate the objectives to the problem definition and to the observed symptoms.)

1. The sales districts must be able to accept 95 percent of all business at the time of the customer's first inquiry, without further recourse to Toledo or plant agencies.
2. On the balance of inquiries, a final response should be made to the customer within 4 business hours, and preferably the same day.
3. The plant should ship orders to customers on or before the scheduled ship date for 95 percent of all orders and 98 percent of all poundage, with the balance delayed not more than 1 week.
4. The system must have over-all compatibility among the basic product lines, including identical paper forms and sales district procedures, identical sales forecasting procedures, and identical plant procedures, except where product and process differences preclude this identity.
5. The system must interface with the balance of VAST, Inc., insofar as use of the company data network and adherence to general accounting rules are concerned.
6. The system must be capable of modification to accept new developments in data processing equipment and new means of making decisions (transfer functions). It must also be sufficiently flexible to accommodate changes in process and product or the expansion of the division into any new product lines.
7. The operating cost of the new system must not exceed by over 10 percent the current operating cost at current sales and production rates.

Note that the objectives stated above are specific to the situation we encountered in this division and that they relate to over-all performance. We also included at this time the general design objectives stated earlier in this chapter, so that we would bear them in mind as we proceeded to the model building and analysis step. We later, in the development stage, refine all these objectives.

Model Building and Analysis

In stating our objectives, we set goals and placed certain limitations on our system. It was then in order to put things together into a model that would meet these objectives. In reading this section, the reader should refer to the definition of problem and statement of objectives listed in the two preceding sections.

In model building, we proceeded from a gross model to a specific one, being careful—since we were still in the survey and concept stage—not to become too engrossed in detail. In this case, we started with an over-all model, which had the following general features:

1. The provision of an availability report to each sales district. This report, to be updated daily, was designed to show, as of the close of the prior day's business, inventory on hand at the plant but uncommitted to a customer and scheduled production for the next 8 weeks not committed to a customer. This report related directly to the first objective and dictated many of the other system features.
2. Direct transmission, by teletype, of all sales orders to the plant. The teletype transmission tape was to be obtained automatically as a by-product of the required typing of the order acknowledgement for the customer. At the plant, the order received by teletype was to be received in machine-readable form, and subsequent plant orders and invoicing was to be done by machine rather than manually. In addition, the data was to be used as primary input to production control.
3. Preparation, at the plant level, of several sets of operating information, including:
   - A daily stock status report, reflecting balance on hand, receipts from production, shipments, transfers, and returns from customers.
   - A daily shipping schedule, showing shipments required for the next 3 days and current shortages, if any.
   - The daily availability report to the sales districts.
   - Certain daily and weekly reports in support of the production control group.
   - Daily summary reports to Toledo production and sales management.
4. The placing of all responsibility for all production planning and control decisions (other than level of operation) at the plant level.
5. Consolidation at the plant level of the above decision functions and associated operating groups into one organizational group. This group was to absorb the present production control, order and invoice, shipping, and materials control groups.

6. The establishment of direct communication between the sales districts and plants on matters relating to availability of products and details of shipping.
7. The referral of exceptions (with a goal of less than 1 percent) to Toledo sales and production management, together with the necessary detailed information, for the decision of management.

Implied in these features was the fact that the two Toledo groups, sales order scheduling (which, we recall, had no real decision function) and production planning, would no longer be needed for routine operations. It was felt that they could be consolidated into a single, smaller group charged with long-range planning and handling the referrals from the plants.

These broad features, upon analysis, seemed to be entirely feasible, considering the current status of art relative to data processing and communications and the capabilities of available personnel. These features also gave rise to several more detailed features, particularly with respect to the production planning and control function at plant level:

1. A statistically-based forecasting technique would be employed, which would use the sales data now to be available at the plant. While no final decision as to the exact transfer function was made at this point, exponential smoothing was felt to be feasible.
2. It appeared feasible for the planning function to use the Gozinto method for the Akron plant, at least. Some reservations were felt relative to this use in the thermoplastic plants, since it might be more economic to handle the very few materials involved in some other way. However, there was little doubt that Gozinto would also apply if necessary.
3. In the area of inventory control, it was felt that a fixed order size system, with safety stock, would be the initial model of choice, due to its simplicity. However, in at least one area, an economic manufactured quantity (noninstantaneous production) would have to be considered as well. We could also foresee the use of multi-item models in the future. It was felt wise *not* to include this transfer function in the original model because of the lack of basic data and the greater difficulty of training involved.
4. The actual scheduling of production at the plant level posed a substantial problem, one which could not be resolved during this survey and concept stage. We only postulated at this point that a usable system could be developed in time and that we would start the system with very simple graphical and tabular methods. A general decision was made to schedule within the production control group, rather than to delegate this function to the foremen. This decision was based on the fact that scheduling was now done this way, even if crudely, and that training time and costs

would be less. Also, it was VAST, Inc., policy to free the foremen from as much of this sort of detail as possible, so that they could devote more time to process problems, personnel, and quality concerns.

5. Closely related, of course, to scheduling was the dispatching and progress control function. Here again was a substantial design problem, particularly since under the present method no formal progress control existed. This problem seemed to dictate that we start with a relatively simple system, until better data could be collected and until the plant groups gained some experience operating in a more structured system. Again, since this portion of the over-all system could not be obtained ready made from existing literature or practice, it was merely postulated that a reasonable system could be developed.

At this point, we felt that we had a workable general model and concept. This model was discussed at the plant, sales districts, and Toledo levels, and there seemed to be general agreement as to its feasibility and desirability. Rough analysis of the several features also seemed to indicate workability. We felt ready for the next step.

Evaluation of Proposed System

At this point, we could state the problem, describe the proposed general system, and present some evidence of technical feasibility. Our next task was the economic analysis and the development of a proposed plan for the development, testing and installation, and early operations stage. In this step, we used many of the usual techniques of engineering economy and prepared and processed a gross PERT network for the work to be done. We do not detail this portion, except to say that it was necessary to estimate both the computer size necessary to handle the system and the volume of transactions to establish system operating costs. Extensive use was made of existing standard data for estimating personnel costs in the proposed system. In general, every effort was made to show costs on the high side and benefits on the low side, in order to be conservative.

We also made plans at this time concerning the organization of the design team, following the concepts discussed earlier in this chapter. Beginning early in the survey, we met periodically with our coordinator, the assistant to the general manager. At most of these meetings, the division accounting manager, director of sales, and one or both of the directors of production were present. We proposed that these men now form a steering committee for the balance of the program. We also proposed a counterpart committee at the plant level, consisting of the assistant plant manager, the manufacturing superintendent, the accounting superintendent, and our newly conceived, but as yet unnamed, superintendent of control functions.

We also had to project manpower needs for the balance of the program. Provision was made to use specialists from the engineering department in such areas as programming and statistics, and an over-all level of eight persons from engineering was established. These eight persons would be joined by nearly 40 others, drawn from the plastics fabrication division. While most of these personnel would be from the plant, several were to be drawn from the sales districts and from Toledo operations.

The presentation to management was made verbally, using simple flip-charts as a visual aid to present the key points. These charts were later reduced in size photographically and used as an appendix to a very brief written report confirming the oral presentation. The presentation followed very closely the outline discussed earlier in this chapter. While it is not feasible to reproduce the entire presentation, the key items were as follows:

1. A brief review of the origin and conduct of the survey. We included the places visited and the amount of time spent in each place.
2. A summary of the problem. Here we used the three statements of definition of problem. These statements were put on flip-charts and amplified verbally by describing typical symptoms which arose from these problems, such as inability to accept customer orders promptly.
3. A description of the proposed general system. Our system description had to include all of the features we have discussed under model building and analysis and preliminary system objectives. To assist in describing the system, we prepared flip-charts of the general flow of information between the sales districts and the plant and within the plant. We also indicated the principal decision points in the system and *briefly* described the type of transfer function to be used at each point.
4. Evidence of technical feasibility was not dealt with as a separate item but intermixed with the system description. In general, we tried to show why a given item was reasonable, or cited similar circumstances where a given feature had been successfully used.
5. The economic evaluation of the proposed system was presented in summary form, but complete details were held available in case of detailed questions from management. The principal points, presented on flip-charts, were:

    - Development and installation would take 30 months for the entire division and would have an out-of-pocket cost (for engineering services) of \$235,000. This cost assumed full engineering responsibility for development, installation, and start-up of the system for the Akron plant, but consultant services only for the balance of the plants.
    - Operating costs of the proposed system after installation would be about the same as the present system. Increased equipment rental costs and

more operating personnel at the plant level would be offset by virtual elimination of the sales order scheduling and production planning groups in Toledo and by some reduction of clerical forces in the sales districts.

- The principal benefits to be derived were improved control and alleviation of the many observed symptoms, such as late deliveries and inability to accept orders promptly. These benefits were summarized as improved customer service and proved to be the most important benefit in management's eyes. No actual cost savings could be foreseen, and profit improvement was virtually impossible to predict. In this case, however, the prospect of profit improvement through improved control was sufficient to justify the risk of investing in a new system.

6. A plan for the balance of the system development was presented. This consisted of the organizational features previously discussed and the time-phased PERT diagram. The start-up date for the Akron plant was set for 12 months after authorization of the program.

Proposed System Approval and Revision

After our presentation, management took our proposal under advisement; that is, they needed some time to make up their minds. In about 3 weeks, we were summoned to the general manager's office, where he signed the necessary authorization. In this case, no major revisions in our plan were necessary.

We have dealt with the survey and concept stage of this example in substantial detail, even though, in terms of time, this stage amounts to only about 10 percent of the entire design effort. We have done this because it is vital that this stage be done well. Later errors in design details can be detected and corrected, but if we have not properly perceived the problem or have established a grossly erroneous concept, the system can never be successful. It might be said that the secret of system design is the ability to think big and yet take care of a million pesky details. We have now disposed of much of the "big thinking" and proceed to the "pesky details."

DEVELOPMENT—VAST, INC.

It is manifestly impossible within the limitations of a book to discuss every detail of the development of the VAST, Inc., system. The documentation for operating purposes, for example, may well run several hundred pages. To discuss the reason for inclusion of each point, to relate the lengthy study and discussion necessary to establish each detail, and to present this in readable form would require as much work as developing the system itself. To read it all would be prohibitively time consuming.

We shall resort, therefore, to an over-all description of activities in each step and to subexamples to illustrate specific points. We hope, in this way, to convey a sense of the detailed care necessary in the development of a real system.

Operating System Objectives

As basic input to this step, we have the preliminary system objectives, developed during the survey and concept phase. We also have full knowledge of the general model concept. From these inputs, we must develop detailed objectives and, in the next step, the operating system description. We also at this point recall to mind the several specific considerations presented earlier in this chapter. Now let us try to "zero in" on some detailed objectives for one small segment of the system.

Our third preliminary objective stated, "The plant should ship orders to customers on or before the scheduled ship date for 95 percent of all orders and 98 percent of all poundage, with the balance delayed not more than 1 week." It is evident to us that if this objective is to be met, we must assure ourselves that shortage of raw materials will rarely occur, since without raw materials shipping schedules clearly cannot be met. We also reflected this fact in our conceptual model, when we postulated the use of the Gozinto method as one basic system feature. Implied here was the fact that a decision must be made, at the plant level, relative to quantities of materials needed for production.

We must now look at this decision process in detail, as it relates to our specific problem. In our design program, one of the systems designers from the engineering department, with prior experience with Gozinto, was assigned to this task. Also assigned was a young man from the Akron plant who had been working in the materials group and who was being groomed to head that function in the new organization.

These two first established in great detail the decisions that must be made to discharge this function. They agreed that there were no decisions of a commitment nature, as the input to this activity of an approved production plan completely implied prior commitment of the plant to manufacture and, therefore to procure the necessary raw materials. Decisions were primarily as to timing and amount, and even here decisions were largely a clerical procedure based on prior decisions made during preparation of the production plan.

Based on this examination of decisions (or lack of them), the two established as the objective for this small segment, "Furnish to plant purchasing a statement of each raw material requirement at least 1 day prior to the day when purchasing must place the order with the vendor, considering the average vendor lead time for the specific item." Note that this detailed objective also met preliminary objective 5, relative to interfacing with other systems. Also note that this objective was far from being the only one relative to materials control, but it is a key one which we can follow into the next step.

Operating System Description

To illustrate this step, we can continue to use the small segment of materials control as an example. Having stated the above objective and concluding that this segment involved low-level decisions of timing and quantity, our designer and his plant counterpart needed to determine detailed data requirements. By examining their particular problem, by reference to material such as that in Chap. 4, and by coordination with the designers concerned with the production plan and plant purchasing, they established a parameter file containing bill of material information, materials descriptions, vendor data (including lead time), and the necessary coding for retrieval. They also established the variable data file as the output from the production planning activity. During this period, they were assisted by a data processing specialist from engineering and by several plant people familiar with basic data sources. A great deal of coordination was needed at this point, as the files had to serve all needs. The detail at this point extended to such things as field lengths in the file, and precise determination of report format.

At this point, it was finally determined to use the Gozinto method as the transfer function for the process. It was established that it could be used without modification.

While this process was taking place, other designers and division personnel were doing similar detailed work in other aspects of the system. We seemed ready to proceed to the next step.

Operating System Analysis

To test the assumptions made relative to the stated objectives, the data requirements, and the use of Gozinto as a transfer function, our two men now obtained a set of test data, partially from records of past operations and partially fabricated, and put them in the format agreed upon for the data file. They then arranged to use some computer time and a Gozinto program already running in a plant in another division of the company. In this respect, they were fortunate since the fact that Gozinto was already being used saved them the time of programming. In a 2-day stay at the other plant, they were able to confirm most of their assumptions, but they did find that the form of the bill of material file was clumsy, and upon return to the Akron plant, they made some minor modifications in the system description.

They were also left with some lingering doubts about some aspects of the production plan input, which they discussed with appropriate people upon return to the plant. Some minor changes were made which helped ease their doubts, and they could now only wait for later testing.

Hardware Specification and Procurement

As the several substudy groups dealt with their individual subsystems, the supervising consultant was carefully coordinating the several developmental aspects, and, with the assistance of a data processing specialist, was exploring alternatives as to hardware, both the plant computer and the data origination and transmission equipment for sales districts and plant use. By the time the system description was complete and the analysis underway, it was possible to reach a firm decision. We do not detail this step. A medium-scale third generation computer was specified and scheduled for delivery early in 1967.

It was also necessary at this point to decide on the programming language for the system. The manufacturer of the equipment offered several options, including all the popular languages. A language already in wide use throughout VAST, Inc., was selected to minimize training requirements and to assure compatibility with other divisions and the comptroller's department.

Programming

As soon as a decision on hardware and language was made, the programming of equipment could begin. Insofar as the Gozinto portion was concerned, this phase was relatively easy. The plant where the test was made used a similar computer and the same language, so a copy of their program was obtained and plugged into the over-all system program with relatively few modifications.

The other aspects of programming, those of writing detailed operating manuals, were more difficult. However, our designer assigned to this aspect, with the help of his plant counterpart, was able to write a very detailed portion of an over-all manual dealing with materials control. In this manual, the designers detailed the day-to-day operation of the materials control supervisor and each clerk and specified how to deal with all anticipated exceptions to routine. The manual also dealt with interpretation of systems reports and contained a section on the activities of the plant buyer upon receipt of the system output. The manual was written both for the training of new personnel and for the daily use of the persons charged with materials control when faced by some nonroutine occurrence.

Training Program Design

The programming activity, and the associated documentation of the equipment programs and the operating procedures, led naturally to the design of a training program. The documentation and manuals would form the texts, at least for the operating personnel. (The section on training program design should be re-read at this point.)

Since district and plant operating personnel had been heavily involved in the system's development, much of the training of operating personnel could be vested in them and done informally on the job. However, it was necessary to conduct more formal sessions for the computer operators and the operators of data input equipment at both the district and plant levels. In these two cases, lecture, demonstration, and practice sessions were planned over a 1-week period.

Management orientation was largely the responsibility of the supervising consultant and his key design engineers. The steering committee had met often and functioned well. This group was asked to assist in arranging a meeting with small groups of management from all levels of the organization. A standard presentation of the basic operation of the system, again on flip-charts, as well as a manual describing the system and showing key output reports were made up.

With the assistance, again, of the steering committee, a 30-minute general presentation was also prepared for use in orienting peripheral personnel. Also assisting in this preparation was the division's director of personnel. It was decided that this presentation would be made about a week prior to start up to all personnel in the division not directly involved in the system. At the sales district and Toledo offices, it would be conducted by appropriate managers during regular hours. At the Akron plant, wage roll employees would be given overtime for the meeting, and the assistant plant manager would make the presentation.

Preliminary Training

The preliminary training embraced only a segment of the operating people —just enough to operate the live testing program. For example, only one computer operator was trained, one girl in each sales district, and two plant operators of data input equipment. Personnel who had been active in the design process were considered to be adequately trained to conduct the live testing.

TESTING AND INSTALLATION—VAST, INC.

The day finally arrived, some 10 months after authorization, when equipment was delivered and in place, programs had been written and debugged, manuals written, and key personnel trained. The development phase could be considered over, and we could proceed to testing and installation.

Live Testing

In our system, compared to the old system, the data content, the data flow, and the location and means of making decisions differed drastically. We could not test the actual decision portions, but we could and had to test carefully the mechanical features of the system. Even then, we could not test the system in its entirety, since the decisions would influence the data content under real operation.

The data processing portions were thoroughly tested, however, including data origination equipment in the sales district and plant, data transmission on the VAST, Inc., data network, plant processing, and preparation of key output reports. First testing was done with sets of concocted transactions, in which the output was predetermined. Later testing sampled from real data, but at no time during the testing was any segment of the new system used to supplant the old. Testing was live, but off-line so far as the existing system was concerned.

During this step, many discrepancies were found and remedied. For example, we found that our fears about one part of the production planning output and its effect on the Gozinto program were well founded, and the production planning portion was modified slightly to correct this discrepancy. The testing took about 2 weeks, and correction overlapped with the next step, the training program.

Full Training Program

With testing complete and the necessary corrections made, the full training program could start. During this step, lasting about 2 weeks, all personnel including peripheral personnel and managers were trained. This step went well, due to careful planning during the development stage, and start up was approached with confidence.

Start Up

It was decided to start up, insofar as possible, using the stepwise parallel method. For this system, this meant starting up the data origination from the sales districts first. Then, in a few days, when it was felt that this process was working smoothly, certain portions of plant data origination were started. Later, successive features of data processing were started. In the case of Gozinto, it was nearly a month after the initial start-up day before the first run was made of the Gozinto program. Even then, the output was ignored for decision purposes for 2 more weeks, until we were sure that the processing was being done correctly. The last thing started up in each subsystem was the making of operating decisions.

When the Gozinto portion started, the designer and his plant counterpart worked closely together and consulted often with the plant buyer and other plant activities. In this way we discovered, for example, that we had failed to round off to carload lots on the high consumption bulk raw materials. This flaw was easily corrected by inserting a program "patch," which saved the plant buyer from computing this adjustment manually.

In general, the start up was smooth, and it was necessary to fall back on the old system only once, in one phase of production reporting. This fault was corrected in a day or two, and by the end of the stipulated 12 months, the system was "go."

Reprogramming

We have already mentioned the need for a program patch in the materials control subsystem. Several other patches were also necessary, both in equipment and procedures, to clean up some awkward points. Manuals had been written with a view to revision, so that no great difficulty was encountered in this step.

OPERATION—VAST, INC.

For the continued operation of the system, and indeed for its installation for the other product lines and plants, the management of the plastics fabrication division very wisely established a group in Toledo. This group was assisted during the installation for the other product lines and in some aspects of the continued operation by retraining some of the engineering personnel as consultants for several months. This group carried on the functions of maintenance, surveillance, and refinement.

It is worthy to note that, because the system was well conceived initially, refinement was relatively easily. For example, the continuing group can now consider using much more sophisticated equipment, such as cathode ray tube displays of availability data in the sales districts, without changing or violating the objective of accepting a large percentage of business at the time of the customer's first inquiry. The group must answer only the question of whether sufficient improvement will be gained to justify the cost of the new equipment. Similarly, if a better transfer function than Gozinto emerges, it can be adopted with minimal changes to the over-all system.

We have tried, in this VAST, Inc., example to illustrate the design process for a realistic and complex system. The fact that it is realistic and complex has limited our efforts, because the amount of detail in a realistic systems development is substantial, but certainly not insurmountable, if approached using the concepts set forth here.

REFERENCES

Glans, T. B.; Grad, B.; Holstein, D.; Meyers, W. E.; and Schmidt, R. N. 1968. *Management Systems.* New York: Holt, Rinehart and Winston, Inc.

Magee, J. F.; and Boodman, D. M. 1967. *Production Planning and Inventory Control.* Second Ed. New York: McGraw-Hill Book Company.

Plossl, G. W.; and Wight, O. W. 1967. *Production and Inventory Control.* Englewood Cliffs, N. J.: Prentice-Hall, Inc.

Starr, M. K. 1964. *Production Management: Systems and Synthesis.* Englewood Cliffs, N. J.: Prentice-Hall, Inc.

Webster's New World Dictionary of the American Language, College Edition, 1959. Cleveland, Ohio, and New York: The World Publishing Company.

OTHER SUGGESTED READINGS

Published examples of total operations planning and control systems are relatively rare. Listed below are a few publications concerned with total control systems.

Elmaghraby, S. E.; and Cole, R. T. 1963. On the Control of Production in Small Job Shops. *The Journal of Industrial Engineering* 14: 186-96.

Ferguson, R. L.; and Jones, C. H. 1969. A Computer Aided Decision System. *Management Science* 15: B-550.

O'Malley, R. L.; Elmaghraby, S. E.; and Jeske, J. W., Jr. 1966. An Operation System for Smoothing Batch-Type Production. *Management Science* 12: B-433-B-449.

Reiter, Stanley. 1966. A System for Managing Job-Shop Production. *The Journal of Business* 34: 371-93.

IBM Reference Manual E20-0280-2. 1968. *The Production Information and Control System.* (White Plains, N. Y.)

IBM Reference Manual E20-8041. 1960. *General Information Manual, Management Operating System for Manufacturing Industries* (White Plains, N. Y.)

The following seven articles appeared as a series in two consecutive issues of *IBM Systems Journal:*

Baker, C. T. 1965. Fabrication and Assembly Operations, Part I: The Outlines of a Control System. *IBM Systems Journal* 4: 86-93.

Calica, A. B. 1965. Fabrication and Assembly Operations, Part II: Long-Range Planning Techniques. *IBM Systems Journal* 4: 94-104.

Loewner, P. G. 1965. Fabrication and Assembly Operations, Part III: Matrix Methods for Processing Configuration Data. *IBM Systems Journal* 4: 105-21.

Dzielinski, B. P. 1965. Fabrication and Assembly Operations, Part IV: Linear Programming in Production Planning. *IBM Systems Journal* 4: 122-35.

Calica, A. B. 1965. Fabrication and Assembly Operations, Part V: Production Order Sequencing. *IBM Systems Journal* 4: 225-90.

Gorenstein, S. 1965. Fabrication and Assembly Operations, Part VI: Parameter Values for Sequencing Control. *IBM Systems Journal* 4: 241-49.

Shapiro, S. 1965. Fabrication and Assembly Operations, Part VII: Adaptive Control in Production Planning. *IBM Systems Journal* 4: 250-55.

9 Concepts for the Future

In Chap. 3, we learned that future demand cannot be predicted with certainty. It is equally true that we cannot predict with certainty what the future holds in store for the field of operations planning and control. Nevertheless, we can attempt to assess current trends and hopefully make some meaningful comments about the directions in which we can expect this important field to move.

The first section of this chapter discusses the applicability of formal control theory to operations control systems. The second section then speculates on the general nature of this field in the future.

Applicability of Formal Control Theory*

The industrial engineer is concerned with controlling a segment of his environment, an industrial system, in order to provide a useful product for society. However, before any system can be controlled, it must be understood. The "new" challenge to industrial engineers is the understanding and control of modern, complex systems such as educational systems, traffic systems, and total industrial systems. Yet some of the systems that are understood by industrial engineers have not yielded completely to adequate control. It appears reasonable then that new techniques must be found to solve some familiar problems before solutions can be found to unfamiliar problems.

The production system is an example of a system which industrial engineers understand but have not been able to control adequately as a total system. Certain of its functions, e.g., inventory and job sequencing, have been researched

*This section is based upon an unpublished paper by Mr. Lavon F. Jordan.

and analytical tools have been developed for their performance. However, no quantitative method for optimizing the performance of the total production system has been offered.

A possible explanation for the lack of a total optimizing procedure is that the commonality of the production process with other processes, such as chemical, mechanical, and electrical processes, has not been researched adequately. Perhaps the same methods used to control these processes can be applied to the production system. This section attempts to put some of the major control theories related to production control into proper perspective.

A production system, in this section, is an enterprise that produces goods on a continuous or batch basis, with profit as a primary objective. The purpose of the control system is to control this operation. Control theory is rather broadly defined in this section to include classical control theory, modern control theory, optimization theory, industrial dynamics, and cybernetics.

BASIC CONTROL SYSTEM CONCEPTS

Before looking at individual control theories, it is beneficial to mention some basic control system concepts. What is a control system? Dorf (1967) states, "A control system is an interconnection of components forming a system configuration which will provide a desired system response [p. 1]." It is assumed that a cause-effect relationship exists for the components of a system.

There are two basic types of control systems, open loop and closed loop. The simplest of the two, the open-loop system, is shown in Fig. 9.1. An open-loop control system utilizes a controller to obtain the desired response. An example of an open-loop system is a house heating system without a thermostat. Regardless of how carefully the on and off time periods are chosen initially, when the outside conditions change, the time periods must be reset. Of course, this resetting requires continual monitoring of outside and inside temperature. Using this type of heating system, the house would be either too cold or too hot much of the time. Other examples of open-loop systems are washing machines, toasters, and electric motors.

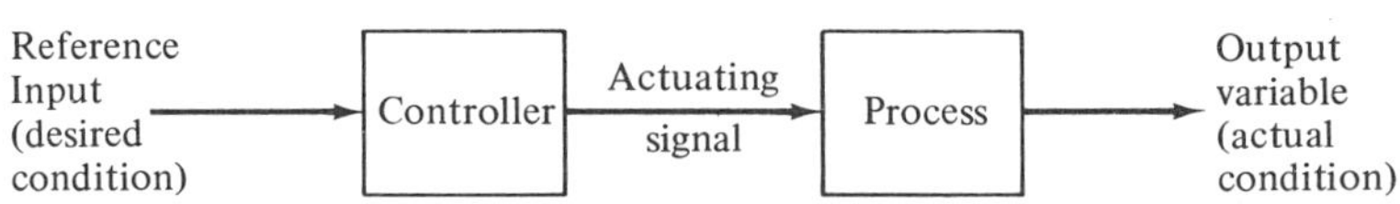

Figure 9.1 Open-loop Control System

The closed-loop feedback system in Fig. 9.2 compares the output variable with the referenced input through the use of feedback. This comparison is used to generate an error signal, which is amplified and fed to the process, which in turn changes the output variable. A definition (AIEE Committee Report 1951, p. 905) of a feedback control system is: "A feedback control system is a control system which tends to maintain a prescribed relationship of one system variable to another by comparing functions of those variables and using the difference as a means of control." A much broader definition of a feedback control system is offered by Forrester (1961, p. 14): "An information-feedback system exists whenever the environment leads to a decision that results in action which affects the environment and thereby influences future decisions."

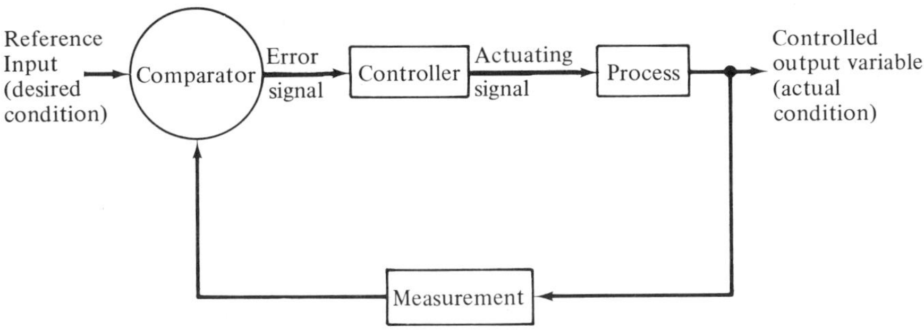

Figure 9.2 Closed-loop Feedback Control System

A satellite orbit control system is a closed-loop feedback system. A target orbit and time are specified, but the actual orbit and time deviate from the specified orbit and time. The actual orbit can be measured by navigational observations and tracking systems to obtain information which can be fed back to the control system. The amount of error is established by comparison, and a correction maneuver is computed based on control laws. The correction maneuver is executed, and the preliminary actual orbit is modified into a corrected actual orbit. This same procedure is repeated at predetermined time intervals or when the control variable exceeds established limits. Conceptually, this type of control system is very similar to a production control system.

CONTROL THEORIES

Classical Control Theory. Conventional linear control theory (Dorf, 1967) is the foundation for most, if not all, other control theories. It offers the concept of a closed-loop feedback control system. Understanding and controlling complex systems begins with a mathematical model of the system. This model forces us to analyze the relationships among the system variables. Since the systems are dynamic in nature, the descriptive equations are usually differential equations. If these equations can be linearized, then the Laplace transform can be used to simplify the solution method.

To be linear, a system must exhibit the properties of superposition and homogeneity. The principle of superposition states that the response to every disturbance runs its course independently of preceding or succeeding inputs to the system; the total result is no more or less than the sum of the separate components of system response.

Homogeneity means that the magnitude scale factor is preserved. Consider a system with an input X and an output Y. It is necessary that the response of a linear system to a constant multiple K of an input X is equal to the response to the input multiplied by the same constant so that the output is equal to KY.

One of the more useful concepts in linear control theory is the transfer function. It is defined as the ratio of the Laplace transform of the output variable to the Laplace transform of the input variable, with all initial conditions assumed to be zero. A transfer function may be defined only for a linear, stationary (constant parameter) system. A nonstationary system, often called a time-varying system, has one or more time-varying parameters, and the Laplace transformation may not be used. The transfer function concept has been generalized in other theories, such as industrial dynamics, to mean output divided by input. In this text, we extend even further the transfer function concept to include any process that converts information to a decision.

In any control system, the sensitivity of the system to variations in parameters is of prime importance. A primary advantage of a closed-loop feedback control system is its ability to reduce the system's sensitivity. System sensitivity can be defined as the ratio of the percentage change in the system transfer function to the percentage change of the process transfer function. The sensitivity of an open-loop system is equal to one. The sensitivity of a closed-loop system can be reduced below that of the open-loop system by increasing the loop gain over the frequency range of interest. In general, a closed-loop system allows the transfer function to be less accurately specified, since the sensitivity to changes or errors in the transfer function is reduced by the loop gain. This advantage is an important one of this system in a production control environment because individual transfer functions are very difficult to define.

Since the purpose of control systems is to provide a desired response, the transient response of control systems often must be adjusted until it is satisfactory. A closed-loop system can often be adjusted to yield the desired response by adjusting the feedback loop parameters. Feedback systems are advantageous in that the effects of distortion, noise, and unwanted disturbances can be effectively reduced.

Feedback is used primarily to reduce the sensitivity of the system to parameter variations and the effects of disturbance inputs. It is noteworthy that the effort to reduce the effects of parameter variations and disturbances is equivalent; they reduce simultaneously. However, these advantages have some cost. Feedback requires additional components, increases the complexity of the system, and causes a loss of gain, exactly the factor that reduces the sensitivity of the system. Usually, it is more advantageous to trade this loss of gain for increased control of the system response. In the case of a production system in which gain is undesired, the use of feedback is not only a means of increasing control but also of reducing the gain inherent in the system. Finally, a cost of feedback is the introduction of the possibility of instability. In summary, the fundamental reasons for using feedback, despite its cost and additional complexity, have caused a multitude of feedback control systems to appear in industry, government, and nature.

Modern Control Theory. Modern control theory is extremely useful, not only for designing a specific optimal control system but also for improving the principle on which the system will operate. This theory, although gaining more and more attention, is limited to linear systems. Most physical systems are nonlinear; but if the deviation from linearity is so small that it is not important for the specific problem considered, we can treat the system as linear.

A reasonably accurate and complete description of a complex process requires several to over a hundred first-order differential equations. By the use of state space concepts (Ogata 1967), these equations may be reduced to the essential equations in a systematic manner with no loss of essential information.

A state of a dynamic system is the smallest collection of numbers which must be specified at an instant of time to be able to predict uniquely the behavior of the system for some time in the future for any input belonging to the input set. This collection of numbers is called state variables. The input set is the set of all possible inputs that can be applied to the system.

The set of N state variables can be considered as N components of a state vector $\mathbf{X}$. State space is defined as an N-dimensional space in which $\mathbf{X}_1, \mathbf{X}_2, \ldots, \mathbf{X}_n$ are coordinates. We deal with a set of N first-order differential equations, rather than with a single Nth-order differential equation. A differential equation of order N can always be reduced to a set of N first-order differential equations.

Ogata (1967, pp. 27–28) compared classical control theory (linear control theory) and modern control theory. A summary of his comparison is:

1. Classical control theory is best suited for design of single-input single-output linear time-invariant systems. Modern control theory can be applied to the design of linear multivariable control systems and linear time-varying control systems. (Both areas leave out the production control system since it is a nonlinear system; however, the amount of error induced by assuming it to be a linear system is, in some cases, tolerable.)
2. It is usually difficult to write a set of differential equations representing the dynamics of a complex system so that modern control theory can be applied directly. Hence, from this viewpoint, classical control theory is more convenient than modern control theory.
3. In classical control theory, algebraic manipulation of the parameters can

be made before numerical values are substituted. This is not the case in modern control theory, because of its extensive use of vector matrix notation.

4. Inclusion of the initial conditions in the system design is impossible in classical control theory but not in modern control theory.

Ogata concludes that, although modern control theory is more general than classical control theory, the latter will not be discarded unless the former becomes better in every respect of control system analysis and synthesis.

Industrial Dynamics. Forrester (1961) and the M. I. T. Industrial Dynamics Group have performed extensive research in the area of linear control theory as applied to complete industrial systems, of which the production system is one entity. Industrial Dynamics is a methodology of studying the behavior of industrial systems to show how policies, decisions, structure, and delays are interrelated to influence growth and stability. It views economic and industrial activities as closed-loop information feedback systems. Conditions are converted to information by use of transfer functions and subsequently used as a basis for decisions that control action to alter the surrounding conditions. Thus, all management decisions are made in the framework of an information feedback system in which the decision ultimately affects the environment that caused the decision.

The Industrial Dynamics approach views management as the process of converting information into action. The conversion process is decision making. A *policy* is a rule that states how the day-to-day operating decisions are made. *Decisions* are the actions taken at any particular time and are a result of applying the policy rules to the particular conditions that prevail at the moment. Management success then depends primarily on what information is chosen and how the conversion is made.

Figure 9.3 shows the system structure as it surrounds the decision point. A decision is based on the state of the system, which is described by the condition of various levels. The dotted lines representing information about the levels are inputs to the decision. Some levels describe the present instantaneous condition of the system and others our presumed knowledge about the system. A level may be inventory, the number of employees, bank balance, etc.

Forrester points out that most industrial systems are not linear because the external effects on the system are not purely additive (refer to the definition of *superposition* given previously). When we no longer insist on obtaining a general solution that describes in one neat package all possible behavior characteristics of the system, the difference in difficulty between linear and nonlinear systems vanishes. Simulation methods that obtain only a particular solution to each separately specified set of circumstances can deal as readily with nonlinear as with linear systems. Industrial Dynamics attempts to sim-

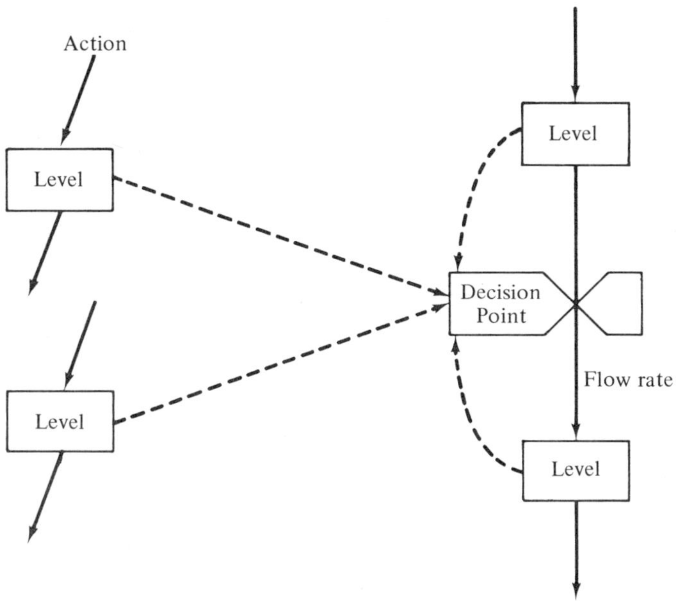

Figure 9.3 System Structure at a Decision Point

ulate the time-sequential operation of mathematical models of dynamic systems, linear or nonlinear, stable or unstable, steadfast or transient, by using difference equations.

Roberts (1963) has presented some examples drawn from Industrial Dynamics research studies of management control systems inadequately designed for the problem. Two of these systems are discussed here. The first area was the control of research and development projects. By using the Industrial Dynamics approach, we can show that many factors other than desired schedule determine the resultant actual schedule of research and development projects. Control systems for research and development, such as PERT, which resort to schedule and effort rate control without full understanding of the system structure of projects, are bound to be ineffective.

A second system inadequate for its problem was a production-inventory-employment control system with large variability in incoming orders. Initially, the traditional approach was used to design a control system, which appeared acceptable. Roberts pointed out that if additional improvement was desired, the traditional control designer would begin by adjusting parameters. He emphasized that this way was not the answer and that the source of error was not due to the fluctuations in incoming customer orders. Instead, another feedback loop was found in the company-customer total system: delivery delay changes affected the customer release rate of new orders, which in turn influenced the company delivery delay.

From this observation, it was concluded that to dampen the fluctuations in customer order rate, the manufacturer must control not inventory, nor backlog, nor employment, but rather he must stabilize the factory lead time for deliveries. This observation is somewhat typical of ones that are possible when a more broadly bounded approach, such as Industrial Dynamics, is employed.

Modern Optimization Theory. Modern optimization theory (Wilde and Beightler 1967) encompasses the differential approach and unifies such diverse elements of optimization theory as unconstrained optimization; nonlinear, linear, geometric, and dynamic programming; Pontryagin's maximum principle; and others. Some of these approaches are highly specialized and not in common usage at the present time.

Development of this approach begins with the definition of a scalar objective function that depends on real scalar independent variables. These variables can be written as an **N**-component column vector or point **X**. When particular numbers are assigned to the components of **X**, the resulting vector is called a policy. A feasible region can then be defined for **X** that consists of all points that satisfy constraints established for the components of **X**. The problem then becomes one of finding the optimum within the established constraints.

In the simpler cases, an indirect method can be used to find the optimum. These methods allow us to proceed directly to the optimum without stopping at points along the way that are less than optimum. The method is called indirect because we solve an equation rather than search for an optimum. The root of the equation locates the optimum.

The familiar differential approach is an indirect method. In many instances, the differentials have to be manipulated to give necessary and sufficient conditions for an optimum. In more restricted cases, a second indirect method, called geometric programming, can be used to locate the optimum by simple inspection of the exponents in the objective function. Instead of seeking the optimal values of the independent variables first, geometric programming finds the optimal way to distribute the total cost among the various terms of the objective function. Then, by inspection of simple linear equations, the optimal cost can be found. If this cost it acceptable, the policy can then be found to obtain it.

When working with equality constraints, we generate decision derivatives to transform boundary optimization problems into interior ones. This step makes it possible to use extensions of classical techniques to find the optimum. By using sensitivity coefficients, we can also evaluate the effect on the value of the optimum of changing constraints.

In more complex cases, we cannot proceed directly to the optimum and must use other methods referred to as direct methods. These important techniques are discussed in Wilde and Beightler (1967).

Cybernetics. The word *cybernetics* has taken on many different meanings. The classic definition given by Norbert Wiener (1948) is "the science of control and communication in the animal and the machine." Klir and Valach (1967) stated: "Cybernetics is a science which investigates, from a single point of view, the problems of the control of various processes in different systems and the transfer or communication of control signals within these systems and between them." Another more limited class of definitions views cybernetics as the science of the processes of transmission, processing, and storage of information. Still another group views cybernetics as the system that studies the methods of the forming, structure, and transformation of algorithms describing processes of control that occur in reality. For our purposes, the definitions offered by Weiner and Klir are accepted, since we are concerned with controlling a production system.

The old philosophy of system analysis is to vary one parameter of the system at a time. This, however, is often impossible in complex systems because they are so dynamic and interconnected that the alteration of one factor immediately causes changes in others. This type of system, which includes production control systems, is the type that cybernetics is designed to solve. However, as is true with most new methodologies, the techniques are still undergoing research and development, and very scanty data are available for study.

In general, cybernetics attempts to bring together and reexamine lines of research that have formerly been pursued in isolation. In many cases, it is assumed that science develops only by the subdivision of each branch into still smaller branches; whereas, in reality, despite all the specialization and subdivision of subjects, there is also a good deal of recombination and regrouping constantly taking place. One example of this phenomenon is the field of statistics. In its early stages of development, simultaneous research was conducted in many different fields with little, if any, exchanging of results. More recently, these separate findings have been combined into one body of knowledge and used by many disciplines. Many authorities believe that cybernetics will follow much the same pattern.

Some of the concepts used in cybernetics are worth mentioning. Cybernetics includes the now familiar concept of feedback in a "self-regulating" role. It also includes the concept of interface, which can be defined as locations where parts of a system (or whole systems) come together and exchange information, components, resources, etc. Cybernetics depends primarily on this interfacing, and to a lesser degree on the characteristics of the parts of the system, for control of the system.

Miscellaneous Control Theories. In 1963, Alcalay and Buffa offered a conceptual framework toward the development of a mathematical model of a generalized production system. They viewed the production system as separate from its control subsystems. They felt that a number of simultaneous differential, balance equations and a set of constraint equations and profit criteria relationships, when solved simultaneously, would lead to an optimum control scheme.

In 1964, Reisman and Buffa co-authored an article which depicted another structural framework of a production system. This framework recognized feedback concepts and attempted to illustrate the analogy of a production system to an electrical circuit. They compared their model to the Industrial Dynamics models developed at M. I. T.

OBSERVATIONS

As was stated in Chap. 7, at the present time there exists no known technique or theory that can be used to optimize the total system. (Optimality in this case refers to the minimization of total costs.) Our consideration of the several control theories is an attempt to determine whether one or more such theories might possibly allow us to progress toward the development of an over-all optimizing procedure. We now make some general observations regarding the applicability of the several control theories discussed to the design of operations planning and control systems.

Regardless of how we view the control process, its basic function is the transformation of information. Through analysis of the individual functions that must be performed in a production control system, we may discover how a subsystem acts upon a set of input variables to obtain a set of output variables. This acting or operating has come to be called a *transfer function.* After these transfer functions are defined, we can vary the independent variables of each subsystem within desired constraints, and, in accordance with our understanding of the dynamics of each subsystem, until we approach an optimum. This technique, although relatively new to industrial engineering, is used extensively in linear control theory. This technique is precisely the balancing or tuning procedure discussed in the last section of Chap. 7.

Another concept mentioned in many of the theories is the concept of closed-loop feedback systems. Feedback is an absolute requirement in present-day production control systems. It is difficult to imagine another system that would contain more distortion, noise, and unwanted disturbances. Forecasts are continually in error, unpredictable production problems occur, and humans err in their judgment, to name only a few of the disturbances a production control system must deal with. In addition, many parameters vary with time or some other variable. Feedback through a closed-loop system allows us to detect these variations and make appropriate corrections, thereby making the system less sensitive to these random disturbances.

Industrial Dynamics is one methodology that includes these fundamental concepts. To date, it has emphasized mainly the total system operation. As applied to only a production control system, it is relatively unproven but definitely shows promise for the immediate future. Because it views a total system, it is ideally suited for establishing the over-all structure of the production control system. After this macro-level analysis has been performed, it appears possible that optimization techniques could be applied at the micro-level to optimize the control system.

Industrial Dynamics appears closer to immediate adaptation for total system representation than do any of the other control theories. It is based on linear control theory but has been expanded to include nonlinear systems as well. It is also much easier to arrive at numerical solutions using Industrial Dynamics. One disadvantage is the amount of detail required in formulating the model, specifically in the area of interrelationships.

Cybernetics shows some promise for the more distant future; however, it still requires a great amount of research. The more mathematical approaches, such as modern control theory and modern optimization theory, appear to be limited in application at the present time. It is usually difficult to define the state space vectors or the differential equations or both for some processes. These approaches appear more useful in finding the optimum for individual functions and do not, at present, appear general enough for total system analysis. However, further research and development could very well lead to useful procedures for optimally controlling total production systems.

In reflection, it seems likely at the present time that no one theory can be adapted for all production control functions, but that general concepts, such as feedback and the transfer function, can be used as a basis for designing production control systems. These concepts have been fundamental in this text. More insight concerning the extent to which the individual theories can be adapted to production control must be gained through evaluation of the more promising theories in a hypothesized production system. To carry out this evaluation, appropriate criteria must be established.

EVALUATION CRITERIA

The evaluation of theories for use in production control systems should involve several major areas. Some major questions that can be answered through research are:

1. How easily can the production control system be abstracted to fit the conceptualization of the system defined by a particular theory? For example, when using optimization theory, can a differential equation be written for each process?
2. After the system has been abstracted in accordance with a given theory, how accurate a representation of the real system is it? For example, how accurately does the differential equation describe the physical process?
3. How easily can this representation of the system be manipulated to give numerical or other types of solutions? Are we forced to make unacceptable approximations in seeking solutions so that our solution does not accurately approximate the true performance of the system or process?

4. Does the theory permit experimentation with various system designs?
5. Is the theory proven, or does it still require research?

Answers to these questions should provide a sound basis for evaluation of the various control theories, and on this basis, some quantitative measures of application to production control can be gained. A great amount of research remains to be performed in this exciting and important area.

The Future

The field of operations planning and control is as old as the Industrial Revolution, whose beginning dates back to 1760. Production and distribution processes have become progressively more complex and sophisticated. While some advances have been made in the development of new techniques for control systems, the science of control has not, in general, kept pace with the science of production and distribution. We are still searching for a general systems theory.

What can we say about the future of operations planning and control systems design? Certainly no general theory is likely to emerge in the near future which would be capable of rendering optimal control system designs.

In lieu of a formal optimizing procedure, it seems reasonable to expect that more attention will be directed toward the refinement of the balancing procedure discussed in Chap. 7. This procedure appears to have many merits, particularly its ability to fine tune a control system in response to occurrences in the dynamic environment. Many chemical and petroleum operations are controlled by such systems.

It is also expected that many concepts from formal control theory will be applied to the design of operations planning and control systems. In this text, we used the concepts of the transfer function and feedback. We also introduced elementary notions of adaptive control systems. Many other concepts of control theory (see Dorf 1967; and Sage 1968) are also expected to find application in designing operations systems. Bhattacharyya (1968) and Hwang and Fan (1966) are two examples of research papers in which Pontryagin's maximum principle is applied to the control of production systems.

Another exciting possibility for the more distant future is the application of the emerging science of cybernetics to the design of management control systems. Some basic concepts of cybernetics and their application to management problems are discussed in two books by Stafford Beer, *Cybernetics and Management* (1959), and *Decision and Control* (1966).

The application of the concepts of Industrial Dynamics (Forrester 1961) to an operations system is another possibility that appears to have considerable merit. This application is done to some extent in PROSIM V (Mize *et al.* 1971a and b; Mize 1969), the production system simulator discussed in Appendix C.

It is evident that much basic research and development is needed in the area of operations planning and control systems design. Such research is likely to be pursued in many diverse directions. We feel confident in making one prediction about the nature of the research: it will be more and more quantitative and will feed generously on the developed and developing theories of system control in other disciplines. The future student, teacher, and practitioner of operations control system design will need a very good background in such mathematical techniques as differential equations, matrix theory, state-space techniques, and optimization theory.

REFERENCES

AIEE Committee Report. Proposed Standards and Terms for Feedback Control Systems, Part 2. *Electrical Engineer* 70: 905-9.

Alcalay Jack A. and Buffa, Elwood S.; 1963. A Proposal for a General Model of a Production System. *International Journal for Production Research.* 3, 73-87.

Ashby, W. Ross. 1961. *An Introduction to Cybernetics.* London: Chapman and Hall, Ltd.

Beer, Stafford. 1959. *Cybernetics and Management.* London: English Universities Press.

Beer, Stafford. 1966. *Decision and Control.* London: John Wiley and Sons, Ltd.

Bhattacharyya, R. K. 1968. Optimization of a Production System Response with Respect to the Input Characteristics. Paper presented at the 1968 National ORSA Conference (San Francisco).

Dorf, R. C. 1967. *Modern Control Systems.* Reading, Mass.: Addison-Wesley.

Forrester, J. W. 1961. *Industrial Dynamics.* Cambridge, Mass.: M. I. T. Press.

Hwang, C. L.; and Fan, L. T. 1966. The Application of the Maximum Principle to Industrial and Management Systems. *Journal of Industrial Engineering* 17: 589-93.

Klir, Jiri; and Valach, Miroslav. 1967. *Cybernetic Modelling.* Princeton, N. J.: D. Van Nostrand Co., Inc.

Mize, J. H., *et al.* 1971a. *PROSIM V Administrator's Manual: Production System Simulator.* Englewood Cliffs, N. J.: Prentice-Hall, Inc.

Mize, J. H., *et al.* 1971b. *Production System Simulator (PROSIM V): A User's Manual.* Englewood Cliffs, N. J.: Prentice-Hall, Inc.

Mize, J. H. 1969. PROSIM V: A Production System Simulator. Paper published in *Proceedings of the Third Conference on Applications of Simulation* (Los Angeles, California).

Ogata, Katsuhiko. 1967. *State Space Analysis of Control Systems.* Englewood Cliffs, N. J.: Prentice-Hall, Inc.

Reisman, Arnold, and Buffa, Elwood S. 1964. A General Model for Production and Operations Systems. *Management Science* 11, 64–79.

Roberts, Edward B. 1963. Industrial Dynamics and the Design of Management Control Systems. *Management Technology* 3: 2, 100–118.

Sage, A. P. 1968. *Optimum Systems Control.* Englewood Cliffs, N. J.: Prentice-Hall, Inc.

Wiener, Norbert. 1948. *Cybernetics.* New York: John Wiley & Sons, Inc.

Wilde, D.: and Beightler, C. 1967. *Foundations of Optimization.* Englewood Cliffs, N. J.: Prentice-Hall, Inc.

Appendices

Appendix A
Elementary Matrix Operations

In our development of the Gozinto procedure in Chap. 4, we made use of some elementary matrix operations. The purpose of this appendix is to explain the mechanics of these elementary matrix operations. The theory underlying these operations and other elements of matrix algebra may be found in texts such as those listed in the references for this appendix.

General Concepts

DEFINITIONS AND NOTATION

A matrix is a group of numbers arranged in a rectangular array of rows and columns. For example, the numbers in the rows and columns of Fig. A.1 form a matrix.

$$
\begin{array}{cc} & \text{Columns } (j) \\ & \begin{array}{cccc} 1 & 2 & 3 & 4 \end{array} \\ \text{Rows } (i) \begin{array}{c} 1 \\ 2 \\ 3 \end{array} & \begin{bmatrix} 2 & 0 & 0 & 1 \\ 0 & -1 & 0 & 2 \\ 1 & 2 & 1 & 0 \end{bmatrix} \end{array}
$$

Figure A.1 An Example of a Matrix

Each number in a matrix is called an *element*. The position of each element is indicated by a pair of numerical subscripts. The first subscript indicates the row number in which an element is found; the second subscript indicates the column number in which the element is found. We denote the matrix in

Fig. A.1 as **A** and each element as a_{ij}, where i and j indicate the row and column numbers, respectively. Using this notation, $a_{31} = 1$, $a_{14} = 1$, $a_{24} = 2$, $a_{21} = 0$, etc.

We often represent a matrix in general form, as in Fig. A.2.

$$\mathbf{A} = \begin{bmatrix} a_{11} & a_{12} \cdots & a_{14} \\ a_{21} & a_{22} \cdots & a_{24} \\ \vdots & & \vdots \\ a_{m1} & a_{m2} \cdots & a_{mn} \end{bmatrix}$$

Figure A.2 General representation of a Matrix

We see in Fig. A.2 that, in general, m is the number of rows and n is the number of columns of a matrix; m and n may not be equal. If $m = n$, the matrix is *square*. The matrix in Fig. A.1 is not a square matrix since it has four columns and only three rows.

TYPES OF MATRICES

In addition to square matrices, three other types of matrices have special importance in our discussion: the *identity matrix*, the *diagonal matrix*, and the *triangular matrix*.

The identity matrix, denoted **I**, is a square matrix with each element on its diagonal having the value unity and all other elements having the value zero. A 4×4 (four rows and four columns) identity matrix is shown in Fig. A.3.

$$\mathbf{I} = \begin{bmatrix} 1 & 0 & 0 & 0 \\ 0 & 1 & 0 & 0 \\ 0 & 0 & 1 & 0 \\ 0 & 0 & 0 & 1 \end{bmatrix}$$

Figure A.3 A 4×4 Identity Matrix

A diagonal matrix, also a square matrix, is similar to the identity matrix in that all elements not on the diagonal have zero values. The elements on the diagonal may have any values. We see, then, that the identity matrix is a special case of the diagonal matrix. An example of 4×4 diagonal matrix is shown in Fig. A.4.

$$\mathbf{A} = \begin{bmatrix} 3 & 0 & 0 & 0 \\ 0 & -1 & 0 & 0 \\ 0 & 0 & 0 & 0 \\ 0 & 0 & 0 & 17 \end{bmatrix}$$

Figure A.4 A 4×4 Diagonal Matrix

A square matrix whose non-zero values all lie on the diagonal and on only one side of the diagonal is called a triangular matrix. If all the elements lie above and to the right of the diagonal, the matrix is "upper triangular" (see Fig. A.5). If all the elements lie below and to the left of the diagonal, the matrix is

$$\mathbf{A} = \begin{bmatrix} 3 & 2 & 4 \\ 0 & 0 & 0 \\ 0 & 0 & 1 \end{bmatrix}$$

Figure A.5 Upper Triangular Matrix

"lower triangular" (see Fig. A.6). If a matrix contains non-zero elements on only one side of the diagonal and no non-zero elements on

$$\mathbf{A} = \begin{bmatrix} 1 & 0 & 0 \\ 2 & 13 & 0 \\ \sqrt{7} & 6 & 4 \end{bmatrix}$$

Figure A.6 Lower Triangular Matrix

the diagonal itself, the matrix is "strictly triangular." The matrix in Fig. A.7 is strictly upper triangular.

$$\mathbf{B} = \begin{bmatrix} 0 & 7 & 9 & 0 \\ 0 & 0 & 4 & 1 \\ 0 & 0 & 0 & 2 \\ 0 & 0 & 0 & 0 \end{bmatrix}$$

Figure A.7 Strictly Upper Triangular Matrix

Matrix Operations

MATRIX ADDITION AND SUBTRACTION

Two matrices may be added and subtracted only if each has the same number of rows and the same number of columns. Two $m \times n$ matrices, $\mathbf{A} = [a_{ij}]$ and $\mathbf{B} = [b_{ij}]$, may be added to determine a third $m \times n$ matrix, $\mathbf{C} = [c_{ij}]$, where the elements c_{ij} are determined from the relation $c_{ij} = a_{ij} + b_{ij}$ for all i and j. $\mathbf{B}$ may be subtracted from $\mathbf{A}$ to determine a third $m \times n$ matrix $\mathbf{C}$, where the elements c_{ij} are determined from the relation $c_{ij} = a_{ij} - b_{ij}$ for all i and j.

EXAMPLE OF MATRIX ADDITION

Suppose we have two matrices, $\mathbf{A}$ and $\mathbf{B}$, as follows:

$$\mathbf{A} = \begin{bmatrix} 2 & -1 & 3 \\ 1 & 0 & 1 \\ 0 & 1 & 3 \end{bmatrix} \qquad \mathbf{B} = \begin{bmatrix} 0 & -4 & 1 \\ 0 & 1 & 0 \\ 2 & 1 & 0 \end{bmatrix}$$

$$\mathbf{C} = \mathbf{A} + \mathbf{B} = \begin{bmatrix} 2 & -5 & 4 \\ 1 & 1 & 1 \\ 2 & 2 & 3 \end{bmatrix}$$

EXAMPLE OF MATRIX SUBTRACTION

Using the same matrices, **A** and **B**, as in the previous example,

$$\mathbf{C} = \mathbf{A} - \mathbf{B} = \begin{bmatrix} 2 & 3 & 2 \\ 1 & -1 & 1 \\ -2 & 0 & 3 \end{bmatrix}$$

Also,

$$\mathbf{C} = \mathbf{B} - \mathbf{A} = \begin{bmatrix} -2 & -3 & -2 \\ -1 & 1 & -1 \\ 2 & 0 & -3 \end{bmatrix}$$

Notice that $\mathbf{A} + \mathbf{B} = \mathbf{B} + \mathbf{A}$ and that $\mathbf{B} - \mathbf{A} = -1(\mathbf{A} - \mathbf{B})$.

MATRIX MULTIPLICATION

Two matrices may be multiplied together only if the number of columns of the first matrix is equal to the number of rows of the second. When this condition is satisfied, the two matrices are said to be conformable. The order in which two matrices are written in multiplication is very important. In general, $\mathbf{AB} \neq \mathbf{BA}$, where **A** and **B** are two matrices and **AB** indicates that they have been multiplied together.

Suppose that we have two matrices, **A** and **B**, and that the number of columns in **A** is equal to the number of rows in **B**: thus, they can be multiplied as **AB**. The result of this multiplication is a third matrix $\mathbf{C}(=\mathbf{AB})$. The number of rows in **C** will equal the number of rows in **A** and the number of columns in **C** will equal the number of columns in **B**. Each element in **C** (each c_{ij}) is the sum of the products obtained by multiplying the kth element in the ith row of **A** by the kth element in the jth column of **B**, where k is an index denoting the element number in the row and column being multiplied.

Let us define **A** as an $m \times p$ matrix and **B** as a $p \times n$ matrix. In general, the elements of $\mathbf{C} = \mathbf{AB}$ may be obtained as follows:

$$c_{ij} = \sum_{k=1}^{p} a_{ik} b_{kj} \qquad \text{for } i = 1, \ldots, m$$
$$j = 1, \ldots, n$$

While this formulation appears complicated, the mechanical operations involved in matrix multiplication are quite easy to grasp. The method can best be illustrated with a detailed example. We have two matrices, **A** and **B**, which we wish to multiply together to form a third matrix **C**.

$$\mathbf{A} = \begin{bmatrix} 0 & -1 & -2 \\ 7 & -1 & 0 \end{bmatrix} \qquad \mathbf{B} = \begin{bmatrix} 2 & 4 \\ 0 & 0 \\ -2 & 1 \end{bmatrix}$$

Note that **A** is 2×3 and **B** is 3×2; thus, **A** and **B** are conformable since the number of columns in **A** is the same as the number of rows in **B**. Note that **A** could have had any other number of rows and **B** could have had any other number of columns and the two matrices would still have been conformable.

The individual elements of **C** are determined from the general equation given above.

$$c_{11} = \sum_{k=1}^{3} a_{1k} b_{k1} = (0)(2) + (-1)(0) + (-2)(-2) = 4$$

$$c_{12} = \sum_{k=1}^{3} a_{1k} b_{k2} = (0)(4) + (-1)(0) + (-2)(1) = -2$$

$$c_{21} = \sum_{k=1}^{3} a_{2k} b_{k1} = (7)(2) + (-1)(0) + (0)(-2) = 14$$

$$c_{22} = \sum_{k=1}^{3} a_{2k} b_{k2} = (7)(4) + (-1)(0) + (0)(1) = 28$$

Thus,

$$\mathbf{AB} = \mathbf{C} = \begin{bmatrix} 4 & -2 \\ 14 & 28 \end{bmatrix}$$

In this case, we can also multiply **B** times **A**.

$$\mathbf{BA} = \mathbf{C} = \begin{bmatrix} 28 & -6 & -4 \\ 0 & 0 & 0 \\ 7 & 1 & 4 \end{bmatrix}$$

This example clearly shows that, in general, $\mathbf{AB} \neq \mathbf{BA}$.

PARTITIONING OF MATRICES

It is sometimes useful to partition a matrix into several smaller matrices called submatrices. Consider, for example, the following matrix **A**:

$$\mathbf{A} = \begin{bmatrix} 0 & 0 & 0 & 0 & 0 & 0 & 0 \\ 0 & 0 & 0 & 0 & 0 & 0 & 0 \\ 4 & -2 & 0 & 0 & 0 & 0 & 0 \\ 0 & 4 & 1 & 0 & 0 & 0 & 0 \\ 3 & 2 & 6 & 0 & 0 & 0 & 0 \\ 0 & 0 & 0 & 1 & 3 & 2 & 0 \\ 0 & 0 & 0 & 0 & 1 & 0 & 0 \\ 0 & 0 & 0 & 0 & 0 & 1 & 0 \end{bmatrix}$$

There are many ways to partition a matrix; the particular partitioning scheme used depends upon the requirements of the problem. One way to partition **A** would be:

$$\mathbf{A} = \left[\begin{array}{ccc:cccc} 0 & 0 & 0 & 0 & 0 & 0 & 0 \\ 0 & 0 & 0 & 0 & 0 & 0 & 0 \\ \hdashline 4 & -2 & 0 & 0 & 0 & 0 & 0 \\ 0 & 4 & 1 & 0 & 0 & 0 & 0 \\ 3 & 2 & 6 & 0 & 0 & 0 & 0 \\ \hdashline 0 & 0 & 0 & 1 & 3 & 2 & 0 \\ 0 & 0 & 0 & 0 & 1 & 0 & 0 \\ 0 & 0 & 0 & 0 & 0 & 1 & 0 \end{array}\right]$$

Each of the six sets of elements constitutes a submatrix. These submatrices may be written as:

$$\mathbf{A}_{11} = \begin{bmatrix} 0 & 0 & 0 \\ 0 & 0 & 0 \end{bmatrix} \qquad \mathbf{A}_{12} = \begin{bmatrix} 0 & 0 & 0 & 0 \\ 0 & 0 & 0 & 0 \end{bmatrix}$$

$$\mathbf{A}_{21} = \begin{bmatrix} 4 & -2 & 0 \\ 0 & 4 & 1 \\ 3 & 2 & 6 \end{bmatrix} \qquad \mathbf{A}_{22} = \begin{bmatrix} 0 & 0 & 0 & 0 \\ 0 & 0 & 0 & 0 \\ 0 & 0 & 0 & 0 \end{bmatrix}$$

$$\mathbf{A}_{31} = \begin{bmatrix} 0 & 0 & 0 \\ 0 & 0 & 0 \\ 0 & 0 & 0 \end{bmatrix} \qquad \mathbf{A}_{32} = \begin{bmatrix} 1 & 3 & 2 & 0 \\ 0 & 1 & 0 & 0 \\ 0 & 0 & 1 & 0 \end{bmatrix}$$

Now, if we denote the partitioned matrix as $\mathbf{A}_p$, we can write:

$$\mathbf{A}_p = \begin{bmatrix} \mathbf{A}_{11} & \mathbf{A}_{12} \\ \mathbf{A}_{21} & \mathbf{A}_{22} \\ \mathbf{A}_{31} & \mathbf{A}_{32} \end{bmatrix}$$

Notice that the submatrices $\mathbf{A}_{11}$, $\mathbf{A}_{31}$, $\mathbf{A}_{12}$, and $\mathbf{A}_{22}$ contain all zeros as elements. Matrices containing nothing but zeros are called null matrices and can be denoted by $\mathbf{0}$. Thus, we can write

$$\mathbf{A}_p = \begin{bmatrix} \mathbf{0} & \mathbf{0} \\ \mathbf{A}_{21} & \mathbf{0} \\ \mathbf{0} & \mathbf{A}_{32} \end{bmatrix}$$

It is shown in linear algebra texts (see Hadley 1961, p. 81) that the rule for addition of partitioned matrices is the same as the rule for addition of ordinary matrices, provided the submatrices are conformable for addition.

In order for two partitioned matrices, $\mathbf{A}_p$ and $\mathbf{B}_p$, to be conformable for multiplication, the columns of $\mathbf{A}_p$ must be partitioned in the same way as the rows of $\mathbf{B}_p$.

When submatrices are appropriately partitioned for addition or multiplication, the submatrices can be treated exactly as ordinary elements of unpartitioned matrices. These features of partitioned matrices were used in developing Eqs. 4.5 through 4.9 in Chap. 4.

VECTORS

We often use the term *vector* to denote a special kind of matrix consisting only of a single row or a single column. For example,

$$\mathbf{x} = \begin{bmatrix} x_1 \\ x_2 \\ \vdots \\ x_n \end{bmatrix}$$

is a column vector, while

$$\mathbf{x} = [x_1, x_2, \ldots, x_n]$$

is a row vector.

With this definition, any row or column of a matrix may be regarded as a vector. Indeed, referring to the concepts of partitioning discussed in the preceding section, a $m \times n$ matrix can be partitioned into either m row vectors or n

column vectors. The notion of a vector was used extensively in developing the "$\mathbf{T} = (\mathbf{I} - \mathbf{N})^{-1}$ Algorithm" in Chap. 4.

MATRIX INVERSION

The inverse of a square matrix **A** is denoted $\mathbf{A}^{-1}$. By definition, if **A** and **B** are square matrices such that $\mathbf{AB} = \mathbf{BA} = \mathbf{I}$, then **B** is called the inverse of **A**, and **A** is the inverse of **B**. Thus, $\mathbf{B} = \mathbf{A}^{-1}$ and $\mathbf{A} = \mathbf{B}^{-1}$. Considering a single matrix **A**, we have $\mathbf{AA}^{-1} = \mathbf{A}^{-1}\mathbf{A} = \mathbf{I}$. This says that the inverse of a matrix, when multiplied by the original matrix, will result in the identity matrix **I**.

A matrix can be inverted in a number of ways. We illustrate here one method very easy to understand for small matrices. Other procedures may be found in Hadley (1961).

Given a matrix **A** for which we wish to find the inverse, we write an identity matrix **I** to the right of **A** and then perform elementary algebraic manipulations on the two matrices **A** and **I** until **A** has been reduced to an identity matrix. By so doing, we change **I** to some form other than the identity matrix; indeed, **I** will be changed to $\mathbf{A}^{-1}$, the inverse of **A**.

As shown in linear algebra texts, the following elementary operations may be performed on the rows of a matrix without changing the mathematical character of the matrix:

1. Any two rows may be interchanged.
2. The values in any row may all be multiplied by a non-zero number.
3. A multiple of one row may be added to another row.

Suppose we have the following matrix;

$$\mathbf{A} = \begin{bmatrix} 3 & 2 \\ 1 & 3 \end{bmatrix}$$

and that we wish to find its inverse. Following our procedure, we first place an identity matrix to the right of **A**.

$$\begin{matrix} \mathbf{A} & & \mathbf{I} \\ \begin{bmatrix} 3 & 2 \\ 1 & 3 \end{bmatrix} & & \begin{bmatrix} 1 & 0 \\ 0 & 1 \end{bmatrix} \end{matrix}$$

We now use the three row operations listed above to change **A** to **I**. In so doing, we change **I** to $\mathbf{A}^{-1}$.

We can help ourselves by using rule 1 and interchanging rows 1 and 2. This action will place a one in the upper left-hand corner of **A**. We have

$$\begin{bmatrix} 1 & 3 \\ 3 & 2 \end{bmatrix} \quad \begin{bmatrix} 0 & 1 \\ 1 & 0 \end{bmatrix}$$

We want a zero below the one in the matrix on the left. We can obtain this zero by using rule 3 and adding (-3)(row 1) to row 2:

$$\begin{bmatrix} 1 & 3 \\ 0 & -7 \end{bmatrix} \quad \begin{bmatrix} 0 & 1 \\ 1 & -3 \end{bmatrix}$$

We also want a zero in the upper right corner of the matrix on the left. We again use rule 3 and add $(3/7)$(row 2) to row 1:

$$\begin{bmatrix} 1 & 0 \\ 0 & -7 \end{bmatrix} \quad \begin{bmatrix} 3/7 & -2/7 \\ 1 & -3 \end{bmatrix}$$

We have only one step remaining to obtain an identity matrix on the left; we must get a one in the lower right-hand corner. We now employ rule 2 and multiply row 2 by $-1/7$:

$$\begin{bmatrix} 1 & 0 \\ 0 & 1 \end{bmatrix} \quad \begin{bmatrix} 3/7 & -2/7 \\ -1/7 & 3/7 \end{bmatrix}$$

We now have an identity matrix on the left and $\mathbf{A}^{-1}$, the inverse of the original matrix $\mathbf{A}$, on the right.

If our calculations are correct,

$$\mathbf{AA}^{-1} = \mathbf{A}^{-1}\mathbf{A} = \mathbf{I}$$

Checking our results,

$$\mathbf{AA}^{-1} = \begin{bmatrix} 3 & 2 \\ 1 & 3 \end{bmatrix} \begin{bmatrix} 3/7 & -2/7 \\ -1/7 & 3/7 \end{bmatrix} = \begin{bmatrix} 1 & 0 \\ 0 & 1 \end{bmatrix}$$

$$\mathbf{A}^{-1}\mathbf{A} = \begin{bmatrix} 3/7 & -2/7 \\ -1/7 & 3/7 \end{bmatrix} \begin{bmatrix} 3 & 2 \\ 1 & 3 \end{bmatrix} = \begin{bmatrix} 1 & 0 \\ 0 & 1 \end{bmatrix}$$

We see that our calculations were correct and that we successfully determined the inverse of $\mathbf{A}$.

It is obvious that matrix inversion is tedious and time consuming, even for small matrices. For this reason, computer routines have been developed for inverting matrices and for performing many other matrix operations. Most computer centers have several such canned routines in their program library.

REFERENCES

1. Birkoff, G.; and MacLane, S. 1956. *A Survey of Modern Algebra.* New York: The Macmillan Company.

2. Chung, An-min. 1963. *Linear Programming*. Columbus, Ohio: Charles E. Merrill Publishing Co.

3. Hadley, G. 1961. *Linear Algebra.* Reading, Mass.: Addison-Wesley Pub. Co.

Appendix B
Introduction to PERT

Many work projects in both the military and industry consist of a large number of interrelated steps, many of which are either undefined or have never been done before. This type of work is usually characterized by a strong interdependency between parts of the total job; that is, some parts cannot be started until other parts are finished. When there are thousands of such parts, it is very difficult for management to perform its normal functions, such as assigning work at specified times and manpower to specific jobs, etc. The job is even more complicated when tight schedules must be met. Fitting together the many thousands of pieces of a complicated puzzle and making everything flow smoothly is a formidable task, even for the most competent manager. The problem quickly becomes too large for the traditional Gantt-chart approach. Thus management had to seek a new approach to planning and controlling large projects.

As mentioned in Chap. 6, many managers have turned to techniques called "critical path analysis" for planning large projects. The two basic methods of critical path analysis are PERT and CPM. There are many variants of these two methods. Initially, each method had certain distinguishing characteristics. However, each basic method has been modified to include features of the other method so that no essential differences remain. We consider only PERT; details of CPM may be found in Moder and Phillips (1964) and Muth and Thompson (1963).

Basic PERT Concepts

PERT was developed to cope with a complex work program with many interrelated events and activities. The first step in applying PERT is to acquire a thorough understanding of the technical aspects of the work program being planned. All events necessary for the successful completion of the project must be identified. Next, the specific work activities necessary to accomplish each event must be determined. It is important that every activity necessary for the accomplishment of the project be listed.

The next step is to consider the precedence relationship among the activities; i.e., which activities *must* precede other activities. It is helpful in performing this step to construct a directed network of activities and events such as that shown in Fig. B.1. Every activity in the work program is represented by an arrow on the network. The tail of an arrow represents the start of the activity and the head represents its completion. The nodes (circles) of the network represent events. Every activity begins and ends with an event. Note that one event may represent the end or beginning of any number of activities.

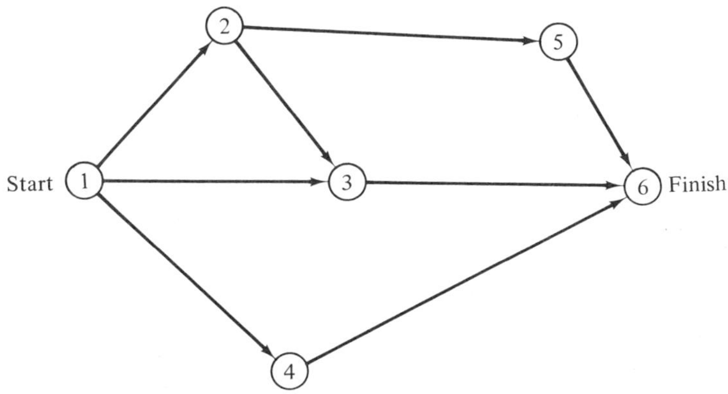

Figure B.1 Project Network

Each event in the network is numbered. Because each activity begins and ends with a numbered event, each activity is also uniquely numbered. Specifically, an activity is designated by the numbers of the two events it connects. We denote the event at the tail of the arrow as i and the event at the head of the arrow as j. Thus, activity (2, 3) in Fig. B.1 is the activity that connects events 2 and 3.

Three definitions become meaningful at this point:

Event: An identifiable point in time (in the course of the work program) at which an activity is started or completed.

Activity: The effort (expressed in units of time) required to accomplish all work leading from one event to the following event.

Network: The graphical representation of the events, activities, and interrelationships which comprise a total work program.

Precedence relationships between events and activities form sequential time paths through the network. Some basic rules helpful in constructing project networks are:

1. Without exception, each activity must have a predecessor and successor event.
2. No event is considered attained until all activities leading to it are completed, and no activity may begin until its predecessor event is completed.
3. Parallel activities between two events, without intervening events, are prohibited. Dummy activities can be used as required to specify intervening events.
4. Independence and dependence of activities must be specified by the network (dummy activities may be used if required).
5. The work performance required by an activity should be associated with a definable organizational entity.
6. Looping is not permitted in a network: that is, if A precedes B and B precedes C, C cannot precede A.
7. All events should be numbered.

At this stage of the PERT procedure, the function of planning has been accomplished. Now, in order to put the plan on a time basis, the function of scheduling must be performed. In scheduling, calendar dates are assigned to the events represented on the project network. In order to make the assignments, time durations must be specified for each activity (each arrow) on the network. A complication arises at this point due to the nature of most project work. It is very difficult to know precisely how much time will be required to perform a particular activity because, in most cases, the activity has never been performed before. Thus, an estimate of the time duration of each activity is required. It is entirely feasible to estimate a single value for this time duration. However, the originators of PERT recognized the high degree of inaccuracy of a single estimate and so, very ingeniously, they devised a scheme for obtaining an approximate probability distribution for each activity duration. The scheme was to obtain three time estimates rather than one. These three time estimates are:

a = optimistic time (an estimate of the time duration under the most favorable conditions).

m = most likely time (an estimate of the time required under normal conditions).

b = pessimistic time (an estimate of the time duration under the least favorable conditions).

These three estimates are then considered parameters of a Beta distribution. a and b are considered the extreme values, and m is considered the modal value. The probability density function of a Beta distribution is shown in Fig. B.2. Using these three estimates, we can compute the expected time and

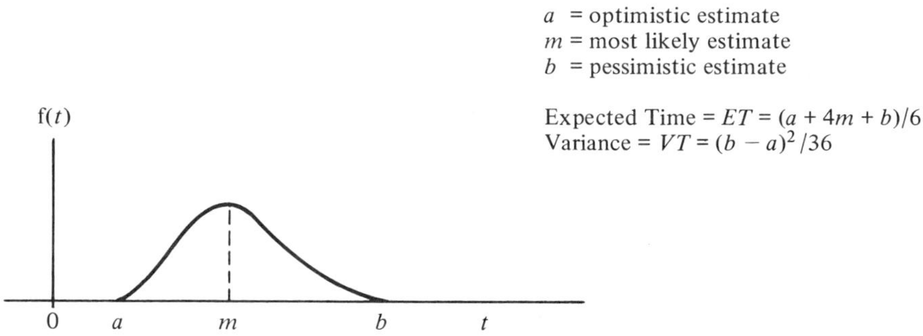

Figure B.2 Beta Distribution

the variance of a particular activity. Three more definitions can now be stated:

Estimate: A technical judgment regarding the amount of time required to perform an activity. This estimate is made in three forms: most likely, optimistic, and pessimistic.

Expected Time: The reduction of the three time estimates to one statistically weighted estimate of the time required to perform the activity.

$$ET = \frac{a + 4m + b}{6}$$

Variance: In the usual statistical sense, the variance is an indication of the reliability of the estimates. For the Beta distribution, the variance is computed as:

$$VT = \frac{(b - a)^2}{36}$$

Using the *ET*'s computed for each activity as the time duration for those activities, we can compute several properties of the project network. These properties, given as definitions, are:

Earliest Starting Time, *TE* (for each event): The summation of all expected times, *ET*, of the activities along the longest path from the first event in the network to the objective event (the event in question).

Latest Starting Time, *TL* (for each event): The latest time, relative to the first event, an event can occur without delaying the scheduled completion time of the entire project.

Slack Time, *TS* $(TL - TE)$: The amount of time a particular event can "slip" without affecting the final completion time.

Critical Path: The longest path from the initial to the terminal event. This path is composed of events with zero slack time; i.e., a delay in the occurrence time of any event on the critical path will result in a corresponding delay in the final completion date. The length of the critical path is the *project duration.*

In addition to assigning calendar dates to each event of the network, this method also establishes rigorous limits to guide operating personnel in the execution of each activity. It tells those responsible for a job when to start worrying about a slippage and to report this fact to higher management, who are responsible for the progress of the over-all project. When this information is considered in conjunction with the project diagram, the project manager can determine when and how to revise the schedule or how to reallocate available resources to assure the schedule will be met. Furthermore, the project manager will know, *in advance*, the probable effect of expediting one activity upon all other activities in the project. This information is not available to the project manager by traditional methods.

A convenient computing algorithm which provides the above information is outlined in the next section. It is important to note that the above discussed procedure provides information which is "event-oriented." The algorithm in the next section provides all the above information plus information which is "activity oriented." This information is essential for assigning jobs to operating departments.

Extension of PERT Model

As mentioned earlier, the original PERT calculations were event oriented. Specifically, the following information was presented about each event:

Earliest start time

Latest start time

Slack time

With this information, we can calculate the critical path and the probability of meeting a specified completion date. This information, though fine for planning purposes, is inadequate for job-loading purposes. To assign jobs to specific facilities, we require the following information for each activity:

Earliest start time

Earliest finish time

Latest start time

Latest finish time

Maximum time available

Total slack

All these quantities can be readily computed from the basic event computations.

Restatement of PERT Algorithm

The original PERT algorithm has been modified to include the above extensions. The new computational algorithm is explained with the help of a simple project network, shown in Fig. B.3. The three time estimates (optimistic, most likely, and pessimistic) are indicated inside the parentheses on each activity.

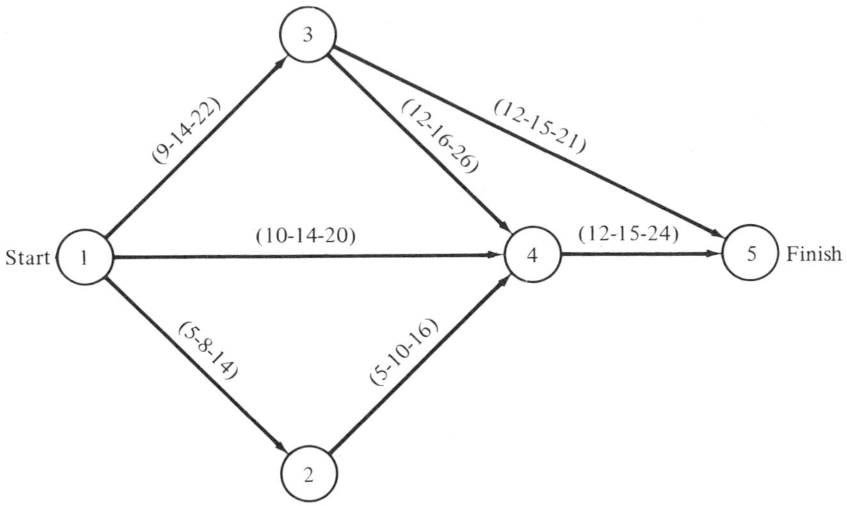

Figure B.3 PERT Network

Given this network, we can perform the computational algorithm as explained below. Table B.1 contains the extended PERT analysis of the network shown in Fig. B.3. Above each column in Table B.1 is a letter used in the following explanations:

Table B.1 PERT Analysis

Event-Oriented Information							Activity-Oriented Information					
A	*B*	*C*	*D*	*E*	*F*	*G*	*H*	*I*	*J*	*K*	*L*	*M*
E	*PE*	*ET*	*VT*	*TE*	*TL*	*TS*	*EST*	*EFT*	*STL*	*FTL*	*TM*	*FT*
5	4	16.0	4.0	47.5	47.5	0	31.5	47.5	31.5	47.5	16.0	0
	3	15.5	2.25				14.5	30.0	32.0	47.5	33.0	17.5
4	3	17.0	5.44	31.5	31.5	0	14.5	31.5	14.5	31.5	17.0	0
	2	10.2	3.36				8.5	18.7	21.3	31.5	23.0	12.8
	1	14.3	2.78				0.0	14.3	17.2	31.5	31.5	17.2
3	1	14.5	4.70	14.5	14.5	0	0.0	14.5	0.0	14.5	14.5	0
2	1	8.5	2.25	8.5	21.3	12.8	0.0	8.5	12.8	21.3	21.3	12.8

Columns A *and* B. The last event (5, in this case) is listed first in Column *A*. Then, in Column *B*, the highest event number leading to the event listed in Column *A* is listed (event 4, in this case). Still in Column *B*, the next highest event leading to the event in *A* is listed. This listing is continued until all preceding events connected to the event in *A* have been listed in descending order. Now, going back to Column *A*, we list the next highest event in the network (4). Switching again to Column *B*, we repeat the above procedure for the new event until all preceding events connected to it are listed in descending order. This entire process is repeated until all events and their predecessors are listed in Columns *A* and *B*, respectively.

Column C. We compute a statistically weighted expected time (*ET*) for each activity by substituting the three time estimates shown on each activity (arrow) in Fig. B.3 into the formula

$$ET = (a + 4m + b)/6$$

For example, *ET* for activity (1, 2) is

$$ET(1, 2) = (5 + 4[8] + 14)/6 = 8.5$$

All other *ET*'s are computed in the same way and entered into Column *C*.

Column D. We find variance (*VT*) for each activity by substituting optimistic and pessimistic estimates for *a* and *b* and solving

$$VT = (b - a)^2/36$$

For example,

$$VT(1, 2) = (14 - 5)^2/36 = 81/36 = 2.25$$

Column E. We can compute the earliest expected time (TE) of accomplishment for each event by adding the elapsed time (ET) of each activity to cumulative total elapsed times through the preceding event, always staying within a single path working from start to finish. When more than one activity leads to an event, that path whose elapsed time (ET) gives the greatest sum up to that event is chosen as the earliest time for that event (because it is the longest path that is being sought).

To illustrate the computations involved in finding TE for an event, we here determine TE for Event 4. There are three paths leading to 4, 1-4, 1-2-4, and 1-3-4. To determine the earliest time Event 4 can occur, the *longest* of these paths must be chosen. We can calculate the following times for the three possible paths:

Path 1-4: Elapsed time $= 14.3$
Path 1-3-4: Elapsed time $= 14.5 + 17.0 = 31.5$
Path 1-2-4: Elapsed time $= 8.5 + 10.2 = 18.7$

Thus, $TE(4)$ is 31.5, this being the earliest possible time that Event 4 can occur. The TE's for the other events are determined in the same way, always choosing the longest path to that event.

Column F. We find the latest time (TL) for a particular event by first fixing the earliest time of the final event as its latest time. (It happens that the latest time and the earliest time for the final event are always identical. Thus, there is no ambiguity in the foregoing statement.) Next, we subtract the objective event's* corresponding elapsed time in Column C to find the latest time of the preceding event, staying within a single path working backward from finish to start. When more than one activity leads from an event, the activity which gives the *least* sum through that event is selected.

TL for each event in Fig. B.3 is calculated below:

$$TL(5) = TE(5) = 47.5 \text{ by definition}$$

$$TL(4) = TL(5) - ET(4, 5) = 47.5 - 16.0 = 31.5$$

$$TL(3) = \min \begin{bmatrix} TL(4) - 17.0 \\ TL(5) - 15.5 \end{bmatrix} = \min \begin{bmatrix} 31.5 - 17.0 \\ 47.5 - 15.5 \end{bmatrix} = 14.5$$

$$TL(2) = TL(4) - ET(2, 4) = 31.5 - 10.2 = 21.3$$

These values are the latest time, relative to the starting time of Event 1, at which each event can occur without delaying the project. In this simple ex-

*The objective event is the event for which TL is presently being sought.

ample, only Event 2 can occur after its calculated ET. Thus, it is said to have "slack," and it would be possible to reallocate resources from activity (1, 2) to one or more of the other activities in which no free or slack time exists. This process is called "compressing" the network.

Column G. We can find the slack time, TS, for each event by subtracting earliest time from latest time ($TL - TE$). The purpose is (1) to locate the critical path through the network (this path is designated by those events having zero slack time) and (2) to determine next-most critical paths, as well as those events having substantial slack.

NOTE: Columns A through G contain the information available through the original PERT algorithm (the event-oriented information). Columns H through M, explained below, contain activity-oriented information and represent an extension of the original PERT technique.

Column H. Earliest start time, EST, for each activity (i, j) is defined as the earliest occurrence time of event i.

Column I. Earliest finish time, EFT, for each activity (i, j) is defined as $EST(i, j) + ET(i, j)$. In Table B.1, we find this time by adding Columns C and H for each activity (i, j).

Column J. We can find the latest start time, STL, for each activity (i, j) by subtracting $ET(i, j)$ from TL of event j. In Table B.1, subtract Column C from Column F; change values in F when values in Column A change.

Column K. Latest finish time, FTL, for each activity (i, j) is defined as the latest time, TL, of event j.

Column L. We can find the maximum time available, TM, for each activity i, j by subtracting $TE(i)$ from $TL(j)$.

Column M. We find total slack, FT, for each activity (i, j) by subtracting $ET(i, j)$ from $TM(i, j)$. In Table B.1, subtract Column C from Column L for each activity (i, j).

Probability of Meeting Project Completion Date

The scheduled objective date is defined as a target date for completing the project. This figure may be quite arbitrarily imposed by management or it may be the result of a contractual arrangement, independent of any PERT calculations. PERT can be of real assistance to a manager in two general ways as far as a scheduled objective date is concerned: PERT can (1) help management specify a realistic schedule date and (2) help determine the lack or excess of resources for accomplishing a scheduled date set by any other considerations.

Recall that the completion time for each activity was assumed to be distributed according to the Beta distribution. The expected value of this time (ET) and the variance of the expected value (VT) are obtained from Columns

C and D. In effect, a probability distribution of activity time for each activity has been specified. If a particular event is chosen, one can speak of the distribution of the time required to reach that event from the starting event. To discuss this distribution, we must refer to an important theorem in statistical theory, the central limit theorem.

Suppose there are k activities along the longest path between the starting event and the event in question. Suppose further that the expected times (ET_1, $ET_2, \ldots ET_k$) for these activities are independent random variables with identical means of ET and variances of VT. According to the central limit theorem, the variable

$$T = ET_1 + ET_2 + \cdots + ET_k$$

is approximately normally distributed with mean

$$T_E = kET$$

and with a variance of

$$VT_E = kVT$$

The conditions of equal means and equal variances are not met in most PERT networks; however, this difficulty is not a major one. In fact, this theorem does an amazingly good job when the time distributions are not identical, provided the variances are not too different. For purposes of the PERT network, it is now assumed that T is normally distributed with mean

$$T_E = ET(1, 2) + ET(2, 3) + \cdots + ET(k - 1, k)$$

and variance

$$VT_E = VT(1, 2) + VT(2, 3) + \cdots + VT(k - 1, k)$$

The activity notation assumes that the events are occurring in ascending numerical order.

The results of the brief development presented above are simply these: PERT assumes (with some validity) that the earliest starting time for an event is normally distributed with the mean simply being T_E and the variance being the sum of the activity variances along the longest path leading to the event in question. Thus, even though activity time was considered to be Beta distributed, the earliest starting time is considered to be normally distributed. This result is extremely convenient for purposes of discussing the uncertainty in T_E because of the nice mathematical properties of the normal distribution.

Such probabilities are obtained from appropriate tables of the area under the standardized normal curve. The standard normal tables reveal immediately that $P(T \leq T_E) = 0.50$ (which is also apparent geometrically). Generally, a probability of reaching a scheduled completion date in the range of 0.35 to 0.65 is considered acceptable; however, each company should decide this figure for itself.

Considering the PERT project analyzed in Table B.1, let us determine the probability that the project will be completed by time period 45. We first determine the mean and variance of the random variable, "project completion time."

$$T_E = ET(1, 3) + ET(3, 4) + ET(4, 5) = 14.5 + 17.0 + 16.0 = 47.5$$

$$VT_E = VT(1, 3) + VT(3, 4) + VT(4, 5) = 4.70 + 5.44 + 4.00 = 14.14$$

The standard deviation of the random variable, SD_T, is just the square root of the variance, or $SD_T = \sqrt{14.14} = 3.75$. Figure B.4 is a graphical representation of the normal distribution which theoretically describes our random variable project completion time, denoted T in the figure.

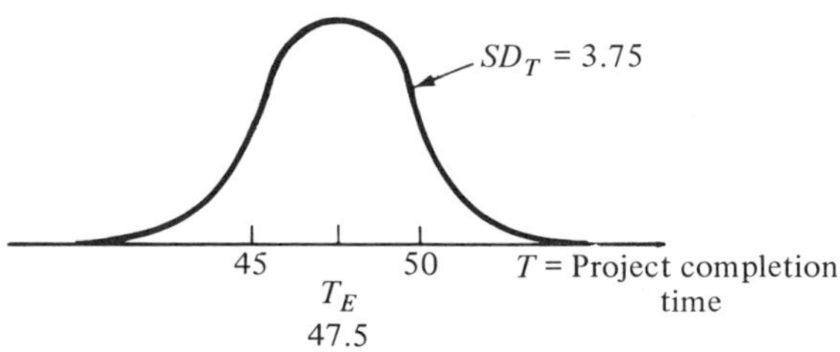

Figure B.4 Normal Distribution for Project Completion Time

The probability that the project will be completed by time 45 is equal to the area under the curve to the left of 45. Reference to a standardized normal curve table (found in most texts on statistics and probability) shows that the area to the left of 45 is approximately equal to 0.25; thus, there is only a 25 percent chance that the project could be completed by time 45. You should be able to show that the probability of completing the project by $T = 50$ is approximately 75 percent.

Replanning

Since PERT directly considers time only (not cost),* a replanning of the work schedule is suggested by the probability of meeting the scheduled com-

*A later version of PERT, called PERT/COST, does consider cost. See *DOD and NASA Guide*, etc. (1962).

pletion date of the project. If this probability is below a certain level (specified by the company, often about 0.35), there is a need for determining how the project duration might be decreased. There are several approaches to replanning in this case, all of which have merit for certain circumstances. These approaches are briefly described here:

Use of Additional Resources. New personnel may be hired or new equipment may be purchased or both. This is one of the simplest ways of replanning. Even though PERT does not consider cost, the manager must consider it, and the cost of additional resources may be prohibitive.

Overtime Use of Present Resources. The overtime use of present resources is always a possibility. In principle, it is essentially the same as acquiring new resources. The lack of easy availability of new and qualified personnel or of special purpose equipment often forces overtime use. Again, cost is extremely important in decisions involving overtime.

Reallocation of Present Resources. One of the advantages of PERT lies in the presentation of information concerning resource use. Slack activities indicate the possibility of additional resources, which might be shifted to the critical activities. Such reallocations may decrease the duration of the critical path. In making these reallocations, new critical paths may come into being.

Re-examine Time Estimates. Sometimes it is very worthwhile to re-examine the time estimates for those activities on the critical path. A more detailed examination may make it possible to decrease some of these estimates. Further, such examinations sometimes point out the possibility that some activities may be performed concurrently instead of in series.

The basic purpose of replanning is to attempt to decide how a scheduled completion date can be reached. Thus, replanning may result in changes in the program or requests for more resources. It is always possible, of course, that replanning would simply convince one of the impossibility of meeting a specified completion date. All such information is of high interest to management, and it should be communicated to appropriate persons.

Even though this discussion has been pointed at project completion, it is apparent that it could be directed at any event in the project.

REFERENCES

Moder, J. J.; and Phillips, C. R. 1964. *Project Management with CPM and PERT.* New York: Reinhold Publishing Corp.

Muth, J. F.; and Thompson, G. L. Eds. 1963. *Industrial Scheduling.* Englewood Cliffs, N.J.: Prentice-Hall, Inc.

DOD and NASA Guide, PERT Cost Systems Design. 1962. By the Office of the Secretary of Defense and the National Aeronautics and Space Administration. Washington, D.C.: U.S. Government Printing Office.

Appendix C
Computer-Aided Instruction in Control System Design

We have intentionally presented many of the concepts in this text from an academic and idealistic point of view. Our purpose has been to emphasize *concepts*, particularly the generality of the information-decision approach to analyzing and designing control systems. We have avoided discussing highly sophisticated transfer functions so that we could direct our attention to the development of a general framework for control system design.

It is difficult to teach principles of system design. Some people feel that design principles cannot be taught; they can only be learned. Nevertheless, systems designers must somehow learn the tools of their trade.

Two basic approaches have been employed in attempting to teach the design of operations planning and control systems. One approach is the ordinary textbook-lecture-homework approach, in which the several control functions are treated individually. Another approach is the case-study approach, in which a real or hypothetical operations system is analyzed.

In teaching the design of a system, it is helpful for the instructor to bring the system into the classroom or laboratory so that pertinent design principles may be discussed. This procedure is common in teaching design principles for electrical systems and mechanical systems.

USE OF COMPUTER SIMULATION

Since it is not generally possible, however, to bring an industrial system into the classroom, a *representation* of the system is the next best thing. Accurate representations are now possible due to the increasing popularity of teaching the principles of control system design through the use of computer simulation. The use of computer simulation is relatively new; most applications have been developed within the past decade. Among the first such applications were the "management games." Perhaps the most widely used management game is the Carnegie Tech Management Game (Cohen *et al.* 1964).

Much interest is currently being shown in the use of computer simulation as an aid in teaching concepts in industrial engineering, production management, and operations research. Porter *et al.* (1966), Vollman (1968), and Whitman and Love (1967) describe such efforts. Computer simulation has also been used for studying higher-order relationships between major components of a firm and between competing firms. The area of Industrial Dynamics, as developed by J.W. Forrester (1961), is concerned with such relationships.

One of the more recent advances in the direct use of computer simulation in the learning process is the development of a production system simulator, called PROSIM V. PROSIM V, developed at Auburn University,* is capable of simulating a variety of operating environments.

The new approach complements the traditional textbook-lecture-homework approach by permitting the student to interact with a dynamic simulated production system. This approach provides the student with a means of gaining experience in controlling a total operating system, of testing his ideas, and of receiving immediate feedback on the results of his decisions. The student literally "closes the loop" in the control system—he is actually part of the adaptive feedback control system. He is both the system designer and manager. The new approach provides the instructor with a means of portraying and manipulating a simulated operations environment and assists him in evaluating student performance in operating and controlling such a system.

PROSIM V is described completely in two manuals, *PROSIM V Administrator's Manual* (Mize *et al.* 1971a) and *PROSIM V User's Manual* (Mize *et al.* 1971b). This approach is summarized briefly in the following paragraphs, which are based in part upon Mize (1969).

Characteristics of the Simulated Environment

PROSIM V is sufficiently flexible to simulate a wide variety of operating environments, including nonmanufacturing operations. It also permits broad flexibility in the type of operations control system that can be designed to control the simulated environment.

Figure C.1 is a typical inventory-production-sales system that PROSIM V is capable of simulating. Systems such as this usually possess the following characteristics:

1. There several finished products, sold in discrete units.
2. Periodic (weekly, monthly, etc.) demand for each product is a random variable and may or may not follow a trend.

*Research supported by a Department of Defense THEMIS Program, Contract Number DAA H01-68-C-0296.

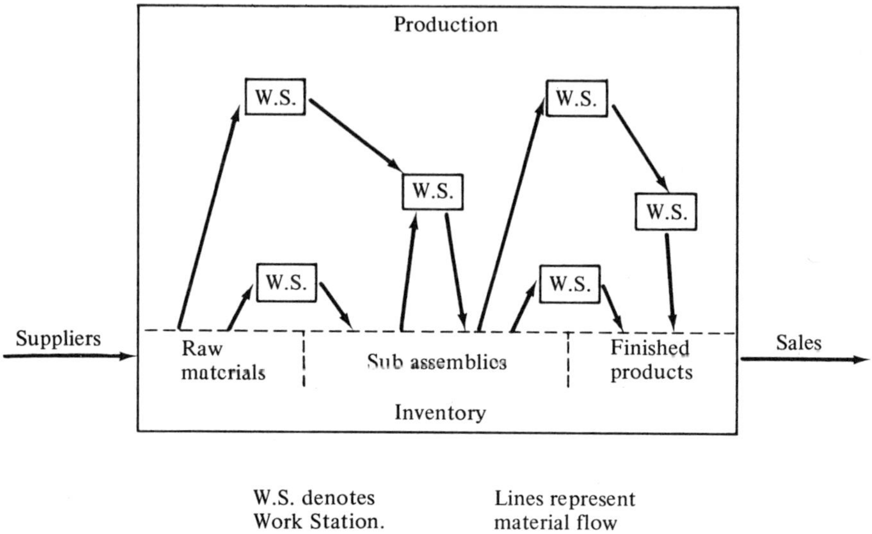

Figure C.1 Typical Inventory-Production-Sales System

3. Each product is composed of assemblies and parts.
4. At least some of the assemblies are composed of subassemblies and purchased items.
5. There are common components and subassemblies among the finished products.
6. Lead time for purchased parts is a random variable.
7. Fabrication and assembly operations are performed at "work stations" (see Fig. C.1).
8. Different assemblies require processing on some of the same work stations.
9. Processing times are essentially deterministic at certain work stations and random variables at others.
10. The quality (percent defective) of incoming raw materials and of manufactured items is a random variable.
11. Machines at work stations experience breakdowns at random intervals.
12. Repair time of machine breakdowns is a random variable.

The above characteristics are typical of many manufacturing systems. A particular system is simulated by initializing pertinent parameters in the computer model. Characteristics 1 through 8 are included in PROSIM V. Characteristics 9 through 12 could easily be added to provide a more realistic simulation. Caution must be used, however, in including more realism and detail than the student is capable of coping with.

Note that the general type of production system just described is essentially the same as discussed in Chap. 1 and around which most developments in this text have been built.

CHARACTERISTICS OF THE SIMULATOR

Size and Speed. Important characteristics of any production simulator are the size of operating system it can simulate and the computer time required for a typical simulation run. Generally, these characteristics depend upon the number of work stations and the number of stock numbers (a unique stock number is assigned to each finished good, assembly, part, and raw material) the simulator is capable of handling. PROSIM V is limited in this respect only by the size of the available computer.* A manufacturing system consisting of 15 work stations, 60 stock numbers, and 3 finished goods requires approximately 100,000 bytes of memory on an IBM 360/50 computer system.

Although there is essentially no limit to the number of periods of simulation that can be run, the computer time required for the simulation of one operating period depends upon the size of the manufacturing system being simulated. For the 15 by 60 problem, one week of simulated operation requires approximately 30 seconds of IBM 360/50 computer time when the time increment used is 1 minute. The computer time can be reduced somewhat by using a larger time increment.

Flexibility. Another important characteristic of this kind of simulator is its flexibility. The flexibility of PROSIM V is shown by the following features:

1. The operating period is variable. Usually a period of 1 week is used.
2. Several periods of simulation can be run on one computer run. There is no limit to the number of periods of simulation that can be run during one academic term.
3. The time increment is variable. If a time increment other than 1 minute is used, all processing times and set-up times must be in multiples of the time increment.
4. The lost size is variable. Items may be produced in lots of any integer size.
5. Certain features, such as random demand and random lead time, can be turned "on or off" with a parameter card.

*If *NWS* is the number of work stations, *NSN* is the number of stock numbers, and *NFG* is the number of finished goods; the dimensioned variables in the computer model require the following number of words of memory: $1000 + (NSN) \times (38) + (NWS) \times (86) + (NFG) \times (15) + (NSN) \times (NWS) \times (4)$. The computer model itself is written in FORTRAN IV and consists of 1000 statements.

6. No more than five work stations may process any one stock number and no more than five raw materials or subassemblies may merge into one stock number. This restriction is overcome easily by specifiying dummy assemblies.

OPERATION OF THE SIMULATOR

Overview. The instructor selects the particular problem he wishes to use for the term. The problem (i.e., a hypothetical manufacturing system) is translated into PROSIM parameters, and the simulator is initialized. The problem and the simulator are described to the class. The class is divided into teams, usually of three members each. The teams are given pertinent data, such as demand histories and costs. Each team then analyzes this data and designs a preliminary production control system.

The simulation now begins. Each team uses its control system to make a set of operating decisions for one operating period (usually 1 week). These decisions are keypunched in a prescribed format and entered into the simulator. PROSIM V then simulates one operating period of plant operation under the control of the student's decisions. The results are printed out and returned to the student teams. The teams analyze their feedback, make appropriate adjustments to the control system, and then make a new set of decisions for the following period of simulated operation. The instructor may interject random disturbances (such as a change in the demand trend) at any time.

The above process is repeated for the desired number of simulated operating periods. More periods can be simulated during a term by requiring several period's decisions on one computer run.

PROSIM V accumulates all costs resulting from each team's sequential decisions. The teams are trying to operate the plant at lowest cost over the entire term. The student's grade in the course is influenced to some extent (usually 30 percent) by the performance of his control system. The above process is shown in Fig. C.2.

Initializing PROSIM V. All variables and parameters pertinent to the manufacturing system being simulated must be initialized. Among the more important are:

1. Work station sequences for each manufactured item.
2. Parameters for demand generators (one demand generator for each finished product).
3. Parameters for lead time generators (one lead time generator for each raw material).
4. Process time and set-up time for each manufactured item at each work station.
5. All costs, parameters, and constraints pertinent to the simulated system.

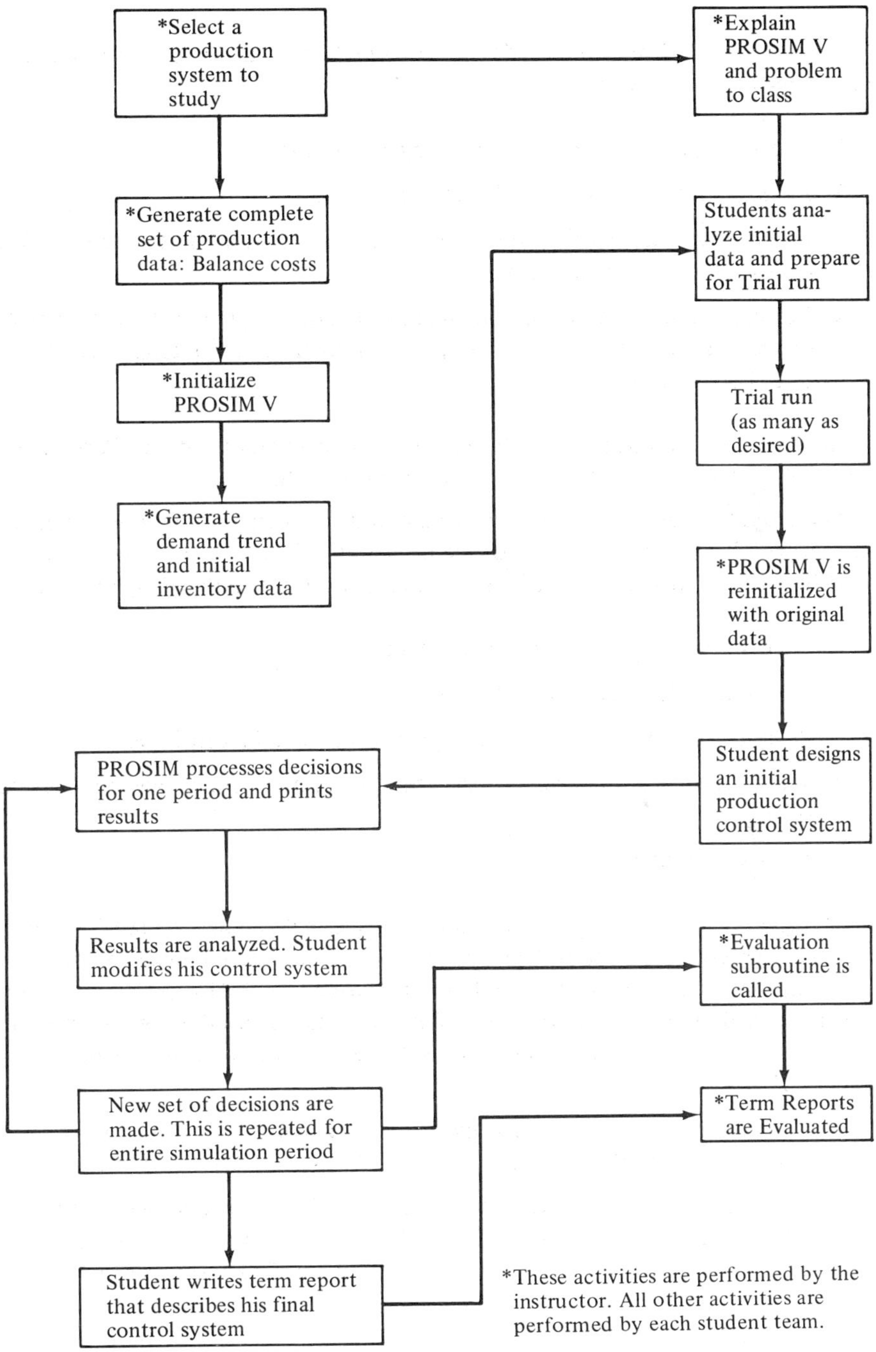

Figure C.2 Overview of the use of PROSIM V as a Teaching Aid

Initial Student Data. The student is provided the following data as a starting point for his analysis:

1. Items 1, 4, and 5 from the preceding paragraph.
2. A demand history for each finished product.
3. Labor and work station rates for each work station: shift change cost; overhead rate; idle time cost.
4. For each stock number: initial stock on hand, carrying cost, reorder cost, discount order quantity, regular price, discount price, average lead time, out-of-stock cost.

In addition, the student may be given a lot of irrelevant data. Part of his problem is to sort out only the data that is meaningful.

Student Decisions. Each student team must design a production control system for the particular operation being simulated. His control system must lead him to the following decisions for each period of simulated operation:

1. Demand forecast for all finished goods.
2. Purchase orders for raw materials.
3. Production orders and desired sequence for manufactured items.
4. Time available for each work station (regular time, overtime, extra shifts).
5. In-process buffer sizes.

In designing a control system to make the above decisions, the student relies on previous courses in industrial engineering or industrial management. In such courses, the student should have acquired a knowledge of inventory theory, scheduling techniques, forecasting techniques, production smoothing, etc. It is entirely up to the student how he uses this knowledge in designing his system.

Ground Rules. PROSIM V acts upon the student's decisions according to a set of ground rules explained thoroughly in PROSIM V User's Manual.

Output. PROSIM V processes the student's decisions for one period according to the ground rules just mentioned. The student then receives feedback in the form of a set of reports. These reports are identified and described briefly below:

1. *Results of Forecast.* This report compares the forecasted demand to the actual demand on a weekly and cumulative basis.

2. *Status of Production System.* This report shows the production orders waiting to be processed at each work station at the end of the simulation.
3. *Idle Time.* This report shows the amount and cost of idle time at each work station.
4. *Inventory Status.* This report shows all inventory transactions (receipts, issues, on order, back ordered, carrying cost, stock on hand) for each stock number.
5. *Total Manufacturing Costs.* This report shows manufacturing costs for the following factors: labor, machines, materials, overhead, and shift change.
6. *Cost Summary Report.* This report summarizes all costs for this period and all previous periods.

The student uses the above information to make any necessary adjustments to the parameters of his control system. He then makes a new set of decisions, and the entire process is repeated.

Student Evaluation Routine. PROSIM V accumulates data on each student team that aids the instructor in evaluating the designed control system. After all simulated periods have been run, the evaluation routine constructs several charts to show the stability of the control system, how well it reacts to disturbances, etc. In this way, the several designs can be compared and evaluated.

The authors have used PROSIM V in conjunction with this text in teaching production control at Arizona State University and at Auburn University. We have found it to be a very effective teaching aid. In comparison with the ordinary textbook-lecture-homework approach, PROSIM V has the following advantages:

1. It emphasizes the design of the control system.
2. It forces the student to consider the dynamic nature of the production environment.
3. It provides a conceptual understanding of the total operations control system and of the interactions among system components.
4. It fosters an appreciation of the concepts of feedback, corrective action, and integrated information systems.

REFERENCES

Cohen, K. J., *et al.* 1940. *The Carnegie Tech Management Game.* Homewood, Illinois: Richard D. Irwin, Inc.

Dill, W. R., *et al.* 1961. Experiences with a Complex Management Game. *California Management Review* 3: 39–51.

Forrester, Jay W. 1961. *Industrial Dynamics.* New York: The MIT Press and John Wiley and Sons.

Mize, J. H. 1969. PROSIM V: A Production System Simulator. Paper published in *Proceedings of the Third Conference on Applications of Simulation* (Los Angeles, California).

Mize, J.H., *et al.* 1971a. *PROSIM V Administrator's Manual: Production System Simulator.* Englewood Cliffs, N.J.: Prentice-Hall, Inc.

Mize, J.H., *et al.* 1971b. *Production System Simulator (PROSIM V): A User's Manual.* Englewood Cliffs, N.J.: Prentice-Hall, Inc.

Porter, J.C., Sasieni, M. W., Marks, E. S. and Ackoff, R. L. 1966. The Use of Simulation as a Pedagogical Device. *Management Science.* 12: B170–B179.

Sasieni, M. W., Marks, E. S. and Ackoff, R. L. 1966. *Instructor's Manual for Course in Production and Inventory Control.* Prepared for IBM under contract with Case Institute of Technology.

Sasieni M. W., and Ackoff, R. L. 1966. *Student's Handbook for Course in Production and Inventory Control.* Prepared for IBM under contract with Case Institute of Technology.

Vollmann, Thomas E. 1968. A Gaming-Monte Carlo Simulation Approach to Teaching Some Fundamental Concepts of Operations Research. Paper presented at the 1968 ORSA-TIMS Joint National Meeting (San Francisco, California).

Whitman, David, and Love, Roy E., Jr. 1967. *The Use of Games in Industrial Engineering Education.* Paper presented to the Annual Meeting of the American Society for Engineering Education (East Lansing, Michigan).

Exercises

EXERCISES FOR CHAPTER 2

2-1. Designate in separate lists: (1) the information components (indicate whether each component is a variable or a parameter) and (2) the decision components (be very explicit) for each of the following systems:

a. A thermostatic temperature control system for a house; the objective is to maintain a particular temperature level.
b. An individual taking a shower; the objective is to maintain a constant temperature of the water striking the person's body. The cold water temperature is constant but the hot water temperature fluctuates.
c. An individual driving a car; the objective is simply to stay on the road.
d. An individual driving a car on a two-lane road on which other vehicles are going in both directions; the objective is to maintain a constant speed.
e. A guided missile (the guidance and control system is on-board the missile) being launched at a target; the objective is to hit as near the target as possible.
f. An air compressor-storage tank combination at a large service station; the objective is to maintain sufficient air pressure for the operation of the station.
g. A carpenter driving a nail; the objective is to drive the nail as straight as possible.
h. A worker guiding an irregular pattern through a band saw; the objective is to follow the pattern.
i. A numerically controlled machine; the objective is to guide the cutting tool in a prescribed pattern.
j. An elevator operator lining up the floor of the elevator with a hallway floor; the objective is to line up the two floor levels as closely as possible.
k. An automatic elevator; the objective is to stop at a desired floor with no up and down oscillations.
l. A policeman directing traffic at a busy intersection; the objective is to move the maximum number of cars through the intersection without causing any line of traffic to wait longer than some maximum time limit.
m. A gas station attendant filling a gas tank; the objective is to fill the tank as full as possible, but to stop on an even $0.05 amount.
n. An individual attempting to cross a crowded room; the objective is to cross the room as quickly as possible without colliding with another person.
o. A student taking a course for credit; the objective is to expend the minimal amount of effort and still pass the course.
p. A student taking a course for credit; the objective is to make the best grade possible in this course, but to also pass the other courses he is taking.
q. A student taking several courses for credit; the objective is to divide his efforts among the courses so that he makes as high a grade as possible in all courses.

2-2. Represent each of the information-decision control systems described in Ex. 2-1 in the context of Fig. 2.4.

2-3. For each information-decision control system described in Ex. 2-1, discuss the problem of information timing.

EXERCISES FOR CHAPTER 3

3-1. Suppose you are trying to forecast the demand for automobile batteries. Your company sells about 10 percent of all batteries included as original equipment on new automobiles and a substantial portion of replacement batteries on old automobiles. Table E3-1-1 lists demand values for new automobiles and the number of replacement batteries your company has sold for the past five years. The data is given as three-month totals, or quarterly demand. The demand values can be regarded as coded for ease of computation and are for illustration purposes only; they are not intended to be consistent with actual automobile demand values for any five-year period.

Table E3-1-1

Quarter	1	2	3	4	5	6	7	8	9	10
New Auto Sales	1300	1280	1270	1310	1370	1380	1340	1360	1520	1540
Replacement Batteries	98	87	82	88	117	98	107	110	108	107

Table E3-1-1 (continued)

Quarter	11	12	13	14	15	16	17	18	19	20
New Auto Sales	1380	1240	1280	1270	1260	1300	1430	1410	1440	1490
Replacement Batteries	131	123	120	120	148	163	126	97	108	96

a. Develop a transfer function for forecasting the demand of batteries for this company.
b. Forecast demand for the next four quarters. Comment about the expected variability in future demand.
c. Economists predict a sharp drop in demand for new automobiles during the next year. How will this affect the demand forecast for this company's batteries?
d. Show your forecasting transfer function in the context of the general transfer function representation of Fig. 2.4. Use Fig. 3.5 as a guideline.

3-2. In the section on combination methods, we developed a forecasting transfer function in which regression analysis and the index method were combined. Note that we can also combine exponential smoothing with the regression-index method to obtain a three-way combination forecasting transfer function. Such a technique is useful only if it improves forecasting accuracy enough to justify the added complexity.

a. Develop a procedure that combines regression analysis, the index method, and EWMA into a forecasting transfer function. It is suggested that the pro-

Table E3-2-1

Month	1	2	3	4	5	6	7	8	9	10	11	12
Demand	180	186	179	170	170	165	155	150	170	192	195	205

Table E3-2-1 (continued)

Month	13	14	15	16	17	18	19	20	21	22	23	24
Demand	215	208	195	200	194	185	180	180	181	205	225	235

Table E3-2-1 (continued)

Month	25	26	27	28	29	30	31	32	33	34	35	36
Demand	225	225	215	225	210	200	204	195	210	220	240	250

cedure first adjust for the trend with regression analysis, then for the cycle with the index method, and finally for "drifts" with EWMA. Write the procedure in a format similar to the one outlined for the regression-index combination method.

b. Determine experimentally (either manually or in a computer simulation model) the best value for the smoothing constant α to use in the above procedure for the demand data in text Table 3.11. Use the procedure to forecast for the third year, and compare the resulting sum of absolute deviations to that of the regression-index method, shown in text Table 3.13.
c. For the data given in Table E3-2-1, apply the above procedure for determining a three-way forecasting transfer function; determine the best value for α; and apply the transfer function to obtain forecasts for the third year. Then use only the regression-index method of forecasting to again determine forecasts for the third year. Compare the results, and comment.
d. Suppose that actual demand for the fourth year is:

Month	37	38	39	40	41	42	43	44	45	46	47	48
Demand	260	255	250	255	240	230	235	220	235	235	270	285

Now apply the regression-index-EWMA method and the regression-index method to the fourth year (use parameters determined from the first three years' data for both methods), and compare the two methods.

e. Show your regression-index-EWMA method in the context of the general transfer function representation of Fig. 2.4 (use Fig. 3.5 as a guideline).

3-3. A company manufactures four different items and sells them through four district warehouses. It also sells a replacement part for Item 2 at each warehouse. Monthly sales records for the years 1968, 1969, and 1970 for the four warehouses are given in Tables E3-3-1 through E3-3-12. In addition to the items sold at the warehouses, special orders occur at irregular times. These are orders given directly to the factory from customers who do not do business with the warehouses.

a. Develop a forecasting transfer function, including control limits, for each of the four items and the replacement part at each warehouse.
b. Develop a forecasting transfer function, including control limits, for each item and the replacement part at the central factory (you must consider composite demands from the four warehouses in conjunction with the special order sales).
c. Generate monthly demand forecasts for 1971 for each item (including the replacement part) for each warehouse and for the total company.
d. Show the forecasting transfer function for Item 1 at the central factory in the context of the general transfer function representation of Fig. 2.4.

TABLE E3-3-1 1968, Warehouse 1 Monthly Sales Records

	Item 1	Item 2		Item 3	Item 4
Month	Number Sold	Number Sold	*R. P. Sold	Number Sold	Number Sold
1	386	412	20	102	1802
2	312	386	15	116	1796
3	352	414	15	127	1888
4	311	450	10	120	1615
5	406	426	22	118	2040
6	387	501	30	110	1822
7	317	487	26	83	1750
8	396	520	27	68	2004
9	415	514	35	75	1672
10	385	583	38	75	1551
11	440	591	38	80	2040
12	425	720	40	97	1886

*R. P. means replacement parts for Item 2.

Table E3-3-2 1969, Warehouse 1

	Item 1	Item 2		Item 3	Item 4
Month	Number Sold	Number Sold	R. P. Sold	Number Sold	Number Sold
1	445	812	40	103	1870
2	502	982	39	125	1980
3	497	1155	41	124	2140
4	485	860	45	134	1907
5	510	702	43	127	2208
6	420	928	50	103	1752
7	486	667	47	92	2004
8	478	1244	50	74	1880
9	465	991	55	68	1863
10	516	1134	58	78	2258
11	510	1014	60	82	1980
12	505	985	70	104	1860

Table E3-3-3 1970, Warehouse 1

	Item 1	Item 2		Item 3	Item 4
Month	Number Sold	Number Sold	R. P. Sold	Number Sold	Number Sold
1	510	1267	80	103	2085
2	540	1123	100	111	2270
3	580	1061	115	126	1890
4	535	1672	86	111	2192
5	570	1336	75	128	2206
6	573	1245	93	112	2214
7	568	1514	67	94	1990
8	585	1337	125	81	2140
9	595	1397	102	69	2406
10	602	1091	115	65	1960
11	590	1614	100	82	1928
12	620	1384	100	92	2267

Table E3-3-4 1968, Warehouse 2

	Item 1	Item 2		Item 3	Item 4
Month	Number Sold	Number Sold	R. P. Sold	Number Sold	Number Sold
1	302	1511	75	122	819
2	400	1485	80	136	804
3	338	1520	60	147	1010
4	547	1523	60	141	670
5	571	1512	39	138	1166
6	228	1561	85	130	919
7	615	1598	60	103	836
8	363	1503	38	88	1005
9	565	1572	60	95	777
10	227	1710	105	95	636
11	498	1695	140	101	1141
12	524	1795	130	117	871

Table E3-3-5 1969, Warehouse 2

	Item 1	Item 2		Item 3	Item 4
Month	Number Sold	Number Sold	R. P. Sold	Number Sold	Number Sold
1	360	1815	151	123	1086
2	410	1889	148	142	1129
3	293	2075	150	144	1232
4	163	1860	150	154	990
5	470	1681	151	147	1300
6	301	1924	158	123	760
7	175	1663	161	112	1020
8	412	2242	150	93	786
9	400	1892	158	88	962
10	368	2136	172	98	1255
11	637	2012	170	100	1085
12	669	1881	180	123	947

Table E3-3-6 1970, Warehouse 2

	Item 1	Item 2		Item 3	Item 4
Month	Number Sold	Number Sold	R. P. Sold	Number Sold	Number Sold
1	542	1968	180	121	1185
2	425	1824	190	131	1417
3	373	1760	210	146	876
4	279	2372	190	132	1198
5	410	2034	160	148	1136
6	340	1943	190	131	1318
7	322	2218	165	113	1044
8	318	2039	225	100	1136
9	402	2095	180	89	1370
10	349	1774	210	85	869
11	614	2315	200	103	922
12	320	2085	190	114	1074

Table E3-3-7 1968, Warehouse 3

	Item 1	Item 2		Item 3	Item 4
Month	Number Sold	Number Sold	R. P. Sold	Number Sold	Number Sold
1	350	285	15	117	548
2	415	293	16	131	520
3	300	320	16	142	623
4	165	309	13	135	515
5	470	338	16	133	702
6	305	330	17	125	590
7	173	362	17	98	540
8	412	346	16	83	673
9	400	349	21	90	516
10	368	389	30	91	435
11	475	401	22	95	698
12	510	410	26	112	560

Table E3-3-8 1969, Warehouse 3

	Item 1	Item 2		Item 3	Item 4
Month	Number Sold	Number Sold	R. P. Sold	Number Sold	Number Sold
1	398	432	29	118	702
2	505	460	30	140	812
3	430	438	32	139	530
4	650	747	30	150	710
5	671	579	30	142	670
6	428	522	33	115	760
7	710	665	36	107	626
8	463	573	32	89	678
9	665	609	32	82	838
10	427	435	39	93	540
11	598	710	40	97	565
12	624	592	41	119	640

Table E3-3-9 1970, Warehouse 3

	Item 1	Item 2		Item 3	Item 4
Month	Number Sold	Number Sold	R. P. Sold	Number Sold	Number Sold
1	742	478	43	117	639
2	625	542	46	126	675
3	573	629	44	141	729
4	479	484	75	127	602
5	610	390	60	143	875
6	540	514	50	127	480
7	622	382	66	109	603
8	518	773	57	95	498
9	602	498	61	84	585
10	549	620	43	82	725
11	690	555	71	96	646
12	620	444	60	107	578

Table E3-3-10 1968, Warehouse 4

	Item 1	Item 2		Item 3	Item 4
Month	Number Sold	Number Sold	R. P. Sold	Number Sold	Number Sold
1	650	1425	100	127	1870
2	700	1375	110	141	1802
3	630	1460	105	152	1688
4	753	1544	111	145	2168
5	771	1622	120	145	1930
6	558	1586	130	135	1848
7	715	1752	110	110	2110
8	563	1775	122	93	1935
9	765	1819	117	100	1980
10	550	1763	123	101	1567
11	698	1821	128	105	2105
12	724	1964	127	122	1973

1969, Warehouse 4

All 1969 sales records for this warehouse were destroyed in an office fire. No reliable duplicate records could be found in the main office at the central factory.

Table E3-3-11 1970, Warehouse 4

	Item 1	Item 2		Item 3	Item 4
Month	Number Sold	Number Sold	R. P. Sold	Number Sold	Number Sold
1	724	1796	190	128	1660
2	698	1693	202	136	1850
3	527	1919	213	151	1440
4	765	1579	190	136	1520
5	563	2069	220	154	1810
6	715	1819	166	137	1625
7	558	1730	192	119	1710
8	771	1962	168	106	1603
9	753	1672	186	100	1528
10	630	1538	216	93	1612
11	700	2044	198	109	1580
12	597	1775	186	119	1570

Table E3-3-12 Special Order Sales

	Number Sold			
Date of Order	Item 1	Item 2	Item 3	Item 4
March 16, 1968			40	
June 14, 1968	350			1410
October 16, 1968		300		1000
January 13, 1969			60	
April 1, 1969		220		
May 28, 1969				1200
July 4, 1969	430		40	
September 10, 1969		780		
December 2, 1969			20	
December 28, 1969				4000
February 14, 1970		180		
March 16, 1970				10,000
July 22, 1970	500		30	
September 19 1970			40	8000
November 4, 1970			80	
December 16, 1970		460	20	

3-4. A company manufactures four different items and sells them through four district warehouses. Sales records for each item for the years 1968, 1969, and 1970 are given in Tables E3-4-1 through E3-4-4.

a. Develop a forecasting transfer function, including control limits, for each item at each warehouse.

b. Develop a forecasting transfer function, including control limits, for each item for the total company.

c. Generate monthly demand forecasts for 1971 for each item for each warehouse and for the total company.

d. Compare the control limits found for each item for the total company with those found for each item for each warehouse. Explain any statistical concept involved.

Table E3-4-1 Item 1

Month	Warehouse 1			Warehouse 2			Warehouse 3			Warehouse 4		
	1968	1969	1970	1968	1969	1970	1968	1969	1970	1968	1969	1970
Jan.	312	712	1167	1401	1710	1868	181	432	378	1319	1885	1696
Feb.	286	882	1023	1387	1885	1724	187	360	442	1260	1912	1593
Mar.	311	1055	961	1420	1975	1660	223	338	529	1366	2034	1819
Apr.	352	760	1572	1413	1760	2272	207	647	384	1473	1792	1479
May	326	602	1236	1422	1581	1934	238	479	290	1516	2148	1969
June	401	828	1145	1461	1824	1843	230	422	414	1486	1561	1719
July	387	567	1414	1498	1563	2118	262	565	282	1652	1821	1630
Aug.	420	1144	1237	1503	2142	1939	246	473	673	1674	1588	1862
Sept.	414	791	1297	1472	1792	1993	249	509	398	1709	1764	1572
Oct.	483	1034	971	1601	2036	1674	289	330	520	1663	2057	1438
Nov.	491	914	1514	1593	1912	2219	301	610	455	1721	1886	1944
Dec.	620	885	1284	1691	1881	1983	310	492	443	1864	1765	1673

Table E3-4-2 Item 2

Month	Warehouse 1			Warehouse 2			Warehouse 3			Warehouse 4		
	1968	1969	1970	1968	1969	1970	1968	1969	1970	1968	1969	1970
Jan.	1452	675	1150	297	360	542	444	1002	1003	327	668	525
Feb.	1003	705	576	400	412	425	972	995	1205	610	603	490
Mar.	1171	269	1374	330	293	373	513	1316	906	304	308	226
Apr.	1105	555	1077	553	163	279	448	561	817	335	405	567
May	831	714	162	571	474	410	600	542	996	314	405	360
June	670	981	1138	228	299	340	335	920	422	327	204	618
July	380	1033	1151	615	173	322	323	668	507	349	304	229
Aug.	1287	549	601	363	412	318	465	714	303	417	472	578
Sep.	1178	1315	444	565	400	402	1072	565	397	271	168	554
Oct.	403	567	517	227	368	349	1370	246	402	370	297	339
Nov.	694	997	973	498	637	614	573	665	489	423	419	441
Dec.	1455	1002	447	524	669	320	1151	926	710	543	363	293

Table E3-4-3 Item 3

Month	Warehouse 1			Warehouse 2			Warehouse 3			Warehouse 4		
	1968	1969	1970	1968	1969	1970	1968	1969	1970	1968	1969	1970
Jan.	721	986	1085	1795	1884	2081	518	669	641	1868	1769	1671
Feb.	704	1029	1317	1800	2011	2280	490	812	680	1720	1885	1880
Mar.	910	1132	776	1913	2135	1775	603	532	727	1664	2053	1433
Apr.	570	894	1098	1610	1897	2093	480	706	599	2250	1769	1569
May	1066	1240	1036	2060	2246	2036	682	670	870	1931	1583	1860
June	819	668	1218	1812	1664	2214	569	763	484	1848	1837	1625
July	736	920	944	1733	1926	1947	510	626	620	2114	1610	1712
Aug.	905	686	1036	1961	1688	2036	633	679	492	1935	2146	1938
Sep.	677	862	1370	1672	1840	2373	486	841	584	1990	1799	1520
Oct.	536	1155	769	1531	2158	1763	422	539	720	1577	2038	1812
Nov.	1041	985	822	2047	1988	1828	671	563	646	2115	1917	1590
Dec.	771	847	974	1776	1845	1967	538	639	578	1983	1887	1694

Table E3-4-4 Item 4

Month	Warehouse 1			Warehouse 2			Warehouse 3			Warehouse 4		
	1968	1969	1970	1968	1969	1970	1968	1969	1970	1968	1969	1970
Jan.	1117	1118	1117	1127	1128	1128	1102	1103	1103	1122	1123	1121
Feb.	1131	1140	1126	1141	1150	1136	1116	1125	1111	1136	1142	1131
Mar.	1142	1139	1141	1152	1149	1151	1127	1124	1126	1147	1144	1146
Apr.	1135	1150	1127	1145	1160	1136	1120	1134	1111	1141	1154	1132
May	1133	1142	1143	1145	1152	1154	1118	1127	1128	1138	1147	1148
June	1125	1115	1127	1135	1128	1137	1110	1103	1112	1130	1123	1131
July	1098	1107	1109	1110	1117	1119	1083	1092	1094	1103	1112	1113
Aug.	1083	1089	1095	1093	1099	1106	1068	1074	1081	1088	1093	1100
Sep.	1090	1082	1084	1100	1093	1100	1075	1068	1069	1095	1088	1089
Oct.	1091	1093	1082	1101	1103	1093	1075	1078	1065	1095	1098	1085
Nov.	1095	1097	1096	1105	1107	1109	1080	1082	1082	1101	1100	1103
Dec.	1112	1119	1107	1122	1129	1119	1097	1104	1092	1117	1123	1114

3-5. For the data given below, answer the same questions as in Ex. 3-4, for the year 1964.

Table E3-5-1 Item 1

Month	Warehouse 1			Warehouse 2			Warehouse 3			Warehouse 4		
	1961	1962	1963	1961	1962	1963	1961	1962	1963	1961	1962	1963
Jan.	72	98	108	179	188	208	51	69	64	186	176	167
Feb.	69	101	131	169	201	231	50	81	65	172	188	194
Mar.	91	113	77	191	213	177	60	53	72	166	205	143
Apr.	57	89	109	157	189	209	44	70	59	227	176	156
May	106	124	103	206	224	203	68	67	87	193	158	186
June	81	66	121	181	166	221	56	76	48	184	183	162
July	73	92	94	173	192	194	52	61	61	211	156	171
Aug.	96	68	103	196	168	203	63	67	49	193	214	193
Sept.	67	86	137	167	186	237	48	84	58	199	179	147
Oct.	53	115	76	153	215	176	42	53	72	157	203	181
Nov.	104	98	82	204	198	182	67	56	64	211	191	159
Dec.	77	84	96	177	184	196	53	63	57	198	188	169

Table E3-5-2 Item 2

Month	Warehouse 1			Warehouse 2			Warehouse 3			Warehouse 4		
	1961	1962	1963	1961	1962	1963	1961	1962	1963	1961	1962	1963
Jan.	67	74	86	167	174	186	33	43	37	198	188	169
Feb.	94	88	72	194	188	172	47	36	44	221	191	159
Mar.	43	105	66	143	205	166	22	33	52	167	203	181
Apr.	57	76	127	157	176	227	28	64	38	199	179	147
May	86	58	93	186	158	193	43	47	29	193	214	196
June	63	82	84	163	182	184	32	42	41	211	156	171
July	71	56	111	171	156	211	36	56	28	184	182	163
Aug.	96	114	93	196	214	193	48	47	67	193	158	186
Sep.	47	79	99	147	179	199	24	50	39	227	176	157
Oct.	81	103	67	181	203	167	40	33	52	166	205	143
Nov.	59	91	121	159	191	221	30	61	45	172	188	194
Dec.	62	88	98	169	188	198	31	49	44	186	176	167

Table E3-5-3 Item 3

Month	Warehouse 1			Warehouse 2			Warehouse 3			Warehouse 4		
	1961	1962	1963	1961	1962	1963	1961	1962	1963	1961	1962	1963
Jan.	145	67	115	29	36	54	44	100	130	32	66	52
Feb.	135	70	57	44	41	42	97	99	70	61	60	49
Mar.	117	26	137	33	29	37	51	131	40	30	30	22
Apr.	110	55	107	55	16	27	44	56	120	33	40	56
May	83	71	16	57	47	41	60	54	130	31	40	36
June	67	98	113	22	29	34	33	92	30	32	20	61
July	29	103	115	61	17	32	22	66	67	34	30	22
Aug.	128	54	60	36	41	31	46	71	83	41	47	57
Sep.	117	131	44	56	40	40	107	56	110	27	16	55
Oct.	40	56	51	22	36	34	137	24	120	37	29	33
Nov.	69	99	97	49	63	61	57	66	130	42	41	44
Dec.	145	100	44	52	66	32	115	92	140	54	36	29

Table E3-5-4 Item 4

Month	Warehouse 1			Warehouse 2			Warehouse 3			Warehouse 4		
	1961	1962	1963	1961	1962	1963	1961	1962	1963	1961	1962	1963
Jan.	102	103	103	122	123	121	117	118	117	127	128	128
Feb.	116	125	111	136	142	131	131	140	126	141	150	136
Mar.	127	124	126	147	144	146	142	139	141	152	149	151
Apr.	120	134	111	141	154	132	135	150	127	145	160	136
May	118	127	128	138	147	148	133	142	143	145	152	154
June	110	103	112	130	123	131	125	115	127	135	128	137
July	83	92	94	103	112	113	98	107	109	110	117	119
Aug.	68	74	81	88	93	100	83	89	95	93	99	106
Sep.	75	68	69	95	88	89	90	82	84	100	93	100
Oct.	75	78	65	95	98	85	91	93	82	101	103	93
Nov.	80	82	82	101	100	103	95	97	96	105	107	109
Dec.	97	104	92	117	123	114	112	119	107	122	129	119

3-6. A company manufactures and sells three products, *A*, *B*, and *C*. Weekly sales records for the two previous years are given in Table E3-6-1.

Table E3-6-1

Week Number	Product *A*	Product *B*	Product *C*	Week Number	Product *A*	Product *B*	Product *C*
1	89	100	188	47	89	124	261
2	99	120	147	48	84	115	192
3	98	112	179	49	83	102	230
4	90	99	188	50	82	109	241
5	80	83	208	51	73	113	235
6	72	110	214	52	73	114	235
7	70	95	162	53	90	126	220
8	62	112	187	54	98	117	266
9	65	117	202	55	91	133	216
10	51	109	196	56	100	139	214
11	70	104	163	57	110	125	216
12	71	119	189	58	104	111	194
13	89	117	200	59	105	143	246
14	87	127	240	60	103	142	244
15	94	115	184	61	97	141	215
16	106	133	190	62	92	136	252
17	105	121	180	63	80	150	250
18	94	124	178	64	84	115	234
19	91	125	176	65	76	112	207
20	82	130	210	66	77	135	261
21	68	117	174	67	89	140	239
22	67	133	187	68	97	141	220
23	61	121	205	69	97	154	256
24	63	123	230	70	117	147	274
25	72	132	208	71	116	142	241
26	83	136	236	72	114	152	260
27	85	125	178	73	113	142	281
28	105	126	223	74	115	147	294
29	104	139	191	75	104	151	202
30	109	132	208	76	98	146	216
31	104	145	184	77	82	141	232
32	100	122	198	78	82	142	240
33	92	139	160	79	85	144	252
34	88	154	200	80	86	135	275
35	72	121	222	81	90	152	240
36	67	141	188	82	101	150	258
37	61	137	182	83	115	140	276
38	68	124	236	84	119	152	235
39	75	128	230	85	117	150	261
40	79	123	246	86	121	175	288
41	100	133	223	87	119	152	261
42	99	118	236	88	114	163	258
43	108	120	221	89	110	123	251
44	107	125	214	90	102	143	268
45	111	139	246	91	100	154	284
46	102	122	245	92	82	142	268

Table E3-6-1 (continued)

Week Number	Product *A*	Product *B*	Product *C*	Week Number	Product *A*	Product *B*	Product *C*
93	84	140	276	99	128	148	302
94	86	138	291	100	130	136	261
95	92	114	245	101	124	137	277
96	107	155	306	102	128	152	267
97	106	154	284	103	108	138	284
98	116	153	286	104	100	148	287

a. Develop a forecasting transfer function, including control limits, for each of the three products.
b. Forecast the demand for each product for the next 52 weeks.
c. If you were forecasting for four weeks at a time rather than for one week, would your forecasts be more or less accurate? Justify your answer with statistical reasoning.
d. Suppose that actual demand for the next 26 weeks is represented in Table E3-6-2.

Table E3-6-2

Week Number	Product *A*	Product *B*	Product *C*	Week Number	Product *A*	Product *B*	Product *C*
105	94	148	275	118	107	173	291
106	99	132	277	119	105	182	285
107	86	144	282	120	101	180	329
108	93	137	284	121	97	180	285
109	102	147	331	122	92	173	318
110	95	158	306	123	102	182	315
111	114	154	281	124	114	180	300
112	125	166	278	125	120	184	317
113	131	161	315	126	122	185	294
114	134	168	325	127	130	179	324
115	131	181	327	128	130	194	276
116	130	170	297	129	131	192	310
117	113	174	296	130	125	186	302

Comment on the adequacy of your transfer functions to forecast demand accurately from week to week. How well does your transfer function respond to disturbances to the equilibrium of the demand system? Discuss the concepts of feedback, corrective action, and parameter adjustments as they are related to this situation.

EXERCISES FOR CHAPTER 4

4-1. Discuss thoroughly the relationships between demand forecasting (Chap. 3) and operations planning (Chap. 4).

4-2. A product is assembled from several components. Ten work elements are involved and 400 minutes are required to assemble one unit of the product. The work elements and their required times are shown in the following table.

Work Element	1	2	3	4	5	6	7	8	9	10
Time (minutes)	45	35	30	80	30	20	40	40	50	30

The precedence relationships between the elements are given in diagram Ex. 4-2-1.

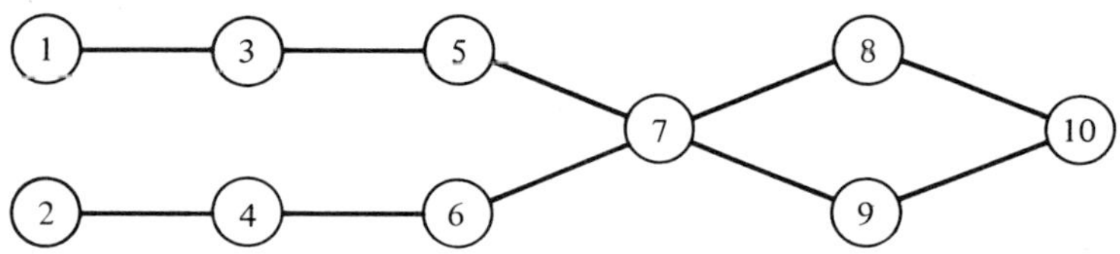

a. Draw a tree diagram of the permissible orderings of the work elements.
b. Use the ranked positional weight technique to obtain a feasible line balance for this product. If possible, improve the solution given by the algorithm.
c. What is the efficiency of the best balance you found in part *b*?
d. Is the best balance you found in part *b* an optimum solution? Why?

4-3. A product is assembled from several components. Eight work elements are involved and 300 minutes are required to assemble one unit of the product. The work elements and their required times are shown in the following table:

Work Element	1	2	3	4	5	6	7	8
Time (minutes)	30	40	30	80	30	20	30	40

The precedence relationships between the elements are shown in diagram Ex. 4-3-1.

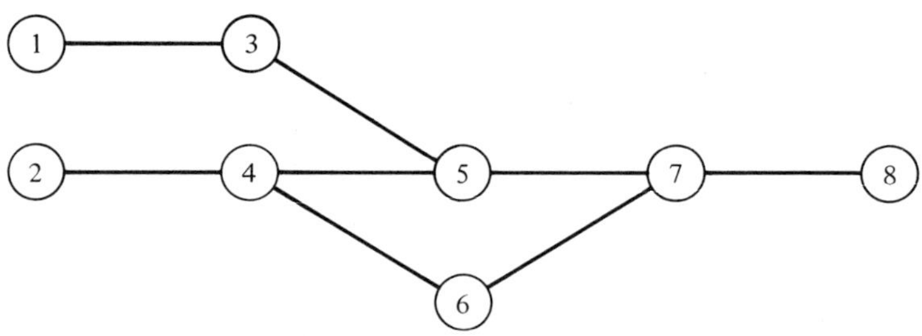

a. Draw a tree diagram of the permissible orderings of the work elements.
b. Use the ranked positional weight technique to obtain a feasible line balance for this product. If possible, improve the solution given by the algorithm.
c. What is the efficiency of the best balance you found in part *b*?
d. Is the best balance you found in part *b* an optimum solution? Why?

e. Would it matter if one or more of the work elements were fabrication operations rather than assembly operations?

f. Suppose that the annual demand for this product is 10,000 units and that the physical dimensions of the plant prevents more than 4 lines to be set up for this product. Assuming 250 working days per year and 8 productive hours per shift per day, what combinations of cycle time and number of shifts would be reasonable (refer to Table 4.4 as a guideline)?

4-4. From the 36 months of demand data given in Table 3.11, develop a 12-month forecast for the fourth year. From the forecast, determine a requirement-time profile such as the one shown in Table 4.9 and Fig. 4.11. Determine three workable production plans for this product. Construct one of your plans such that there will be 100 units of the product remaining in inventory at the end of the year. Show your plans on a graph such as Fig. 4.11. Are any of your plans optimal? Discuss the concept of optimality for problems such as this one.

4-5. The product for which you determined a requirement-time profile in Ex. 4-4 is manufactured from five purchased components, *P*1, *P*2, *P*3, *P*4, and *P*5, and three manufactured components, *M*1, *M*2, and *M*3. The explosion chart of the product is given in diagram Ex. 4-5-1.

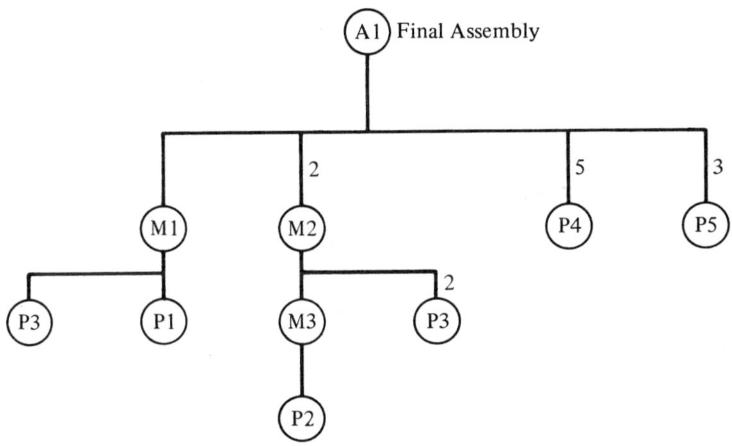

As usual, the number of units of a component required for one unit of the next higher level of assembly is shown just above each bubble. When no number appears, only one unit of the component is required in the next higher level of assembly. This product is assembled according to the work elements and precedence relationships given in Ex. 4-3.

a. Determine a requirement-time profile (such as Table 4.9) for each of the eight components for each of the three production plans you determined in Ex. 4-4.

b. Determine a requirement-time profile for assembly line production time for each of the three production plans (use the work element times in Ex. 4-3).

c. The work element times given in Ex. 4-3 pertain only to the assembly of the product from components *M*1, *M*2, *P*4, and *P*5; they do not reflect the time required to fabricate components *M*1, *M*2, and *M*3. Fabrication occurs in

Departments 1, 2, and 3. The assembly line will be called Department 4. Per unit fabrication time (in minutes) for each fabricated component in each department is indicated in the next table.

Department	*M*1	*M*2	*M*3
1	10		15
2	10	20	
3		10	20

Determine a requirement-time profile for each department for each of your three production plans.

d. Construct the next assembly matrix **N**, as in Eq. 4.4. Determine the total requirements matrix **T**, as in Eq. 4.12.
e. An order is received for 200 units of the product and 600 units of the component *M*2. Use the concepts associated with Eq. 4.24 to obtain component requirements for this order.
f. Use your forecast for this product for the first 3 months of the fourth year to obtain the gross time-phased component requirements. Assume no outstanding purchase orders or production orders. Assume also that initial inventory of all items is zero. It is desired to end the 3-month period with an inventory for the product and each component equal to 20 per cent of the requirements for the third month.
g. For Departments 1, 2, and 3, and the cycle time for Department 4 use the fabrication times given in part *c* above, to determine the total amount of work generated for each department by the requirements stated in part *f* above.

4-6. A company manufactures three products which contain several common components. Explosion charts for the three products are presented in diagram Ex. 4-6-1.

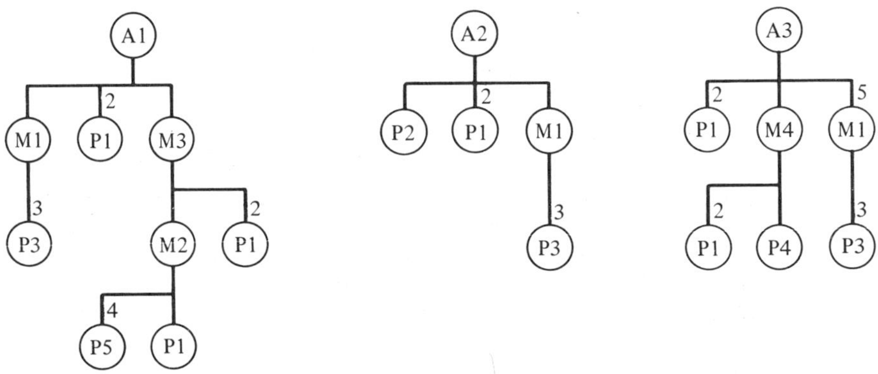

The manufactured components *M*1, *M*2, *M*3, and *M*4 are fabricated in Departments 1, 2 and 3; the three products *A*1, *A*2, and *A*3 are assembled in Department 4. Per unit process times (in minutes) and processing costs (in dollars) for each department are given in the next table. A department is capable of working on only one product or component at a time.

Unit	Department 1		Department 2		Department 3		Department 4	
	Time	Cost	Time	Cost	Time	Cost	Time	Cost
*A*1							5	$0.50
*A*2							4	0.40
*A*3							5	0.50
*M*1	1.0	$0.10	1.5	$0.30				
*M*2			4.0	0.80	20.0	$2.00		
*M*3	5.0	0.50	4.0	0.80				
*M*4			4.0	0.80	1.0	0.10		

Experience has shown that 4, 5, 5, and 6 percent of the manufactured items in Departments 1, 2, 3, and 4, respectively, are defective. No rework is allowed. The purchased items *P*1, *P*2, *P*3, *P*4, and *P*5 also have a percent defective factor. These factors, along with the reorder level, cost, and lead time (in weeks) for each item are:

Unit	Percent Defective	Cost Per Unit	Reorder Level	Lead Time
*P*1	6	$0.04	2500	1.0
*P*2	3	0.80	400	2.0
*P*3	4	0.06	6000	1.0
*P*4	4	0.20	700	2.0
*P*5	5	0.30	1000	2.0

The following table shows current stock on hand and outstanding production and purchase orders for all items. In the case of outstanding orders, the week number (relative to "now") in which they are scheduled to be available is also shown.

Unit	Stock on Hand	Outstanding Orders	Scheduled for Availability
*A*1	50		
*A*2	30		
*A*3	20		
*M*1	110	200	1
*M*2	20	100	1
*M*3	0		
*M*4	40	200	1
*P*1	2000	300	1
*P*2	300	200	2
*P*3	6000	1200	1
*P*4	800	400	2
*P*5	1000		

Weekly demand for $A1$, $A2$, and $A3$ is estimated at 110, 150, and 300 units, respectively.

a. Construct the **N** and **T** matrices for this problem.

b. Determine the total cost of each product, $A1$, $A2$, and $A3$. (Remember to include the quality factor.)

c. Determine the time-phased parts requirements (netted against stock on hand, outstanding orders, and desired ending inventory) for a 4-week period. Use weekly demand values of 110, 150 and 300 units for $A1$, $A2$, and $A3$, respectively. It is desired to end the 4-week period with 1 week's requirements of all items: $A1$, $A2$, $A3$, $M1$, $M2$, $M3$, $M4$, $P1$, $P2$, $P3$, $P4$, and $P5$. Remember to include the quality factor.

d. Determine the weekly workload on each department generated by the results of part *c* above. Specify the number of shifts and the amount of overtime each department should work.

e. Determine when purchase orders should be placed for $P1$, $P2$, $P3$, $P4$, and $P5$ if the amount ordered each time for each item is equal to 2 week's requirements.

4-7. Use the information in Table 4.15 to obtain one smoothed production plan for Department 2 and another for Department 3. Are these two plans compatible with each other and with the plan for Department 1 shown in Fig. 4.12? If not, revise the three plans so that they are mutually compatible. How much confidence do you have, after the revisions, that the resulting set of plans is optimal for the three departments?

4-8. Forecasted demand for a single product for twelve months is as follows:

Month	1	2	3	4	5	6	7	8	9	10	11	12
Forecast	150	147	142	136	130	123	121	122	130	145	167	173

Each unit of product produced generates 10 man-hours of work in Department 1 and 5 man-hours of work in Department 2. Workers in Department 1 receive \$4.00 per hour; those in Department 2 receive \$3.00 per hour. Overtime pay is 1.5 times regular pay in both departments. There is a carrying charge of 10¢ per month for each hour's production carried over to following months. The maximum amount of overtime hours permitted in a month is one fourth of the regular time hours available for that month.

a. Use the plant calendar in the first three columns of Fig. 4.15, and the above information to determine for each department a smoothed production plan that employs the optimum number of men. Are the plans for the two departments compatible? If not, revise the plans so that they are compatible.

b. Would the results have been different if the carrying cost in Department 1 had been 5¢ per month and in Department 2, 30¢ per month, for each hour's production carried over to following months?

4-9. Forecasted demand for three products, *A*1, *A*2, and *A*3, for 12 months is as follows:

Month	1	2	3	4	5	6	7	8	9	10	11	12
*A*1	500	510	480	600	600	660	590	700	680	740	790	760
*A*2	430	380	420	370	410	380	440	380	420	370	410	390
*A*3	110	160	150	100	130	70	120	80	80	120	40	90

The three products are manufactured in Departments 1, 2, and 3. The following table shows the number of man-hours of work generated in each department by the manufacture of one unit of each of the three products.

Unit	Department 1	Department 2	Department 3
*A*1	1		3
*A*2		1	1
*A*3	3	2	2

Units are inspected after all processing has been completed. Inspection records show that 5 percent of all items manufactured are defective. Regular hourly pay rates in Departments 1, 2, and 3 are, respectively, \$3.00, \$4.00, and \$6.00 Overtime pay is 1.5 times regular pay. The maximum amount of overtime hours permitted in a month is one fifth of the regular time hours available for that month. There is a carrying charge of 5¢ per month in Departments 1 and 2 and 10¢ per month in Department 3 for each hour's production carried over to following months.

Use the plant calendar in the first three columns of Fig. 4.15 and the above information to determine for each department a smoothed production plan that employs the optimum number of men. Make any revisions necessary for compatibility between departments.

4-10. Develop a computer model* that will perform the basic calculations for the Gozinto precedure. The program should be written such that it can handle any number of items in the matrices up to some reasonable maximum, e.g., 50.

4-11. Develop a computer model* such that it will perform the calculations for the production spread sheet; include incremental costs.

EXERCISES FOR CHAPTER 5

5-1. Annual usage of a purchased item is 1000 units. Ordering cost is \$10.00 per order and carrying cost is \$0.50 per unit per year. There are 250 working days in the year. Lead time is 10 days.

*It is recommended that an algorithmic programming language such as FORTRAN or ALGOL be used if such a language is available on the computer to which you have access.

a. Determine the optimal order quantity for this item for a fixed order size system. How many orders will be placed during a year? How many days will there be between consecutive orders? How much safety stock is needed? Determine RL, the reorder level. State the complete transfer function for this situation.
b. How would the analysis for part *a* be changed if a fixed order interval system were being used?
c. If we decided to manufacture this item in our own plant, what would our annual production rate PR have to be if we wanted MQ_o (optimal manufactured quantity) to be only 10 percent larger than Q_o (optimal order quantity)?

5-2. Determine the transfer function for the situation described in part *a* of Ex. 5-1 if annual demand is a normally distributed random variable having a mean of 1000 units and a standard deviation of 50 units. Assume $\alpha = 0.01$; i.e., that we are willing to be out of stock in only 1 of each 100 times we place an order.

5-3. The following information is given for an item manufactured for inventory:

$PC = \$5.00$ per set up
$D = 900$ units per year
$CC = \$1.00$ per unit per year
$PR = 9{,}000$ units per year

Holding charges are 10 percent of material cost.

a. Determine the most economical lot size to manufacture.
b. Suppose it is not feasible to schedule our production process to produce the number of units calculated in part *a*. We have a choice of producing either 10 percent more or 10 percent less than the amount determined in part *a*. Which amount should we produce? Justify your answer with appropriate calculations.
c. How much money do we lose by producing the amount determined in part *b*? Do you consider this loss very serious?

5-4. The following information is provided for an inventory situation in which it is desired to minimize total annual cost associated with inventory:

$M =$ number of month's supply to be ordered at one time
$D =$ total annual
$PC =$ ordering cost (for one order)
$CC =$ annual holding cost for one unit
$TC =$ total annual cost

Derive directly (avoid substitution into other formulas) an expression for M that will minimize TC. Note that M is in number of month's supply, not in units of product. Assume instantaneous replenishment.

5-5. Consider Eqs. 5.4 and 5.15.

a. In general, is Q_o larger or smaller than MQ_o for the same cost and demand values? Why?
b. What condition, if any, would result in Q_o and MQ_o being equal, for all practical purposes?

c. In Eq. 5.15, what would be the effect of D being greater than PR? What is the effect of D being equal to PR? Discuss these two conditions from both their mathematical and practical aspects.
d. Make any necessary modifications to Eq. 5.4 if D is expressed as monthly (rather than annual) demand and CC remains an annual cost.
e. How would Eq. 5.4 be changed if we denote the material cost of each unit as Z and change CC to be a percentage of material cost?
f. Which terms in Eqs. 5.4 and 5.15 must be consistent in the units in which they are expressed?

5-6. The following information is provided for an inventory situation in which it is desired to minimize total annual cost:

TC = total annual cost
PC = set-up cost (for one order)
CC = annual holding cost for one unit
D = total annual demand
PR = annual production rate
X = number of production runs per year

Derive directly an expression for X that will minimize TC. Assume 250 working days per year.

5-7. Weekly demand for an item is a normally distributed random variable having a mean of 200 units and a standard deviation of 20 units. Ordering cost is \$25.00 per order, carrying cost is \$5.20 per unit per year, and lead time is 4 weeks. Management has specified an α value of 0.05.

a. Determine the complete transfer function for a fixed order size system.
b. Determine the complete transfer function for a fixed order interval system.
c. Compare the total expected costs of the two systems. Explain the difference.
d. Suppose that weekly demands for this item for the next 20 weeks are as follows:

Week	1	2	3	4	5	6	7	8	9	10
Demand	216	224	163	241	206	182	164	218	247	198

Week	11	12	13	14	15	16	17	18	19	20
Demand	181	203	232	173	184	198	221	204	162	183

Initial inventory is at 300 units (at relative time zero), and previously placed orders, each of size 350, are scheduled to arrive at times 1.0, 2.6 and 4.2. (Times are in weeks; a week is considered to be 5 days.) Operate this system with the transfer function determined in part *a*. Convert weekly demand values to daily demand values by dividing the weekly demand by 5 and rounding to the nearest integer. Draw a graphical representation of inventory level over the 20-week period (similar to the graph in Fig. 5.9). Compute total cost for the 20-week period (use a stock out cost of \$10 per unit for each sale lost due to being out of stock; no backorders are permitted).

e. Repeat part *d*, but use the transfer function determined in part *b*. Round off OI_0 to the nearest full day. The graphical representation for this case will be similar to Fig. 5.13.

5-8. For the item considered in Ex. 5-7, the supplier has offered to give you a discount if you will buy 500 units on each order, rather than the amount determined in part *a* of Ex. 5-7. The price per unit without the discount is \$25.00. By how much must the supplier reduce the price per unit in order for the discount price to be attractive to you? (*Hint:* Modify Eq. 5.1 to include price per unit U as a factor:

$$TC = (PC)/(D/Q) + (CC)(Q/2) + (U)(D) + (CC)(SS)$$

Find TC for Q_o found in part *a* of Ex. 5-7 and $U = \$25.00$. Then solve the same equation for U using TC as just found and $Q = 500$ units).

5-9 Consider again the inventory item in Ex. 5-7. It is possible to manufacture this item in our own plant at the rate (PR) of 35,000 units per year and at a cost of \$25.00 per unit. Set-up cost would be \$50.00 for each production run; manufacturing lead time would be 1 week.

a. Use Eqs. 5.15 and 5.16 to determine a complete transfer function for this situation.
b. Would it be more economical to manufacture the item ourselves than to continue buying it for \$25.00 each?
c. If the answer to part *b* is *yes*, what price per unit must the supplier quote to be competitive? If the answer to part *b* is *no*, what cost per unit must we achieve in order to manufacture the item without losing money?

5-10. A process in our plant produces four different products. Pertinent inventory data is given in the next table.

Item	Annual Demand	Set-Up Cost	Carry Cost	Production Rate
1	8,000	\$15.00	\$0.48	40,000
2	12,000	5.00	3.60	36,000
3	4,000	20.00	0.20	20,000
4	10,000	10.00	1.00	50,000

a. Determine MO_o for each item individually. Calculate total cost for each item and add these to obtain total annual cost for the four items.
b. Determine the optimal fraction of annual demand to produce during one production run (use Eq. 5.19). Find the total annual cost for this solution. Compare this total cost to that found in part *a*, and comment.
c. Specify the complete transfer function for the solution found in part *b*.
d. Find the number of units of each item produced each cycle and the number of days in each cycle devoted to the production of each item (see Table 5.2).

5-11. Answer the same questions given in Ex. 5-10 for the following product items and inventory data.

Item	Annual Demand	Set-Up Cost	Carry Cost	Production Rate
1	400	$80.00	$20.00	4,000
2	19,000	4.00	0.30	300,000
3	34,000	6.00	0.15	290,000
4	520	20.00	5.00	10,000
5	15,000	2.00	0.60	215,000
6	390	100.00	35.00	3,000
7	29,000	5.00	1.00	400,000

5-12. In part a of Ex. 5-10, MQ_o for each of the four items was determined individually. Find production ranges for the four items using Eq. 5.23 for the following permissible percent increases in total costs: 1 percent, 5 percent, and 10 percent. Discuss how these results could be used in adjusting MQ_o values for production scheduling.

5-13. Repeat Ex. 5-12 for the seven items shown in Ex. 5-11.

5-14. In the development of the production range transfer function, we used a nondimensional ratio $p = TC'/TC_0$. If we define another nondimensional ratio $q = MQ'/MQ_0$, we can show that the relationship between p and q is:

$$p = \tfrac{1}{2}(1/q + q)$$

from which

$$q = p \pm \sqrt{p^2 - 1} \qquad \text{for } p > 1$$

Plot this relation on a graph in which q is the horizontal axis and p is the vertical axis. Note for each value of p that there are two values of q, q_L and q_U. These two values of q can then be used in a modified form of Eq. 5.23 to obtain the production range:

$$MQ_L = MQ_o(q_L)$$

$$MQ_U = MQ_o(q_U)$$

Verify the results of Exs. 5-12 and 5-13 using the graph.

5-15. The two stages of a two-stage production process have identical Poisson service rate distributions. Per unit carrying cost is $1.00 per 8-hour shift and idle machine cost is $20.00 per hour.

a. Determine the optimum buffer size for the process.
b. Determine the efficiency of the process. What physical meaning does this value have?
c. Determine the expected total cost of operating this system (carrying cost plus idle machine cost) for 40 hours.

5-16. Construct a graph of Eq. 5.27 for several values of the ratio C_2/C_1. Plot this ratio on the horizontal axis and the corresponding J_o from Eq. 5.27 on the vertical axis.

5-17. A production process consists of eight stages having identical Poisson service rates. Per unit carrying cost is \$0.25 per hour and idle machine cost is \$0.50 per minute. Determine the optimum buffer size to provide between successive stages.

5-18. A production process consists of two stages. The production rate of the first stage is a Poisson random variable having a mean of 50 units per hour. The production rate of the second stage is also a Poisson random variable and its mean is adjustable. Management has specified that it is willing to permit the first stage to be idle 10 percent of the time, but that the second stage must not be idle more than 5 percent of the time. Determine the mean production rate of the second stage and the required buffer size between stages.

5-19. Repeat Ex. 5-18 with the mean of the second stage equal to 50 and the mean of the first stage adjustable. All other conditions and requirements are the same.

5-20. A production process consists of two stages. The production rate at each stage is a Poisson random variable with an adjustable mean. Each machine is built such that its mean production rate can be varied from 45 to 55 units per hour. Management will permit each machine to be idle only 10 percent of the time.

a. Determine the permissible ranges of mean production rate values for the two machines. (That is, determine all possible combinations of μ_A and μ_B that satisfy Eq. 5.30 for $\alpha_B = 0.10$ and that also satisfy each machine's 45 to 55 units per hour range.)

b. Assume that we want maximum production through this two-stage process; determine the required buffer size and the mean production rate of each machine.

5-21. Construct a table and graph similar to Table 5.4 and Fig. 5.23 for a ratio of *PC* to *CC* of 10.

5-22. Write a computer program that will calculate and print tables such as Table 5.4 for any values of *PC* and *CC* read in and for any value of *D* desired.

5-23. A company carries three items in inventory. Characteristics of the three items are given in the accompanying table where k is the cubic foot requirement for storing one unit of each item:

Item	D	PC	CC	k
1	8000	\$5.00	\$0.50	2 ft^3
2	800	10.00	1.00	4 ft^3
3	4000	7.50	0.20	3 ft^3

a. If total storage space K is limited to 2000 cubic feet, use the Lagrangian multiplier method to determine the adjusted order quantities which will result in minimum total cost within the space limitation constraint.

b. Look at this problem from a practical point of view. Arrange a staggered order arrival schedule for the three items such that Q_o (determined individually) is used for ordering each item and that total available storage space is never exceeded. Assume a linear usage rate for each item throughout the year.

c. Suppose that total storage space is 1500 cubic feet for the problem above. What ordering policy would you follow for this situation, if you can control order arrival time for each product?

5-24. Consider the three items in Ex. 5-23.

a. Suppose that there is no limitation on storage space, but that the average amount of money that we can have invested in inventory at one time is limited to \$6,000. The three items cost \$10.00, \$5.00, and \$20.00, respectively. Determine the adjusted order quantities which will result in minimum total cost within the budgetary constraint of \$6,000.

b. Determine order quantities for the three items that meet both the space limitation of 2000 cubic feet and the budgetary limitation of \$6,000.

5-25. Write a computer program that will determine adjusted order quantities for space limitations. The program should read as input data the demand (D), the procurement cost (PC), the carrying cost (CC), and the cubic foot storage requirement per unit (k) of each item in the analysis. It should also read as input data the total storage space available. The program should first compute Q_o for each item, print it out, and then perform the iterative procedure for minimizing total cost. A starting value for the Lagrangian multiplier, λ, must also be read in. The program should print out the adjusted values of the order quantities, the final value of λ, and the total cost of the new policy.

5-26. Write a computer program similar to the one in Ex. 5-25, but for budgetary limitations.

5-27. Write a general computer program that will handle either space limitations or budgetary limitations.

5-28. Consider Ex. 4-4, in which you were asked to develop three workable production plans for a single product. A particular purchased part is used in the production of this item. Ten units of the purchased part are needed for the manufacture of one unit of the finished product. A lead time of one-half month is required to obtain an order of this purchased item. The cost of placing an order is \$5.00 and the cost of carrying one unit of the purchased part for one year is \$0.24.

a. For each of the three production plans determined in Ex. 4-4, develop an inventory transfer function for this purchased part.

b. Determine the total annual inventory cost of the purchased part for each production plan. Discuss how these costs could be used in helping to arrive at a decision as to which production plan to use.

5-29. Consider again Ex. 4-4, and the additional information given in Ex. 5-28 above. One workable production plan for the manufactured product would be to produce at a rate of 200 units per month for the first 6 months and at 100 units per month for the last 6 months.

a. Develop a complete inventory transfer function for the purchased part used in the manufacture of this product. Which type of inventory system would be better for this situation, a fixed order size system or a fixed order interval system?

b. Represent your transfer function in a manner similar to Fig. 5.14 or Fig. 5.15, depending upon which type system you selected. Be sure to include a parameter that indicates whether we are in the first 6 months or last 6 months of the year.

5-30. Devise a computerized fixed order size inventory system that will: (1) keep all inventory records updated, (2) automatically place purchase orders when the reorder level has been reached, (3) print out a message when the emergency level has been reached, and (4) keep a running total of all costs (ordering cost, set-up cost, out of stock cost, and total cost). Devise your system so that it will handle up to 100 inventory items.

Note that this is an *operating* system and, as such, an inventory analysis for each item must be performed external to the computer program. Such things as Q_o, RL, LT and α are parameters that are an integral part of the program; they are not determined within the program.

Design your program so that it operates on a daily basis; i.e., total usage and total receipts for each item are read in at the end of each day's operation.

Your transfer function is of the form, "whenever $SOH + SOO \leq RL$, place an order for Q_o units." There will be a unique transfer function for each inventory item. The expected arrival time for each order can be determined by adding the lead time for the item to the current time.

Include in your program the ability to read in at any time new parameters computed externally. Also, build into your system an automatic monitoring feature (see the feedback loop in Fig. 5.15). Establish certain measures of effectiveness, such as the percent of stock out conditions that occur, average stock on hand when an order arrives, etc. Compare on each computer run these computed values to the appropriate "standard" values, and print out the results. The manager of the inventory system then has an opportunity to modify certain parameters if he wishes to do so. For example, if the number of out-of-stock conditions exceeds α, the acceptable proportion of such conditions, the inventory manager should increase RL. He should recognize, however, that in so doing he increases the average stock on hand, thereby increasing total cost. He must literally "fine tune" the controllable parameter RL until he has balanced out-of-stock costs against carrying costs. He should also avoid over-reacting to short-term occurrences. After all, α is usually stated in a manner similar to this: "We will permit an out-of-stock condition only once in 100 replenishments." We see, then, that several hundred replenishments are necessary for a certain RL value in order for us to really know the proportion of time this value is causing us to be out of stock.

Also include in your system a means for indicating when an emergency level has been reached for each item. The emergency level for each item is an externally determined parameter. Whenever an emergency level is reached, have your system print out a warning message.

Design your system so that it will accumulate various costs for each item and for the system as a whole. Have the system prepare a detailed cost report for each month and a cost summary report for the entire year.

5-31. Repeat Ex. 5-30 for a fixed order interval system.

5-32. a. Prepare a detailed flow diagram for the program in Ex. 5-30.
b. Prepare a detailed flow diagram for the program in Ex. 5-31.

EXERCISES FOR CHAPTER 6

6-1. A company has just received four production orders, to which it has assigned the following identification numbers: 41, 42, 43, and 44. Each production order consists of several operations. The operation sequences, the process times in hours, and the facilities to which the several operations are assigned are given in the table below:

Production Order Number	Operation Sequence	Process Time	Facility Number
41	*A*	2	2
	C	5	1
	B	3	3
42	*C*	1	2
	A	2	1
	D	1	2
	B	4	3
	E	3	1
43	*B*	8	3
	A	4	2
	C	5	3
	D	2	1
44	*B*	1	3
	C	1	1
	A	2	2

Previously placed production orders have resulted in existing work loads at the three facilities as follows:

Facility 1: Operation 29*G* scheduled for time 0 to 3. Operation 24*F* scheduled for time 3 to 6.

Facility 2: Operation 24*C* scheduled for time 0 to 2. Operation 37*B* scheduled for time 4 to 8.

Facility 3: Operation 37*D* scheduled for time 0 to 4.

a. Construct a Gantt bar chart similar to Fig. 6.1 to show the existing workload at time zero.
b. Place the 15 new operations onto the Gantt chart, as in Fig. 6.2.
c. Manually manipulate the Gantt chart until a feasible solution is found. Compute the following measures of effectiveness of the schedule you have

generated: (1) total lateness of the four production orders and (2) total amount of idle facility time.

d. How many combinations of operation sequences are theoretically possible for this problem?

e. Apply the "shortest operation first" and the "first come first served" decision rules to this problem. Compare the results with each other and with the solution you found in part *c* above.

6-2. A company has just received three production orders, to which it has assigned the following identification numbers: 45, 46 and 47. Each production order consists of several operations. The operation sequences, the process times in hours, and the facilities to which the several operations are assigned are given in the table below:

Production Order Number	Operation Sequence	Process Time	Facility Number
45	*C*	6	3
	A	10	2
	B	3	3
	D	8	1
46	*A*	12	2
	C	6	3
	B	10	1
	E	5	2
	D	7	1
47	*B*	12	1
	A	10	3
	C	3	2
	D	9	1

There are no previously placed production orders in the three facilities.

a. Construct a Gantt chart for the 13 operations shown, creating an "initial load" for each facility.

b. Find a feasible solution by visual inspection and manual manipulation of the Gantt chart. Compute the following measures of effectiveness of the schedule you generated: (1) total lateness of the three production orders and (2) total amount of idle facility time.

c. How many combinations of operation sequences are theoretically possible for this problem?

d. Apply the "shortest operation first" and the "first come first served" decision rules to this problem. Compare the results with each other and with the solution you found in part *b* above.

6-3. Consider the seven production orders, numbered 41, 42, 43, 44, 45, 46, and 47 in Exs. 6-1 and 6-2 as one problem. Use the previously placed production orders listed in Ex. 6-1 as the existing workload at time zero.

a. Construct a Gantt chart for the 33 operations (5 of which are already on the facilities at time zero).
b. Find a feasible schedule by visual and manual means, and compute the two measures of effectiveness: (1) total lateness of the 7 production orders and (2) total amount of idle facility time.
c. How many combinations of operation sequences are theoretically possible for this problem?
d. Apply the "shortest operation first" and the "first come first served" decision rules to this problem. Compare the results with each other and with the solution you found in part *b* above.
e. Compare the results for parts *c* and *d* above to the comparable results of Exs. 6-1 and 6-2. In particular, discuss whether the results of the combined problem are comparable to the combined results of the individual problems.
f. Discuss the shortest operation first rule in terms of its tendency to delay the longer jobs. Imagine and describe situations in which this tendency would be desirable and others in which it would be undesirable.
g. Suppose that the seven orders came from customers of varying importance to us. In particular, we can classify the seven orders into the following categories:

   Relatively high importance: orders 42 and 46.
   Medium importance: orders 41, 43, and 47.
   Relatively low importance: orders 44 and 45.
   Develop a decision rule that resolves schedule conflicts according to the above classifications. Apply your decision rule to the combined problem. Discuss the results in terms of the objective of giving priority to orders from customers of high importance.

6-4. A job shop receives five orders. There are four machines capable of performing some or all of the orders. Processing times (in hours) for the several possible combinations are given in Table E6-4-1.

Table E6-4-1.

Order Number	$M1$	$M2$	$M3$	$M4$
1	50	40	30	45
2	—	110	80	60
3	60	80	40	—
4	100	90	70	30
5	90	50	60	65
Hours Available	110	100	80	120

Use the index method for loading to assign orders to machines. Comment on the resulting assignment.

6-5. Use the Index Method for loading to assign the ten orders given in Table E6-5-1 to the available five facilities:

Table E6-5-1.

Order Number	$M1$	$M2$	$M3$	$M4$	$M5$
1	90	50	60	60	60
2	70	50	90	40	—
3	80	100	—	120	90
4	—	80	95	60	75
5	120	160	140	—	100
6	30	30	40	20	30
7	75	75	60	50	90
8	60	—	80	40	70
9	60	30	60	50	40
10	40	50	30	40	50
Hours Available	150	70	90	120	140

6-6. A job shop receives four orders for the following quantities:

Order	Quantity
1	100
2	200
3	150
4	150

Three machines could be used to process these orders. The hours available on each machine for the scheduling period are as follows:

Machine	Hours Available
$M1$	150
$M2$	125
$M3$	100

The number of units per hour that each machine is capable of processing for each order are as follows:

Order	$M1$	$M2$	$M3$
1	2	0.5	1
2	1	2.5	2
3	1.5	2.0	1
4	2.0	3.0	1.5

Use the Index Method for loading to assign orders to machines. Can you improve the solution by visual inspection? (Attempt to obtain an improved solution for which the total hours of assigned production is less than that found by the initial application of the Index Method.)

6-7. Consider again the order assignment problem of Ex. 6-4. Let us now modify the problem to allow order splitting; i.e., one order can be split and produced on more than one machine. If we let c_{ij} be the cost of producing one unit of order i on machine j, then we can easily set the problem up for solution by the simplex method of linear programming. Do this by following the procedure described in the subsection, "Linear Programming for Loading." Compare the results to those you obtained using the Index Method.

6-8. Repeat Ex. 6-7 for the order assignment problem of Ex. 6-5.

6-9. Repeat Ex. 6-7 for the order assignment problem of Ex. 6-6.

6-10. Seven operations are listed in the table below, all of which must be processed first on *M*1 and second on *M*2. The processing time on each machine is shown in the next table.

Operation Number	*M*1	*M*2
1	3	5
2	6	3
3	2	12
4	8	3
5	4	4
6	7	6
7	5	9

a. Apply Johnson's sequencing model to this set of operations to determine the optimal sequence. Draw a schedule bar chart for this sequence, as in Fig. 6.12a.

b. Apply the first come first served decision rule to this problem. Draw a schedule bar chart for the resulting sequence, as in Fig. 6.12b. Compare the two sequences in terms of total elapsed time and idle facility time.

NOTE: Exercises 6-11 and 6-12 require the use of the PERT procedure outlined in Appendix B.

6-11. From the information given in Table E6-11-1, construct the project network and perform a complete PERT analysis. Determine the probability of completing this work program in 80 days. Activity (i, j) signifies the activity whose beginning event is i and whose ending event is j. Activity times are given in days.

Table E6-11-1

Activity	Time Estimates		
	Optimistic	Most Likely	Pessimistic
(1, 2)	3	6	10
(1, 3)	7	11	14
(2, 4)	13	19	23
(2, 5)	17	24	32
(3, 6)	16	22	34
(4, 7)	4	9	13
(5, 6)	16	19	24
(5, 7)	8	12	17
(6, 8)	23	29	34
(7, 8)	13	16	20

6-12. Information for a PERT type project is given in Table E6-12-1. In this case, the activities are assigned a number. Activity times are expressed in weeks.

Table E6-12-1

Activity Number	Time Estimates		
	Optimistic	Most Likely	Pessimistic
1	11	16	23
2	36	40	50
3	5	12	17
4	2	4	7
5	7	12	15
6	7	9	12
7	10	13	17
8	8	13	20
9	21	29	40
10	14	19	23
11	16	22	30
12	4	10	14
13	8	13	16
14	32	38	48

A "must precede" matrix is given next. A 1 in a cell indicates that the activity i must precede activity j, where i is the row number of the cell and j is the column number. For example, activity 1 must precede activities 4 and 5.

i \ j	1	2	3	4	5	6	7	8	9	10	11	12	13	14
1				1	1									
2						1	1							
3								1	1					
4						1	1							
5										1				
6										1				
7											1	1		
8											1	1		
9														1
10													1	
11													1	
12														1
13														
14														

a. Construct the project network, and perform a complete PERT analysis.
b. Determine the probabilities of completing the projects in 100 weeks; 108 weeks; and 90 weeks.
c. What completion date should we quote if we want to be 75 percent certain of meeting it?
d. Suppose that each activity is to be performed by a different crew and that the company has no other projects under way. This would mean that crews working on noncritical activities will be idle a considerable amount of the time. Suppose that resources can be shifted from noncritical activities to critical activities at 50 percent effectiveness; i.e., a crew working on an activity having 10 weeks of slack can be used on a critical activity and reduce the expected duration time by 5 weeks. Reallocate resources until the shortest possible project duration time has been achieved.

EXERCISES FOR CHAPTER 7

7-1. Consider the production schedule you developed in Ex. 6-1, part *e*, using the shortest operation first decision rule. Suppose this was the production schedule actually implemented for a 40-hour week consisting of 5 eight-hour days. The process times used in generating the schedule were only estimates. Suppose the actual process times were the same as the estimated process times with the following exceptions:

Operation	Actual Time
41*C*	4
42*C*	2
42*D*	3
43*B*	7
43*A*	6
44*C*	4
44*A*	4

Also, suppose these events occurred during the week:

(1) Department 2 was out of service from noon to 5:00 p.m. on the third day.

(2) An emergency order was received at noon on the fourth day that consisted of one 4-hour operation in Department 1 and then a 2-hour operation in Department 3. Company policy with regard to emergency orders is not to interrupt on-going operations, but to begin processing the emergency order immediately after completion of the operation now in process.

(3) Department 1 did not begin processing on the second day until 1 hour after regular starting time.

a. Show on a Gantt bar chart the actual occurrences for this 40-hour period of operation, as well as the current status of the shop for any operations carried over.
b. Analyze the results of last week's operations, and see if you can determine any trends or symptoms of undesirable behavior that should be investigated.

7-2. Consider the production schedule you developed in Ex. 6-2, part *d*, using the shortest operation first decision rule. Suppose this production schedule was actually implemented for a 40-hour week consisting of 5 eight-hour days. The process times used in generating the schedule were only estimates. Suppose that actual process times were the same as the estimated process times with the following exceptions:

Operation	Actual Time
45*A*	12
45*D*	7
46*A*	15
46*B*	12
46*E*	6
47*C*	5

Also, suppose the following events occurred during the week:

(1) A machine malfunction in Department 1 caused that department to shut down for the last 2 hours on the second day.

(2) Department 1 was authorized to work as many as 4 hours overtime on the third day and 2 hours overtime on the fourth day, if there was that much work ready for it to perform.

(3) Operation 47*B* was subcontracted to an outside firm on the first day. It is returned to the company and ready for the second operation, 47*A*, at the beginning of the second day.

(4) An emergency order was received at the beginning of the fourth day that required 2 hours processing in Department 3. (Company policy for emergency orders is explained in occurrence 2 in Ex. 7-1.)

a. Show on a Gantt bar chart the actual occurrences for this 40-hour period of operation, as well as the current status of the shop for any operations carried over.

b. Analyze the results of last week's operations, and see if you can determine any trends or symptoms of undesirable behavior that should be investigated. (In particular, determine how close each department's time estimates are to actual process times).

c. Given the additional production orders listed in Table E7-2-1, construct a new schedule for the coming week. (You may adjust the estimated process times if you can justify this from your analysis in part *b* above.) Use the shortest operation first rule in scheduling the operations; however, any operations remaining from production order number 46 are to be given top priority.

Table E7-2-1

Production Order No.	Operation Sequence	Process Time	Facility Number
48	*B*	9	2
	C	12	3
	A	8	1
49	*A*	7	1
	D	8	2
	B	5	1
	E	10	3
	C	12	2
50	*C*	6	3
	A	8	1
	B	4	2
	D	8	3

d. Recommend a different policy for scheduling emergency production orders that would improve over-all efficiency of the shop. Also, is there any way to make provisions in the schedule for breakdowns and other work stoppages that might occur?

7-3. The following information is concerned with three manufactured items that are maintained in inventory:

Item	Reorder Level	Manufacturing Lead Time, in days
1	500	20
2	400	10
3	100	10

A production order has been released for each of these three items. Current status is a follows:

Item	Current Stock on Hand	Number of Days Since Order was Placed
1	350	5
2	150	7
3	30	5

a. Determine the "critical ratio" for each item using the method presented in the text. If two or more of these items formed a schedule conflict on a facility, how would you use the critical ratios to resolve the conflict?

b. What assumption in the critical ratio method makes it unrealistic for many manufacturing situations?

7-4. Construct an example involving several items in which the "sum of digits" method is used to resolve schedule conflicts at several departments. Carry the example through several days operation; show how the priorities change each day. Input new orders into your system intermittently.

7-5. Develop a computerized method of using the Gozinto procedure at the operating level. Use as a guideline the procedure outlined in Chap. 4 for the planning level. This method would be quite similar to the program requested in Ex. 4-10.

7-6. In Ex. 4-9, the forecasted demand for three products for 12 months was given and you were asked to determine a smoothed production plan for each of three departments engaged in manufacturing the three products. Suppose that actual demand for the three products for each of the 12 months is as given in Table E7-6-1:

Table E7-6-1

Product	Month											
	1	2	3	4	5	6	7	8	9	10	11	12
*A*1	493	470	504	540	560	590	580	620	610	650	690	710
*A*2	409	412	412	382	431	350	340	390	330	410	420	380
*A*3	120	150	160	140	143	109	132	107	116	114	85	70

In addition to these actual demand values, two other types of feedback information are noted. First, at the very end of the third month, a new machine was installed in Department 3 which reduced labor requirements on $A1$ to 2 man-hours and on $A3$ to 1 man-hour. Secondly, the actual percent defective of all items manufactured was 6 percent for all 12 months.

Imagine that you are production control manager and that you receive the above information at the end of each month, *after the events that had created it occurred.* Labor costs and carrying costs are given in Ex. 4-9. Also, an out-of-stock cost of $10.00 per item is charged for each unit the company is out of stock. Workers who are laid off must be paid for 80 extra hours at regular pay. The cost of training a new employee is $100.00.

How would you react to the above occurrences if you were production control manager? Discuss explicit ways you would measure the performance of the system, and state the actions you would take at the end of each month. Remember to base your actions only on past occurrences, since you do not know what is going to happen in the future.

7-7. In Ex. 4-6, the estimated weekly demand for $A1$, $A2$, and $A3$ was 110, 150, and 300 units, respectively. Based on this estimate and other data given in the problem, you determined several items of information to be used in formulating a production plan.

You are now asked to develop an adaptive production control system based upon the concepts presented in Chap. 7. You are to select a set of transfer functions which will lead to operating desisions on a weekly basis. Your total system should also include meaningful measures of effectiveness, and associated control limits, so that system effectiveness may be monitored on a longterm basis. The additional data needed for this exercise is given on the following page.

You are to run your system for a period of 20 weeks. Actual occurrences are described below. Your system should respond to these occurrences appropriately in terms of modifying the production plan, changing system parameters, etc. Your objective is to operate the system for 20 weeks at lowest total cost. You are encouraged to change any parameter that your feel should be changed, such as order quantities, reorder levels, etc. In running this system, however, you must remember that the actual occurrences are not known to you until *after* they have occurred. You must avoid the temptation to look ahead to see what the actual occurrences will be. This is an opportunity for you to participate in an exercise that is very similar to the real world, one in which a control system must respond to actual occurrences and monitor its own performance.

It is suggested that you design a set of forms to assist you in running this production system. Design the forms so that they simplify your job of record keeping and calculations from 1 week's run to the next.

Additional Data

Order cost ($P1$, $P2$, $P3$, $P4$, $P5$)	$ 5.00
Set-up cost ($M1$, $M2$, $M3$, $M4$)	10.00
Set-up cost ($A1$, $A2$, $A3$)	20.00
Out-of-stock cost (per unit short)	10.00

(No backorders permitted.)

Overtime cost is 1.5 times regular cost.

Manufacturing sequence:

- $M1$ Department 1; Department 2
- $M2$ Department 3; Department 2
- $M3$ Department 1; Department 2
- $M4$ Department 2; Department 3

Carrying cost (dollars per week)

$P1$	$0.001
$P2$	0.015
$P3$	0.0005
$P4$	0.008
$P5$	0.006
$M1$	0.005
$M2$	0.010
$M3$	0.015
$M4$	0.015
$A1$	0.100
$A2$	0.050
$A3$	0.200

Actual Weekly Demand

Week Number	Weekly Demand			Week Number	Weekly Demand		
	$A1$	$A2$	$A3$		$A1$	$A2$	$A3$
1	110	140	310	11	110	140	280
2	95	155	300	12	120	155	265
3	90	150	280	13	110	140	320
4	105	150	295	14	140	145	300
5	100	140	320	15	130	160	280
6	110	145	305	16	135	145	290
7	90	210	285	17	160	150	330
8	95	155	300	18	145	155	295
9	110	160	310	19	160	150	290
10	105	140	305	20	175	145	300

Manufactured Items: Actual Percent Defective and Downtime*

Week Number	Dep't 1		Dep't 2		Dep't 3		Dep't 4	
	Percent Defective	Down-time	Percent Defective	Down-time	Percent Defective	Down-time	Percent Defective	Down-time
1	4.2	0	4.8	0	3.8	4	5.7	2
2	4.5	2	5.1	0	4.0	0	6.2	2
3	3.6	0	5.4	0	1.2	0	6.3	4
4	4.0	0	4.6	0	3.5	2	6.1	0
5	4.1	4	7.3	0	4.4	3	5.5	2
6	3.8	0	4.4	1	4.5	0	6.0	4
7	4.4	1	4.8	0	3.6	0	5.8	3
8	1.1	0	4.6	2	4.4	0	5.7	0
9	3.5	2	4.3	0	3.4	0	6.3	1
10	3.2	3	4.5	0	3.8	4	5.4	1
11	3.8	0	4.1	0	3.7	0	5.7	1
12	4.2	0	3.8	0	4.2	1	5.1	3
13	3.7	0	4.4	0	4.1	0	4.8	4
14	4.4	0	4.1	3	3.4	0	3.5	2
15	4.0	8	3.6	0	4.6	0	6.2	0
16	3.0	0	4.3	0	4.1	0	6.3	1
17	4.5	0	3.9	0	4.1	2	5.7	1
18	3.9	1	4.6	0	4.2	0	5.9	1
19	4.2	0	4.2	1	3.7	2	6.1	0
20	4.0	2	4.3	0	3.9	2	5.7	1

*Downtime is expressed in hours.

Purchased Items: Actual Percent Defective and Lead Times

(Use as many of the following purchase orders as you need for each item for the 20 weeks operation. Use them in sequential order.)

Purchase Order Number	*P*1		*P*2		*P*3		*P*4		*P*5	
	Percent Defective	Lead Time	Percent Defective	Lead Time	Percent Defective	Lead Time	Percent Defective	Lead Time	Percent Defective	Lead Time
1	5	1.0	3	2.0	1	2.0	4	2.0	6	2.0
2	6	1.0	3	2.0	6	1.0	3	3.0	4	2.0
3	6	2.0	3	1.0	3	1.0	4	1.0	5	2.0
4	10	1.0	5	3.0	1	2.0	5	2.0	8	2.0
5	4	1.0	4	3.0	4	1.0	4	1.0	1	2.0
6	7	1.0	3	1.0	6	2.0	3	4.0	3	2.0
7	5	2.0	3	2.0	5	1.0	3	2.0	4	2.0
8	5	2.0	3	2.0	2	1.0	5	1.0	3	2.0
9	6	1.0	2	3.0	5	1.0	4	3.0	2	1.0
10	5	1.0	3	1.0	5	2.0	6	2.0	1	2.0
11	6	2.0	3	1.0	1	2.0	5	1.0	3	2.0
12	6	1.0	4	2.0	1	1.0	6	1.0	1	2.0
13	5	1.0	3	2.0	4	1.0	7	4.0	1	2.0
14	6	1.0	3	2.0	3	1.0	6	2.0	2	2.0
15	5	2.0	3	3.0	3	2.0	6	1.0	4	2.0

Index

Q

R

S

T

U

V

W

Y

Z